EROSION

EROSION

American Environments and
the Anxiety of Disappearance

GINA CAISON

DUKE UNIVERSITY PRESS
Durham and London 2024

© 2024 DUKE UNIVERSITY PRESS
All rights reserved
Project Editor: Lisa Lawley
Designed by A. Mattson Gallagher
Typeset in TheSans C4s by Westchester Publishing Services

Library of Congress Cataloging-in-Publication Data
Names: Caison, Gina, [date] author.
Title: Erosion : American environments and the anxiety of disappearance / Gina Caison.
Description: Durham : Duke University Press, 2024. | Includes bibliographical references and index.
Identifiers: LCCN 2024008007 (print)
LCCN 2024008008 (ebook)
ISBN 9781478031161 (paperback)
ISBN 9781478026914 (hardcover)
ISBN 9781478060147 (ebook)
Subjects: LCSH: Ecocriticism. | American literature—Themes, motives. | Ecology in literature. | Soil erosion. | BISAC: LITERARY CRITICISM / American / General | SCIENCE / Environmental Science (see also Chemistry / Environmental)
Classification: LCC PN98.E36 C357 2024 (print) | LCC PN98.E36 (ebook) | DDC 810.9/36—dc23/eng/20240719
LC record available at https://lccn.loc.gov/2024008007
LC ebook record available at https://lccn.loc.gov/2024008008

Cover art: Tamsin van Essen, *Viral III* (porcelain), from the *Erosion* series. Courtesy of the artist.

For
Fannie Mae, who loved poetry

and
Anchor, who loved the soil

Contents

Acknowledgments ix

Notes 221
Bibliography 251
Index 271

Introduction:
Erosion
1

1

Landslides and
Horizons
of the West
29

2

Surfaces and
Allotments
of the Heartland
68

3

Disappearing Grounds
and Backgrounds of
the Gulf
107

4

Gullies and Removals
of the Plantation
South
141

5

Littoral Cells and
Literal Sells of the
Atlantic
177

Conclusion: What We
Talk About When
We Talk About Erosion
209

Acknowledgments

I've heard it said that one can always tell an author's first book from her second because in the first, the author thanks everyone she has ever met, and in the second, the acknowledgments list is a much tidier, direct affair. I find myself somewhat at odds with this trend. As I age, I see that my list of debts has only grown longer and my gratitude wider. There is no solitary scholarly life. The collaborations and conversations are what make this work both possible and worthwhile. The best parts of this book are owed to the following people and organizations; the shortcomings are mine alone.

To begin, this work would not exist without my time as a junior core fellow (2020–21) at the Institute for Advanced Study (IAS) at Central European University in Budapest. Director Nadia Al-Bagdadi offered enthusiastic and generous support for this project from the beginning, and it made all the difference. The IAS staff—Agnes Bendik, Krisztina Domján, and Andrey Demidov—deserve more thanks than a person can give for the way they held the institute together in unprecedented times. Similarly, Ágnes Forgó and Éva Gelei worked tirelessly to make sure our home spaces could double as our work spaces, and for that they deserve a hearty *köszönöm szépen*.

The community of IAS fellows during the 2020–21 academic year managed to make the surreal experience of writing a book during a pandemic lockdown probably better than it deserved to be. I am grateful for each of them. In particular, the scholarly work of Central European University faculty fellows Constantin Iordachi and Viktor Lagutov helped me to coalesce several key themes that run throughout this book about political and ecological empires. Zsuzsa Hetényi, Raluca Iacob, Frances Kneupper, Oksana Maksymchuk, Lorenzo Sala, Róbert Somos, and Ildikó Zakariás all contributed to an intellectual community that far exceeded the confines of our virtual spaces. The long walks and even longer talks and laughs shared with István Pál Ádám (and Carlos), Larissa Buchholz (and Aimy), Mary Cox, Tyrell Haberkorn, Mostafa Minawi, and Somogy Varga all shaped this project in ways I never expected and much for the better. They each kept my spirits afloat with their own generous spirits and ears. Perhaps by far, though, the greatest gift this project brought me was the opportunity to meet two of my now very best friends—Tanja Šljivar and Petr Vašát. Their careful brilliance improved my thinking at every turn as I researched and wrote *Erosion*.

Much of this project's inception owes itself to the seminar "The Geological Turn," organized by Ruoji Trang and Christopher Walker at the American Comparative Literature Association in 2018, where I was able to workshop my initial ideas about Providence Canyon and the geological registers of *Gone with the Wind*. Comparatively, the book found its final footing in the seminar "Venice Is Leaking: Interventions in the Lagoon-City Continuum," led by Heather Contant, Ifor Duncan, and Sasha Gora at the Anthropocene Campus Venice in 2021, where I began to coalesce the concluding thoughts I had started in Budapest. I am grateful for the scholarly exchanges with Léa Perraudin and Oliver Völker that emerged from each of these opportunities and helped to shape this project. *Erosion* was also supported by funding from the Wilson Special Collections Library at the University of North Carolina at Chapel Hill. Matthew Turi and the entire staff at the Wilson Library deserve special thanks for their dedication to archival scholarship. Although I began research on a different (and ultimately unsuccessful) project at the American Antiquarian Society, my time as a Jay and Deborah Last fellow at the society is perhaps where this book began. Funding from the Kenneth M. England Fund for Southern American Literature at Georgia State University (GSU) helped carry this project across the finish line, and I am thankful to the GSU Foundation for their continued support of my scholarship.

The Department of English at GSU has provided additional support for this project. My department chairs, Lynée Gaillet and Audrey Goodman, have been instrumental in helping me to find the time to give this project the attention it deserves. My faculty mentor, Matthew Roudané, continues to offer the best advice I didn't even know I needed. My colleagues Shannon Finck, Melissa McLeod, Mark Noble, and Paul Schmidt all offered incredibly helpful suggestions to make this book better. In particular, Paul's insight on key parts of the textual analysis in chapter 4 fundamentally shaped my thinking. Melissa and Mark similarly offered helpful suggestions for revision, and Shannon read every word of this manuscript near its completion, improving it by leaps and bounds. Gracie Whitney worked as an undergraduate researcher on this project, and I cannot thank her enough for her attention to detail and her contagious energy.

Many other people have contributed to this work in concrete ways. Lindsey Eckert served as my initial reader for every chapter, and her astute (and often humorous) feedback kickstarted my motivation to push my research to new places. A visit to Providence Canyon with geologist Josh Poole sparked a lot of my initial curiosity about how erosion is narrated. A series of delightful email exchanges about Lynn Riggs and *Green Grow the Lilacs* with David Word and Chris Zahniser helped me to write out several of the initial ideas for chapter 2. I owe enormous thanks to Jo Handelsman and Kayla Cohen, who arranged for us to have a productive and enjoyable interdisciplinary conversation about the ongoing erosion crises across the country and world. I would also like to thank Marc Farinella for sharing his thoughts about the history of Shell Island outlined in chapter 5. Along the way, advice from friends and colleagues Eric Gary Anderson, Dana Arter, Eric Bennett, Michael Bibler, Martyn Bone, Vanessa Esquivido, Ben Frey, John Garrison, Allison Harris, Lisa Hinrichsen, Lauren LaFauci, Malinda Maynor Lowery, Stephanie Rountree, Kirstin Squint, Mandy Suhr-Sytsma, and Bryan Yazell also improved my research and writing process. Cutcha Risling Baldy and Brook Colley continue to be sisters and sounding boards, always knowing the questions to ask that allow me to see the world anew.

Despite being a writer, I cannot capture in words the joy it has been to work with my editor at Duke University Press, Benjamin Kossak. His care, precise eye, and enthusiasm for what this project could be are humbling. Elizabeth Ault and the two anonymous readers of the manuscript each offered suggestions that helped refine the project into something so much better than I could have achieved alone. I am also grateful to Duke University Press's production team for their dedication to transforming rather bland Word documents into *books*—what magic. I also owe thanks to my now longtime indexer, Victoria Baker. People often think I am joking when I say my favorite thing about something I've published is the index; I am not. I would also like to thank Kelly Vines for helping me compose the alt text for images (and for traveling with me to both Providence Canyon and the southernmost tip of Louisiana). I extend enormous gratitude to the Wilson Library and Monique Verdin, along with the Neighborhood Story Project, for their permission to share some of the images in chapters 5 and 3, respectively.

I remain grateful for the education I received from Auburn University, the University of Wisconsin–Milwaukee, and the University of California, Davis. In particular, Mark Jerng and Michael Wilson formed my fundamental approach to many of the questions that I ask in this project. I will owe them thanks forever. Additionally, Margaret Ferguson, Martha Macri, and Hilary Wyss continue to improve my thinking through their intelligence, good humor, and superior example. I always strive to make them proud.

The lasting support of family and friends sustains me and my work. In addition to those named above, Layla Amar, June Bishop, Emily Bloom, Randi Byrd, Billie Caison, Ashley Capel, Krishna Chapatwala, Adjoa Danso, Danielle Cadena Deulen, Sarah Dyne, Sally Hawkins, Kevin Hayden, Fay Heath, Ashley Holmes, Kimmy Kellett, Debbie Kozuch, Molly McGehee, Jessica Parker, Tracey Powell, Lynette Rimmer, Nicola Sharratt, Dustin Stewart, Kris Townsend, and Reanna Ursin have all contributed the necessary encouragement and community to see this endeavor through. Christopher Hollis and Mercedes Williams went far above and beyond the call of friendship (and housesitting) in ensuring I had a house to come home to following my fellowship in Hungary. In addition to my eternal gratitude for my parents, Becky and Ken Caison, I am thankful for the support of the entire Caison and Byrd families, all of whom have contributed in some way to how I know and understand the soil. As always, Scott Heath has had the words when I am at a loss, the energy when I am tired, and the faith when my confidence falters. He is my first and last audience. Even though I have written this book among so much joy, it has also been marked by a time of loss. I hope in some small way that my words here honor the memories of Richard Byrd, Ralph Byrd, John Lowe, Earl Heath, and Brian Horton—each of whom brought a surplus of happiness to the world.

Last, none of the words I have written here matters if I do not honor in some material way the work of those Indigenous peoples, communities, and nations who are fighting for a better way to live on their homelands and waters. Therefore, my royalties from *Erosion* will be donated equally to three Indigenous-run organizations and projects doing this work: the Rou Dalagurr Food Sovereignty Lab and Traditional Ecological Knowledges Institute in northern California, the Land Memory Bank and Seed Exchange in southern Louisiana, and the Waccamaw Siouan Tribe's Tribal Healing Garden in southeastern North Carolina. Indeed, academic works rarely make anyone a lot of money, but I hope that in some small way this book contributes to more than simply a conversation. Thank you for reading and for considering your own way you might do the same.

EROSION

Introduction: Erosion

THE IDEA FOR THIS BOOK resulted from an encounter with a map. When I first saw the US National Resources Board map of soil erosion assessment from 1934, while researching literature set in southern Georgia, I was struck by how the locations marked as high-concern areas aligned nearly perfectly with areas closely associated with specific conflicts over theft of Indigenous homelands (figure I.1). In several cases, these areas marked in black, indicating "Severe Sheet Erosion with Gullying," are familiar to those who study the bookended nineteenth-century policies of Removal and Allotment: the US Southeast of Georgia and Alabama, eastern Oklahoma, and farmlands of the upper Mountain West. Notably, however, the places marked in white, where erosion was deemed "unimportant," are all spaces that are often cited today for their acute erosion concerns: southern Louisiana, the Chesapeake Bay and

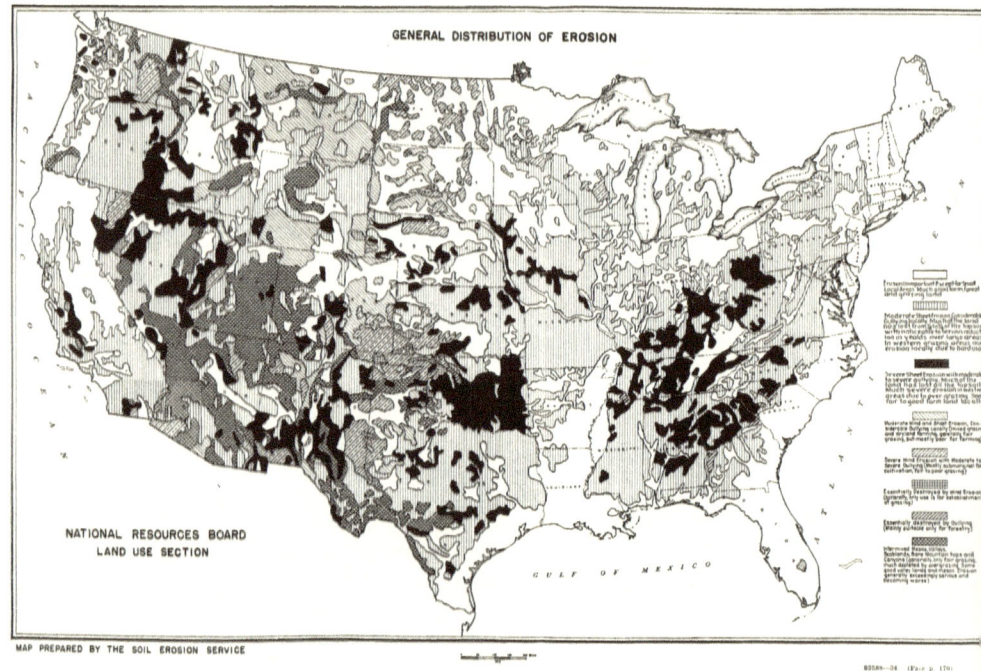

I.1 *General Distribution of Erosion*, 1934. From National Resources Board, *Report on National Planning*.

eastern North Carolina, portions of northern California, and low-lying lands in South Florida. Interestingly, another factor that many of these "erosion-unimportant" locations share is that they are spaces where Indigenous peoples ranging from the Hupa to the Seminole and from the Houma to the Lumbee had managed to hold on to their homelands either collectively or individually. Therefore, this 1934 map became instructive in at least two ways.[1]

The first and perhaps most obvious conclusion is that places where settler agriculture moved into a region saw extensive topsoil erosion soon afterward. For example, as federal and state governments carried out southeastern Indigenous Removal and as white settlers cleared forestland for cash monocrop agriculture, the soil washed away. Following the allotment of tribal lands in Indian Territory and the creation of Oklahoma statehood, a similar phenomenon occurred. As this book demonstrates, these connections have been both directly and tacitly acknowledged by soil scientists since at least the 1930s and in some cases even as far back as the first half of the nineteenth century. The second conclusion this map allows us to draw relates to how one defines *unimportant*. The marking of erosion as unimportant in certain locations does not necessarily indicate that erosion was not occurring in those places. Rather, it's possible to infer that the unimportance of erosion in these spaces had much more to do with the question of importance *to whom*. For example, we know that sites such as southern Louisiana and the Eastern Seaboard are experiencing profound battles with erosion and sea-level rise today. However, in 1934, if the people affected by these erosion challenges were largely Indigenous people in rural eastern North Carolina or Native people living at the end of the bayous, it seems likely that the federal government did not consider these concerns to be of immediate importance for their considerations of agricultural productivity in the 1930s. The perceived erosive threats, then, reflect not only values in place but also values in time. Together, these two conclusions indicate the idea that frames this book's larger concern: a careful analysis of when and how some people evoke ideas of erosion and what that means for a nation built on stolen Indigenous homelands.

This Natural Resources Board map from 1934 illustrates how fundamentally questions of erosion are tied to settler colonialism in both geological and narrative arenas. As a geological phenomenon, erosion is both mapped and lived—a geological process with a narrative footprint. It is limited neither to oceanfront towns (just ask US Great Lakes residents) nor to shores at all (think of the Dust Bowl). Importantly, it is not wholly a positive or negative process. Geologically speaking, some sites need to erode to produce sediment in others.[2] While the humanities have seen a recent investment in narratives of climate change and the grand scales of the Anthropocene, this project zooms in to examine a smaller scale: the individual particles of soil. Although a handful of popular scientific texts have engaged questions of erosion, these texts—written largely by geologists—tend to neglect the very real humanities questions of land attachment and narrative affect that undergird much of colonial history and contemporary nationalism.[3] In other words, erosion narratives are almost always about who claims the material earth to what ends and whether or not those claims are recognized.

Additionally, erosion narratives often merge with erasure narratives. Many non-Indigenous narratives of ecological crises simultaneously draw from and erase the Indigenous dimensions of the story or subsume Indigenous loss into white pathos. Conversely, many Indigenous-authored texts about erosion disrupt previous discussions of land loss that are tied to anxieties about the disappearance of whiteness and white supremacy. These works also challenge a romantic logic of incorporation. To demonstrate this difference in perspective is not to supplant one holistic myth with another but rather to engage

Stephen Nathan Haymes's concept of "small ecologies," which LeiLani Nishime and Kim Hester Williams describe as a concept that attempts to account "for multiple ways of being and multiple scales for conceptualizing the ecological."[4] This attention to small ecologies establishes this book's organizational logic of focusing on a specific region in each chapter as it faces an erosion event. Furthermore, this attention to the "small" joins the conversation of Elizabeth DeLoughrey in *Allegories of the Anthropocene*, where she asserts, "To parochialize the Anthropocene is to uncover its place-based allegories."[5] Via critical regional studies, *Erosion* attaches these analyses to the humanities' geological turn to illuminate moments when historical and agricultural contingencies quickly calcify into naturalized narratives of place. Thus, these erosion narratives become "sense of place" formulations within regional studies that produce troubling affinities with settler colonial practices. These practices include what Kevin Bruyneel calls the *"work of settler memory,"* whereby settler society alternatingly remembers and disavows "Indigenous political agency, colonialist dispossession, and violence toward Indigenous peoples" in order to maintain its own narratives across the multiple ideological registers of a colonialist nation-state.[6] In this way, I argue that erosion texts help us think critically about the *conserve* in *conservatism* and *conservation*—two words that might evoke different US political alignments—in deeply local contexts.

To pursue this argument across the following chapters, I center three guiding questions. The first question is, How much of contemporary humanities-based Anthropocene rhetoric is born of settler colonial anxiety about the physical disappearance of stolen Indigenous homelands? In considering this question, it's useful to turn toward work from scholars such as Kathryn Yusoff, who explains in her preface to *A Billion Black Anthropocenes or None*, "The Anthropocene might seem to offer a dystopic future that laments the end of the world, but imperialism and ongoing (settler) colonialisms have been ending worlds for as long as they have been in existence. The Anthropocene as a politically infused geology and scientific/popular discourse is just now noticing the extinction it has chosen to continually overlook in the making of its modernity and freedom."[7] This critique from Yusoff rightly disjoints a temporal fantasy in which all communities are experiencing the ongoing climate crisis in the same way at the same time. Or as put more simply by Eric Gary Anderson and Melanie Benson Taylor, "The Anthropocene unsettles but does not decolonize."[8] When these Anthropocene concerns are placed in the context of conversations about erosion, the stakes of material land claims come to the forefront all the more vividly. As Métis/Michif scholar Max Liboiron outlines, settlers need Land (capitalized as Liboiron renders it) available in reserve for settler futures, making erosion a clear threat to such futurity.[9] In other words, the apocalyptic Anthropocene may only be on academia's tongue because it is finally on settler colonialism's doorstep. Up until now, it was in the blank white spaces on the map where erosion remained unimportant.

These blank white spaces are related to part of Bruyneel's concept of the work of settler memory as it buttresses a posture of settler distress toward crises of the Anthropocene. These blank white spaces call up a concern similar to that of Métis scholar Zoe Todd, who calls for "an examination of art and the Anthropocene as variations of 'white public space' — space in which Indigenous ideas and experiences are appropriated, or obscured, by non-Indigenous practitioners."[10] Repeatedly, the discussions of erosion follow this model, where the specific problems facing Indigenous peoples and homelands become subsumed into a flattened global community that takes no stock of the material history of the settler state's anti-Indigenous attitudes and policies. An erosion crisis exposes the instability of a capitalist settler state dependent on the myth of the individual property owner. If the rendering of Indigenous homelands into property serves as the bedrock of the settler state's wealth and thus its very existence, then erosion becomes not only a material crisis but a slippery metaphor.[11] Put another way, the erosion of land is not the erosion of democracy. However, in the ubiquity of these phrases, an uncomfortable lurking tenet of environmentalism in the settler state comes forward whereby the preservation of lands (often rendered as the property of the individual citizen or the national "public") works alongside the preservation of the settler colonial nation as a political entity. It demonstrates what Mario Blaser and Marisol de la Cadena argue when they state that "now the colonizers are as threatened as the worlds they displaced and destroyed when they took over what they called *terra nullius*."[12] In this way, the work of settler memory leaves its footprints in the work of the Anthropocene discourse. As DeLoughrey articulates, "The lack of engagement with postcolonial and Indigenous perspectives has shaped Anthropocene discourse to claim the *novelty* of crisis rather than being attentive to the historical *continuity* of dispossession and disaster caused by empire."[13] For my part, this book seeks to keep the concepts of novelty and continuity in view in order to demonstrate how a settler anxiety of disappearance often subtends critical and cultural conversations about erosion, particularly within the US context.[14]

The next guiding question of this book is, How does literary regionalism shape popular discourse around the geological processes of the earth and the ways humans respond to these changes? As with the first question, much of the answer may be found in considerations of temporal scale and orientation as they relate to narratives of place. As Rob Nixon queries, "Beyond the optical façade of immediate peril, what demons lurk in the penumbral realms of the *longue durée*? What forces distract or discourage us from maintaining the double gaze across time? And what forces—imaginative, scientific, and activist—can help extend the temporal horizons of our gaze not just retrospectively but prospectively as well?"[15] This looking beyond asks us to think about deep pasts as well as possible emergent or even deep futures. As Nixon continues, "How, in other words, do we subject that shadow kingdom to a temporal optic that might allow us to see—and foresee—the lineaments of slow terror behind the façade of sudden spectacle?"[16] Narrative form is but one imaginative—and occasionally activist—force that helps jostle assumptions of homogeneous empty time as well as the limits of only understanding events by delineations of "crisis." When writers attempt to narrativize specific regions, they often use the constriction of space to move more capaciously across time, drawing from historical narratives in order to predict futures for the places they center. Moreover, over and above local color, literary regionalism also allows writers to show repeating historical patterns and the ways these prescribe various future options. These options have long been complicit in the settler futurity that Liboiron outlines, but texts that center place as conceit can also be used for imagining productive futures, particularly in the work toward Indigenous homelands reclamation. This possibility for place echoes Kristina Lyons's reflection on how the campesinos and Indigenous communities that partnered in her work on soil science oriented her toward the idea of "knowing where one is standing."[17] In this way, literary regionalism across multiple registers complicates and extends Nixon's structure of time, which remains relatively linear in his theorization, by showing cycles, repetitions, and recurrences across a spiralic time grounded in a specific location.[18] Together, these varying registers engage audiences to ask themselves if they know where they are standing.

Moreover, the question of literary regionalism (an approach that might appear to some as a bit retro for the present moment) animates a productive return to critical grounds that some might assume have been well covered. It allows critics to create a framework for analysis that challenges tacit assumptions around spaces of exceptionalism and fantasies of the local and its supposed healing power. While my previous work has focused more directly on southeastern Indigenous contexts, the methodologies of interrogating the residue of regional fantasies across the nation offer a broader application that undoes narratives of a "we" who experience crisis in temporal lockstep and, further, conceptions of an "out there" where those facing crisis are not "us" but rather a "them" quarantined safely at a remove, brought under by their own supposed folly of location choice. To this critical extent, this endeavor is in conversation with that of scholars such as Anderson and Taylor who work to understand the "challenges of regionalizing a global crisis" as well as with that of DeLoughrey, who argues that even though "global climate, a planetary phenomenon, is not reducible to local weather," nonetheless "scalar telescoping follows a long tradition in postcolonial studies in which universalizing narratives are troubled, contested, and provincialized."[19] A focus on literary regionalism, then, can challenge narratives of a universalizing "we" of the crisis while still holding space for Scott Romine's argument about how invented or "fake" regional spaces such as the South *become* "real . . . through the intervention of narrative."[20] As I demonstrate in this book, these narrative interventions work at the level of locale, especially when literary authors attempt to grapple with narratives of land and belonging in which the work of settler memory buttresses the nation through nostalgia for regional particularity across the mainstream political spectrum.[21] In other words, regions delineated and invented through narratives of settler colonialism, with their various exceptionalist paradigms, often serve to undergird fantasies of a coherent settler national identity and a universalizing "anthropo" identity within the crisis of the Anthropocene.

The final question I consider traverses more metaphysical territory even as it appeals to a literal sense of groundedness: To what extent does the material disappearance of the land following the removal of Indigenous peoples force a recognition of Indigenous knowledges that have long asserted a physical connection between the presence of the people and the presence of their homelands? The popular hashtag #LandBack has gained traction across media platforms ranging from Twitter (X) to the FX series by Sterlin Harjo, *Reservation Dogs* (2021–23), and from the Peacock comedy *Rutherford Falls* (2021–22) to TikTok. Indeed, the return of homelands to Indigenous control is central to the movement for Indigenous sovereignty on this continent. At the same time, when land disappears due to sea-level rise, sediment depletion from damming and river engineering projects, and harmful agricultural practices, what is left to return? This is not to say that control over waterways and shorelines is insignificant but rather to create a more exacting inventory of the myriad ways that settler colonialism has created damage that cannot be simply undone with a transfer of legal title. Moreover, many Indigenous nations and communities across the continent have long asserted the core relationship between the people and the earth given to them by their creator. While some might be tempted to read these knowledge traditions with a skepticism of spirituality born from Western Enlightenment science, a study of erosion tells another story. I argue that when, following Indigenous Removal, erosion and erosive practices have caused the material earth to disappear, it behooves us to pause and think about what Indigenous knowledges have been trying to tell the rest of the world about landed connections. In addition to offering a core lesson about spiritual and religious responsibilities, it is just as likely that Indigenous lessons about connections between lands and people are also about the human stewardship of the material soil in highly particular places. In the absence of these knowledges developed for thousands of years in a specific place with specific soil and water conditions, settler colonialism has removed the physical land in addition to removing Indigenous peoples.

This slippage between land as title and land as soil is one of the core tenets that undergirds this entire project. In many ways, it follows the conceptual recursivity laid out by Robert Nichols in *Theft Is Property!* As Nichols explains in his critique of colonialism and capitalism, "Dispossession can rightly be said to exhibit a 'recursive' structure because it produces what it presupposes."[22] He continues, "In a standard formulation one would assume that 'property' is logically, chronologically, and normatively prior to 'theft.' However, in this (colonial) context, theft is the mechanism and means by which property is generated: hence its recursivity."[23] In other words, the idea of property is *made* by the act of colonial theft whereby Indigenous peoples become merely "prior owners" of territories, imbricating them politically into a structure of perpetual acknowledgment of the property terms of colonial capitalist exploitation. In this structure any attachment to territory or land becomes one of title and valuation within a Western capitalist context. In this work I attempt to acknowledge this critically slippery territory while also not according it absolute prominence in the totality of an analysis of homelands' loss and return. I find the phenomenon of erosion one among perhaps many key ways to think both within and without this recursivity. Because material erosion exposes the fiction of land title as coterminous with land existence, it likewise allows some space for critics to loosen the concept of lands as colonial property to acknowledge land as Indigenous homelands.

Therefore, I frequently use the term *homelands* to signal a variety of connections and kinships that an Indigenous worldview might animate via what LeAnne Howe calls *tribalography*, where "Native stories, no matter what form they take (novel, poem, drama, memoir, film, history) seem to pull all the elements together of the storyteller's tribe, meaning the people, the land, multiple characters and all their manifestations and revelations, and connect these in past, present, and future milieu[s]."[24] In this formulation Howe articulates an Indigenous narrative vision similar to Nixon's call to think both retrospectively and prospectively. These lands of Howe's tribalography are not simply reducible to the fiction of paper land title. However, the land as soil is not immune to the manifold desecrations brought about by erosion via settler land mismanagement. Together, these considerations encourage grappling with the stakes of material erosion for those who claim land as property and those who claim it as home.

Together, these three core questions of the book bring into dialogue ongoing concerns about climate change, lost-cause pathos born of white supremacy, the use of narrative to inspire large-scale policy, and the role of Indigenous studies in the traditional disciplinary divides of the humanities. *Erosion* contributes to an understanding of the way that narrative construction and visual framing shape our understanding of planetary crises, including climate change. In many ways, narrative allows humans to experience the necessary affective investment required for action. Alternatively, as I illustrate in the following chapters, narrative can also limit our possible responses as well-worn assumptions of declension narratives involving settler colonialism and lost cause–ism map all too easily onto stories about the demise of the earth.[25] Because erosion events often generate a visual archive, I consider each literary text within its larger visual context, as many famous works that I examine—for instance, *The Grapes of Wrath* (1939) and *Gone with the Wind* (1936)—exist alongside iconic imagery in the American imagination. Therefore, each chapter places its central texts in dialogue with a visual archive. I make use of the formal features of this visual archive (e.g., sfumato, horizon line, background composition, aerial imagery) to elucidate a reading of the written works. Combined, the visual and literary archives offer a way to understand the legacy of settler colonialism for the soil that takes into account the affective bridges humans have constructed to make sense of both how the earth has been actively changed and how it has refused attempts at human-imposed stasis.

To pursue this analysis, *Erosion* employs two core methodological frameworks: critical Indigenous studies and critical regional studies. While the project speaks to larger theoretical conversations in environmental humanities and American studies, it does so in a way that always foregrounds the methods of Indigenous studies and regional studies together. This allows the book to address the long history of place, as demonstrated by the chapter organization, which zooms in on the literature of particular spaces as they have encountered erosive events over time. I pull together threads from the humanities and earth sciences to demonstrate that how we talk about the dirt matters for how we regard our role as a human species on the planet. At the sixth meeting of the International Soil Conservation Organization, geographer Piers Blaikie called for "an interdisciplinary and multi-level explanation of the problems of land degradation, allowing for a variety of different perceptions of what the problems are."²⁶ This project seeks to extend this interdisciplinary conversation into the realm of literary and cultural texts. Importantly, my previous scholarship queries the limitations of traditional periodization for organizing our understanding of American literatures, particularly as that approach engages colonial and Indigenous texts.²⁷ Therefore, this book is equally recursive in its temporal scope in order to more fully grapple with the question of what regionally specific literatures can add to discussions of large-scale planetary events.

Erosion is grounded within my background as a non-Native scholar trained in a Native American studies department. This fact influences my approach to erosion texts in at least three key ways. The first is that on the basis of work in Indigenous storytelling traditions, I view any given text as a continually informed, dynamic exchange between creator and receiver. Texts are indeed of their moment of telling and/or publication, but they also exist for receivers in their moment of reception. Neither tells the entire truth of the story.²⁸ Together, these two simultaneous truths build a spiralic time of engagement. As Greg Sarris explains of his work with Pomo storyteller and knowledge keeper Mabel McKay, "Remember that when you hear and tell my stories there is more to me and you and that *is* the story."²⁹ Thus, when I examine an image or a novel, for example, I place the object in its own present and in the ever-shifting present of various receivers across spirals of time. Such an approach paradoxically values and displaces both (new) historicism and affect theory to help us understand how texts make meaning recurrently and recursively over time.

This project's investment in questions of temporality aligns with work on what Mark Rifkin calls *settler time* and what Kyle Powys Whyte has theorized as *kinship time*. Rifkin's theories of settler time demonstrate how, in his words, "'natural' time appears as if it were a singular, neutral medium into which to transpose varied experiences of becoming, such that they all can be measured and related through reference to an underlying, 'real' continuity—a linear, integrated, universal unfolding."³⁰ The investment of settler societies in creating a singular narrative of unfolding time affects not only how settlers attempt to control and contain Indigenous epistemologies but also how settlers perceive their own existence within the anxiety-inducing crises of the Anthropocene. As Whyte offers, "When people relate to climate change through *linear time*—that is, as a ticking clock—they feel peril and seek ways to stop the worst impacts of climate change immediately."³¹ This peril generates anxiety. This book argues that this temporal anxiety merges with settler anxieties about whiteness, nation, and property. As an alternative, Whyte offers kinship time, which in an Indigenous worldview focuses on "establishing and repairing shared responsibilities" and views time as a web of connections, not unlike Howe's tribalography.[32] In order to think with both theories of settler time and kinship time, this project works with a methodology of layering texts across temporal spirals of meaning as well as considering how these appeals to temporality work within the selected texts. Each of these approaches is consistent with my focus on erosion, for as soil scientist Jo Handelsman explains, "Time influences soil, varying across an enormous scale, from seconds to billions of years."[33] A focus on erosion doesn't simply benefit from complicating received wisdom about the structure of time; it necessitates it.

This adjustment of temporality is not simply about thinking with a deep time of the Anthropocene that can, in some instances, shuttle too quickly from the settler-state contemporary moment to a universalizing deep global crisis defined by a geological epoch. This risks speeding past the significant time of Indigenous preeminence on the American continent, becoming what Eve Tuck and K. Wayne Yang call "settler moves toward innocence": "those strategies or positionings that attempt to relieve the settler of feelings of guilt or responsibility without giving up land or power or privilege, without having to change much at all."[34] Thus, when discussing the problems of erosion, I often turn to a "deeper" Indigenous history of the continent that seeks to avoid the erasure performed by the universalizing Anthropocenic "deep."

The second key value I bring from my background in Indigenous studies is that Indigenous studies is not merely an object-focused discipline where Western-based interpretations are applied to Indigenous texts and objects. Rather, Indigenous studies encompasses a set of epistemological and methodological practices for engaging all facets of the world around everyone, settlers such as me included. As complex epistemologies, Indigenous studies methods have the capacity to undertake and support readings of settler-produced texts and objects. Indigenous methodologies can offer rich interpretations and analyses of settler objects, cultures, and practices not because they were designed to or chose to but because the survival of Indigenous peoples and communities often was at stake. Put another way, John Steinbeck cannot offer the tools to understand the Cherokee diaspora of the Dust Bowl. But Cherokee epistemologies offer profound insight into Steinbeck's misguided vision of 1930s Oklahoma. My commitment to Indigenous-generated theories and methods of interpretation—combined with my allegiance to seeing time and literary periodization beyond a linear, settler-imposed Gregorian calendar—structures this book's anticolonial approach to understanding narratives of erosion.[35] I do not, however, seek to appropriate these epistemologies for my own claiming. Rather, at every turn I seek to engage these epistemologies on their own terms, in their own contexts, and in ways that respect the relationships and responsibilities they necessitate among their own past, present, and future milieus.

The third way that my education in critical Indigenous studies informs this study is in my consistent interrogation of my own relationship to the concepts of land and soil that I address. Knowing where one stands includes understanding the stakes of being in that place at that time. Throughout the book I attempt to keep my own complicity in settler colonialism in view without making this interrogation simply about myself. I do not wish to recenter my own identity through a constant recursive nod to my role as settler. Nonetheless, I also work carefully not to collapse my specific response to texts, ideas, and concepts as totalizing. Rather, as a settler I interrogate the mechanisms by which settler colonialism attempts to avoid calling attention to itself. I also foreground the fact that "deconstructing coloniality is *not* the same as decolonization."[36] While some of the authors and artists that I examine here are themselves engaged in the active work of decolonialization, my own work in and of itself cannot be considered work in decolonialization, as it emerges from my perspective as a settler scholar. Nonetheless, I am invested in pulling apart the mechanisms by which the settler colonial project attempts to steal material ground. I take the liberty and opportunity to call attention to the way supposedly good white liberal politics (a structure I myself am certainly implicated in, as perhaps are many of my readers) is also buttressed by investments in ongoing settler colonialism. In other words, this project about the soil follows Astrida Neimanis's work on the water to navigate the space between the narcissistically inflected refuge of the "I" and the troubling collapse of the "we" of the Anthropocene. As Neimanis articulates in *Bodies of Water*, "From a feminist perspective, [bodies of water] insist that I find a way of challenging the myth of the 'we' within a *nonetheless mutually implicating* ontology."[37] She continues, "As bodies of water, 'we' are all in this together . . . but 'we' are not all the same, nor are we all 'in this' in the same way."[38] Settler scholars have the responsibility to work against both a universalizing "we" and a mythically isolated self-referential "I." Instead, as Zoe Todd cautions, settler scholars must interrogate their own efforts at establishing the responsibilities Whyte calls for in his articulation of kinship time without appropriating Indigenous frameworks.

In addition to my training in Indigenous studies, much of my career has engaged with critical regional studies via southern studies. This methodological commitment means that I am consistently skeptical of any regional exceptionalism, as I alluded to earlier in the discussion of literary regionalism. While southern exceptionalism may be the first and foremost exceptionalism people imagine outside of American exceptionalism, critical regional studies challenges *all* settler-delineated regions as particular fictions generated by and for specific affective uses. Rightly, very few contemporary scholars hold on to Cold War–era narratives of American exceptionalism; however, I argue that taking a close look at considerations of Indigenous histories of place reveals how non-Native scholars can accidentally lapse back into exceptionalist narratives about the United States via southern exceptionalism and an incomplete consideration of Native American literature and history. This insidious creep of American exceptionalism appears today in casual discourse as social media users regularly evoke ideas of "this isn't my country" in response to Trumpism or suggest that "red state" politics thrives only in southeastern states. In other words, from an Indigenous studies perspective, this *has always been* America, and that America *is everywhere*.

Indeed, American exceptionalism still creeps into our assumptions about literature, and it is buttressed by southern exceptionalism—the belief that the US South represents a distinct region that serves as the exclusive repository of the nation's worst ideologies and simultaneously its downhome authenticity and hospitality. Most recognize that the idea that the United States is somehow a unique nation founded on the equality of persons is quickly undermined by the fact of the enslavement of Africans and African Americans. However, many times the problems of the nation were and are imaginatively quarantined to the southern states, as Jennifer Rae Greeson demonstrates in *Our South*. Both within the field of US literary studies and more broadly in US culture, the region of the US South is imagined as holding all of the nation's ills and moral failings, allowing the rest of the nation to float free of its material and psychological investments in enslavement economies and white supremacy. Although scholars may no longer structure their analyses around tenets of progressive fantasies based on American exceptionalism, the idea that the US South and its enslavement system signify an aberration from American ideals unwittingly reinforces exceptionalist narratives. In other words, unexamined southern exceptionalism continually bolsters American exceptionalism even for those who may claim to denounce the latter.

Regional exceptionalism does not, however, stop at the imagined borders of the US South. For example, Curt Gentry narrates an apocalyptic California landscape as it breaks off from the rest of the United States in his novel *The Last Days of the Late, Great State of California* (1968). Despite its clever geological premise that California floats away in a political-continental drift, Gentry attempts to critique the California fantasy while erasing the material history of settler colonial land management that disproportionately affected Indigenous communities in the state. In other words, the visual and literary erosion narratives of a California dream are not so much about the loss of the land as about the loss of settler colonial prosperity. Thus, even though the erosive mudslides of the California coast may seem a continent away from the vanishing Outer Banks in North Carolina, and though these regionalized communities may have vast differences in wealth and political ideology, the way each space engages questions of erosion and land loss reveals a consistent anxiety about white American disappearance. Put simply, no region is exceptional when it comes to the politics of erosion.

Popular work in geology and the natural sciences, while informative about the physical processes of erosion, frequently relies on clichéd colonial narratives of pathos that reveal more about the cultural assumptions of Euro-American civilization than about geological thought. A prime example of this is David Montgomery's *Dirt*, which, although informative about how Western cultures have engaged questions of the soil, offers little regard for the historical details or stakes of settler agriculture for Indigenous peoples. In more local contexts, Paul Sutter's *Let Us Now Praise Famous Gullies* and Mike Tidwell's *Bayou Farewell* also foreground white pathos and regional exceptionalism in attempting to discuss land degradation. Rather than challenging the agrarian fantasies of settler colonialism, these works merely extend them to evoke concern for a vanishing earth via assumptions of white pathos. This troubling trend continues across popular work in soil science, even from respected erosion specialists such as Orrin Pilkey and Rob Young (*The Rising Sea*); climate change writers, including Roy Scranton (*Learning to Die in the Anthropocene*); and legal scholars such as Jedediah Purdy (*This Land Is Our Land*). Scranton and Purdy in particular never pause to consider in their above-named works exactly whose land is "ours" and whose "civilization" is supposedly now tragically coming to an end. The conflation of anxieties over climate change and anxieties over the end of empire and white settler colonialism often leans on the powerful metaphor of erosion, which links land to nation.

Rather than spend too much time outlining the shortcomings of the natural and social sciences for addressing humanities concerns, I would prefer to offer that it's best to follow the lead of Tuck and Yang in the assertion that "decolonization is not a metaphor."[39] Tuck and Yang offer, "Decolonization, which we assert is a distinct project from other civil and human rights-based social justice projects, is far too often subsumed into the directives of these projects, with no regard for how decolonization wants something different than those forms of justice. Settler scholars swap out prior civil and human rights based terms, seemingly to signal both an awareness of the significance of Indigenous and decolonizing theorizations of schooling and educational research, and to include Indigenous peoples on the list of considerations—as an additional special (ethnic) group or class."[40] I am not arguing that we decolonize Western-based soil science. Present-day Western soil science—as a field with a genealogy emerging from figures such as Charles Lyell and Hugh Hammond Bennett (as critical as he was of Euro-American agricultural practices)—is a colonial science.[41] This is not to pass judgment on Bennett, who led an immediately necessary agricultural reform movement during the first half of the twentieth century, but rather to acknowledge the incongruity of decolonization with "reform."

Instead, my cue from Tuck and Yang comes in the second half of their often-quoted phrase. I assert that erosion is not a metaphor. Despite what I prove to be its near-ubiquitous appearance in metaphorical language for the demise of entities ranging from democracy to public health, I want to maintain that this book considers real land that is imperiled by settler colonial practices and real people who are likewise imperiled by that fact. It would be understandable and even tempting as a literature scholar to lean on the vehicle and tenor of erosion as others have done. At times this may seem like an exceptionally small needle that I am trying to thread—how to talk about literature, a medium that relies heavily on metaphor, without making a metaphor of material land abuse. To do so, I pay extra attention to the geological settings, events, and historical contexts of the texts I examine. Additionally, I work to show how metaphors within the text often have a residual relationship to material erosion events just outside the edges of their universe. In other words, I make every attempt to keep the real ground in view.

Despite this imperative, *Erosion* is not necessarily an entry into discussions of the new materialism. As Chadwick Allen describes in *Earthworks Rising*, those trained in Indigenous studies come to discussions of physical matter and objects from perspectives not overdetermined by trends in Western philosophical movements. He writes, "Sustaining the fiction of the 'newness' of the so-called new materialisms depends on the foregrounding of European and Anglo-American epistemologies and perspectives and on the continued erasure of relevant Indigenous epistemologies and perspectives."[42] Similarly, to some readers the framework I use for understanding questions of erosion may seem akin to concepts of hyperobjects that are nearly invisible in their quotidian ubiquity. While I do not deny anyone's connective entry point into understanding the material stakes of my analysis, *Erosion* emerges from Indigenous studies scholarship that has a long tradition of acknowledging the materiality and material agency of objects and forces that exist within various kinship relationships that are included in Howe's "past, present, and future milieu."

In my attempt to keep the real ground and the real people who live on it or who were removed from it in view, I have found myself turning time and again to the visual archive that often follows erosion narratives. This was not an area I intended to explore when I initially conceived of this project. My training in Indigenous studies has long encouraged me to listen to the archive rather than speaking my own plans into it. When I looked to the West Coast, Lewis Baltz's landscape photography came up time and again. Similarly, the Dust Bowl is dominated by Dorothea Lange's portraits of migrant farmworkers. If we look back in time, the visual register of the Mississippi Gulf Coast was largely shaped by John James Audubon. And the frequency with which contemporary Indigenous women writers such as Deborah Miranda and Monique Verdin include photographic archives in their written works also began to stand out. Even as I researched geological sites in Georgia such as Providence Canyon, I realized that the same awe and terror that inspired my own literary investigations had also captivated visual artists such as Elizabeth Webb. In this realization, I began to follow a line of inquiry informed by Audrey Goodman's work when she asserts, "Women writers have used photographs to revise and reimagine landscape, identity, and history in the U.S. West. Looking beyond the ideologies of wilderness, migration, and progress that have shaped settler and popular conceptions of the region ... women's photo-texts disrupt colonial geographies as they tell stories of contact with the land, of encounters between cultures, and of environmental destruction and change."[43] Although not all the visual artists I examine are women, I began to think about the interrelationship between image and text as it considers the physical earth within a history of colonialism. As Nicholas Mirzoeff explains in *White Sight*, settler colonial whiteness maintained itself both through the use of the visual as surveillance and through the creation of whiteness as naturalized background.[44] This framework appears across the archive to buttress the work of settler memory while some Indigenous creators challenge these sight lines to offer a different view.

In the following chapters, I examine visual and written archives from different creators that exist side by side while often presenting profoundly different perspectives on the erosion crises before them. Rather than try to draw one-to-one connections between these visual and written bodies of work, I instead began to use each form to allow me to read the other. In other words, how does narrative analysis shape an understanding of the visual archive? And how do the formal tools of the visual arts, such as photography, give literary scholars productive language for thinking about written narrative forms? Taking these together, I ask throughout this work how the crisis of erosion draws creators across media forms to try to "capture" disappearing lands.

When this archive of literary and visual texts is read through the logic of settler colonial systems, an uneasy tension emerges between white pathos and what we might think of as the "liberal" politics of earth preservation. As Liboiron says, "Colonial relations are reproduced through even well-intentioned environmental science and activism," and as K. Wayne Yang argues, "Sustainability is the present story of the settler colonial future."[45] To illustrate this, I turn to a book that shares a title with my own: Terry Tempest Williams's 2019 collection of essays, *Erosion*. In this work Williams subsumes Indigenous knowledges into her own anxiety about the death of American democracy amid Trumpism and the destruction of "public lands." Crucially, Williams rarely pauses to interrogate the fact that American democracy is, for Indigenous nations, a colonial government and that "public lands" have been stolen from Native nations and held in a collective fantasy of the white possessive *our*.[46] Leaning heavily on erosion as a metaphor, she writes, "This is a collection of essays written from 2012 to 2019, a seven-year cycle exploring the idea of erosion: the erosion of land; the erosion of home; the erosion of self; the erosion of the body and the body politic. It is a book of competing dreams and actions—the arc between protecting lands and exploiting them and, for many, not seeing them at all; between engaging politics and bypassing them; and the spectrum between succumbing to fear and choosing courage."[47] On its face, this is a powerful meditation. Although she occasionally references Indigenous thinkers, Williams subsumes the material erosion of Indigenous homelands into a pathos of vanishing American democracy, as the above description suggests. The conflation of these various losses strikes me as American exceptionalism by simply another, more poetic name.

This general conflation continues across the collection as Williams writes:

> People often ask how we can stay buoyant in the face of loss, and I don't know what to say except the world is so beautiful even as it burns, even as those we love leave us, even as we witness the ravaging of land and species, especially as we witness the brutal injustices and deep divisions in this country—as exemplified by the separation of families seeking asylum at the southern border; the blatant racism

exposed in Charlottesville; or the students' encounter with a Native elder drumming at the Indigenous March in Washington, D.C. The erosion of democracy and decency feels like a widening crack on the face of Liberty.[48]

There are number of liberally coded "good politics" gestures in this passage, and yet at its core, this sentiment is awash in a pathos of a lost cause; a quarantining of national problems to the geography of the US South (where notably all of her examples occurred); and a valorization of an America endowed with an unquestioning allegiance to democracy, decency, and liberty.[49] Williams's erosion metaphor belies the role of the settler state in upholding the very things that this passage suggests are consequences of its demise.

If it seems as if I am beating up on one of the good guys here, it is perhaps because I am. However, I find few things as dangerous as unquestioned "good" politics that rests on the same core foundation that buttresses settler colonialism. So while I agree in many ways with Williams's hope—"May justice rule on the side of deep time and a storied landscape that has been held in place by the roots of indigenous wisdom"—my affirmative nod is stalled by the assertion a few lines earlier:

> But so many of us in the United States of America are suffering from amnesia, both forgetting and losing track of who we are and what we stand for as a nation. The American landscape *becomes us*. If we see our natural heritage only as a quarry of building blocks instead of the bedrock of our integrity and cultural identities, we will indeed find ourselves made not only homeless but rootless by the impoverishment of our own imaginations. The cruelty of our ambitions and the ruthlessness of our priorities are undermining the ground beneath our feet. We are in a societal rockslide. Our democracy is collapsing.[50]

The problem here is in the parts of speech. Who is the "our" whose natural heritage she laments? And who is the "us" that the "American landscape becomes"? When authors fail to interrogate the specificity of their language when considering climate issues, they seek reconciliation ahead of accountability, replicating a settler move toward innocence that cannot approach the actual solidarity and relationality needed to address these global problems.

The conflation of an Indigenous wisdom with a settler becoming one with the land, as Williams forwards, reads less like a lesson in conservation than like yet another appropriation of Indigenous knowledge to salve the tacit, lurking anxiety that the white settler state should not exist. It illustrates what Yang, writing as La Paperson, describes as "settler environmentalism," which refers to "efforts to redeem the settler as ecological, often focusing on settler identity and belonging through tropes of Indigenous appropriations—returning to the wild man or demigoddess, claiming of one's natural or 'native' self and thus the land, again."[51] In other words, less than a lament for the land, Williams's work is a lament for the disappearance of America. Her work illustrates the problem of making erosion a metaphor of (tacitly white) American loss. She asserts, "As westerners, we take our public lands seriously. We know they are our birthright as American citizens."[52] This claim to a birthright reveals the disturbing tenet on which much of the contemporary American ecological conservation movement is built: a conservative understanding of American exceptionalism where public lands are a settler birthright. I imagine this point is uncomfortable for many readers, and some might even feel that this argument undermines what at its core is surely a good thing: preservation of land, away from privatized mining and logging companies. I don't make this critique of Williams's work lightly. However, as a fellow settler, I argue that there is an ethical imperative to interrogate the root of our core anxieties beyond the facade of the "good" and ask ourselves when we're fighting for the earth and when we're simply fighting for the settler state that has made our place on this earth more comfortable for the past four hundred or more years.

Amitav Ghosh counsels on this issue in his 2016 book *The Great Derangement* when he notes, "In accounts of the Anthropocene, and of the present climate crisis, capitalism is very often the pivot on which the narrative turns."[53] While he offers that he has "no quarrel" with this line of analysis, he continues, "This narrative often overlooks an aspect of the Anthropocene that is of equal importance: empire and imperialism. While capitalism and empire are certainly dual aspects of a single reality, the relationship between them is not, and has never been, a simple one."[54] What Ghosh endorses here is the same point that Williams alludes to in superficial ways but never fully interrogates: imperialism—and in the case of the United States, its cousin settler colonialism—is a key component in understanding how the Anthropocene and its effects have been created and absorbed by people around the globe in profoundly unequal ways based on their relationship to empire. Indigenous peoples have experienced far different effects of and entanglements with climate change as they have borne the brunt of various iterations of the global imperial project.

I want to be clear, however, that this project is not about simply romanticizing a mythical Indigenous past where earth preservation and conservation came "naturally" to Native peoples. It is counterproductive to fall into myths of the so-called "ecological Indian," which serve only to burnish settler sensibilities about contemporary conservation.[55] However, a clear distinction needs to be made between Indigenous knowledges built for thousands of years about a specific place and generic stereotypes about Native ecological warriors. Anderson and Taylor outline how a "renewed appreciation of the deep time . . . offered up by Indigenous memory" is "distinct from non-Native notions of Indians as paragons and icons of ecological wisdom."[56] Conservation and land-responsive agriculture did not just come naturally or mystically to Indigenous peoples, then or now. These knowledges were and continue to be built through careful observation as well as attempts *and* failures at survival in varying climatic conditions. As Handelsman argues, these practices are learned, refined sciences developed by dedicated practitioners focused on longevity in place.[57] This longevity in place can allow a misreading of traditional ecological knowledges as *static* ecological knowledges, when they are anything but. James Scott highlights this problem in *Seeing Like a State*, demonstrating how "'traditional peoples' will embrace techniques that solve vital problems."[58] As Scott explains, these knowledges are developed in highly specified, local spaces that make them "unassimilable for purposes of [settler] statecraft."[59] In other words, the romantic ecological Native depends on the projection of flattening stasis onto Indigenous epistemologies that are ever working against the work of settler memory.

However, in the effort not to romanticize the ecological practices of Indigenous societies, some settler scientists and popular narratives of science can lean too far toward various ecocide models that rely on "the philosophical argument that humans are inherently destructive to the environment."[60] These ecocide models have been applied to Indigenous societies to explain why various cities and towns were "abandoned" before colonial invasion. While in some cases these models may have some basis, they have also been overapplied and overextended in a way that gives cover to an anti-Indigenous relativism that justifies continued colonial erasure, dismissal, and damage to Indigenous homelands and peoples.

One place where an ecocide model has been overextended without supporting physical data is the large Indigenous "Mound City" of Cahokia, whose collapse has been popularly attributed to erosion. In 1993 two researchers forwarded the "wood-overuse hypothesis" to explain the supposed abandonment and collapse of the site. As Caitlin Rankin and her coauthors explain, "The wood-overuse hypothesis suggests that tree clearance in the uplands surrounding Cahokia led upstream erosion, causing increasingly frequent and unpredictable floods of the local creek drainages in the floodplain where Cahokia Mounds was constructed)."[61] However, this hypothesis, which depended on deforestation resulting in large-scale erosion, was only ever intended to be just that: a hypothesis. In fact, as Rankin and her cowriters explain, "[Neal] Lopinot and [William] Woods ... made it clear in their chapter on the wood-overuse hypothesis that there were insufficient data to move their narrative from hypothesis to a probable cause for collapse at Cahokia."[62] Nonetheless, "despite the lack of data to support this hypothesis, the ecocide narrative has been maintained in the literature as a potential contributor to Cahokia's abandonment ... Since the publication of the wood-overuse hypothesis, no attempts have been made to evaluate if erosion in the uplands and/or increased flooding in the floodplain did indeed occur during Cahokia's occupation."[63] In 2021 Rankin and her colleagues attempted to test the hypothesis through geoarchaeological methods at the site. Rather than evidence of Indigenous wood overuse, they found evidence of stable soil conditions throughout Cahokia's inhabitation. In fact, on the basis of the core and soil samples from their research, destabilizing soil events did not appear at the site until the opening of industrialized coal mines near the area in the 1800s. Thus, for several decades the myth persisted that Indigenous peoples at Cahokia contributed to their own demise through deforestation, and subsequently erosion, when, in reality, it's more likely erosion appeared at Cahokia with settler extractive practices.

Rankin and her fellow researchers offer some speculation about the persistence of this narrative via the work of Jane Mt. Pleasant, who "argued that archaeologists tend to underestimate and/or ignore conservation strategies employed by North American pre-Columbian people in agricultural and arboricultural activities)."[64] They continue, "Perhaps, in attempt to push away from the pristine myth of the pre-Columbian landscape, we have ignored the capabilities of these people as purposeful conservationists of their landscape and resources."[65] I do not disagree that many Western-based academics have done important work to shed dehumanizing myths of romanticized Indigenous peoples and societies. However, I also think narratives that fault Indigenous peoples for land mismanagement that results in erosion events find an audience with a mainstream settler culture that simply wants evidence that justifies, or at the very least relativizes, colonialism. While land mismanagement narratives are not the same as colonial relativisms that seek to show how "all peoples" have been colonized, as Tuck and Yang explain, they do run in a similar vein in an ecological context, allowing settler conservationists to imagine themselves as better land stewards than Indigenous peoples and justify their own investment in stolen property. Thus, while I eschew clichéd, generic eco-Native discourse in this work, I also reject the settler fantasy that Native societies were *just as* implicated in ecocide erosion events as the settler state has been since its inception.[66]

Importantly, the concept of erosion in the English language begins with denotations related to the body and temporality. The oldest English usage of *erode*, moving from the Latin via French, appears in 1612 and denotes "of the action of acids, canker, ulceration, etc.: To destroy by slow consumption."[67] This eating away at the body by slow processes eventually evolved to a geological usage, which is often credited to the so-called father of geology, Charles Lyell, who uses the words *eroded* and *eroding* in his *Principles of Geology* in 1830.[68] Despite Lyell frequently being credited with the geological usage of *erosion*, Oliver Goldsmith uses the noun form *erosion* in 1774, in volume 1 of his *History of the Earth and Animated Nature*.[69] While in later chapters I consider the colonial contours of the verb *erode*, whose usage coincided with peak colonial dates, including the 1830 signing of the Indian Removal Act, I want to pause in this introductory moment to consider how the denotations of the verb *erode* and the noun *erosion* force us to adjust our temporal frame. Destruction by slow consumption challenges the immediacy we often give to our understanding of what constitutes an event. The event of colonial conquest is not discrete. In fact, as Liboiron argues via Tiffany King, it "is not an event, not an intent."[70] For as King puts it, "conquest is not even a structure, but a milieu or active set of relations that we can push on, move around in, and redo from moment to moment."[71] It works through slow consumption, and likewise, as Nixon posits, the resistance to colonial conquest also can occur slowly, even imperceptibly to those who don't know what to look for.[72]

As I explain above, in order to understand erosion, we have to let go of a linear calendar time that moves in an assumed predictable fashion and instead think with a timescale that adjusts with outside factors and challenges received wisdom about how "history" supposedly moves across the continent. Therefore, the following chapters are deliberately arranged to undo typical colonial teleology that pushes narratives across the continent from east to west. Instead, the analysis here begins in present-day California and moves east, a method that Indigenous studies scholar Jack Forbes often encouraged his students at the University of California, Davis, to think with in order to jostle a lifetime of the American education system's narrative structure of "history." Together, these chapters engage regional particularity to complicate studies in literary regionalism and to query how lost-cause ideologies and romantic ideas of "vanishing Natives" may very well subtend present-day, well-meaning ecological calls to action. In each chapter I set up a kind of balanced tension where, in some cases, I begin with an analysis of how Western-based science narratives have addressed or cataloged an erosion event and then show how an Indigenous artist or thinker has complicated that understanding. In this way, the chapters do not simply begin with Indigenous voices and then move on to colonial texts, a re-creation of the American education system's own obsessive love of making all Indigenous history a prequel to "America." However, in some cases, I set up an Indigenous thinker's understanding of place to then show how it can offer a productive methodology for interrogating later settler creators' interpretation of events. In these cases, I always return to the Indigenous works later in the chapter to mirror my contention that settler colonial states may very well be a discrete moment preceded by thousands of years of Indigenous control *and* ultimately outlasted by Indigenous landed sovereignty on this continent.

The first chapter, "Landslides and Horizons of the West," takes on literary and visual narratives of exceptionalism amid the retrospective and prospective imagination of California. This chapter, then, works from the call I quote earlier from Nixon that we examine what "forces—imaginative, scientific, and activist—can help extend the temporal *horizons* of our gaze not just retrospectively but prospectively as well."[73] In examining this space of the horizon, I take up Lewis Baltz's landscape photography and Octavia Butler's novels of the speculative near future alongside Indigenous studies sources about traditional Indigenous land management in the region. This constellation of Baltz's and Butler's contemporaneous careers allows me to demonstrate how even progressively minded discourses of erosion and climate change often rely on the erasure of Indigenous homelands claims and on California exceptionalism. Baltz and Butler both theorize the limits of futurity in their work, and they each cast a critical eye toward fantasies of the West Coast landscape that promises fulfillment of so-called American promises. In this analysis I examine one of the most significant formal features of Baltz's photography, which is his use of the horizon line to complicate the viewer's relationship to the American landscape tradition. The horizon line becomes the place one cannot see past regardless of one's own possible well-meaning intentions. For many in conservation and land-preservation movements, this horizon line is the return of homelands to Indigenous peoples. Even for Butler, who is known for her prescient work, particularly the *Parable* texts I analyze in chapter 1, there is a particular horizon line around what role Indigenous peoples play in California's past, present, and future. By taking two artists invested in the prospective imagination encouraged by Nixon and then reconsidering their creations with an understanding of Bruyneel's concept of the work of settler memory, I demonstrate the profound difficulty of building a decolonized future when settler memory attempts to cut off such horizons of possibility. In grappling with this tension between prospective imagination and settler memory, I close this chapter with an analysis of Deborah Miranda's *Bad Indians* to demonstrate how an Indigenous understanding of California complicates received wisdom around the state's paradoxical status as a space beleaguered by climate issues and simultaneously invested in its own progressive fantasies of exceptionalism.

Chapter 2, "Surfaces and Allotments of the Heartland," places the history of surface erosion alongside the Indigenous history of Allotment in the region. I begin by placing Dorothea Lange's most famous image of the Dust Bowl, *Migrant Mother*, firmly within the historical conditions of its subject, Florence Thompson. Lange did not name Thompson and did not realize or learn that she was a Cherokee woman originally from just outside Tahlequah, Oklahoma. I analyze how this nonrecognition of the Indigenous woman by Lange perpetuates a surface reading of the Dust Bowl that reifies abject whiteness. I then take this nonrecognition as a cue to examine how previous considerations of the Dust Bowl have not engaged the Indigenous contours of the story, specifically in how it relates to Indigenous women and their traditional roles of agricultural management in southeastern Native nations. I begin with a brief overview of how specific policies related to Indigenous oppression and settler expansion, such as the General Allotment Act of 1887, created the environmental conditions of the Dust Bowl. I then trace how these concerns resonate across the popular texts regarding the event. I offer a reading of Steinbeck's *The Grapes of Wrath* (1939) that accounts for his conceiving of the Joads as from eastern Oklahoma (Sallisaw, Oklahoma, which notably is in the Cherokee Nation) rather than the western part of the state, which was most affected by the Dust Bowl. I outline how Steinbeck continuously places moments of anxiety regarding disappearing whiteness alongside mentions of Indigeneity throughout the novel. I go on to explicate Lange's (with husband and intellectual partner Paul Taylor) *An American Exodus* (1939) for how it also traffics and attempts to refuse an anxiety of loss around whiteness. However, because of what Lange and Taylor could not see—Indigeneity—their work manages to cement erosion as part of a white pathos. I then turn to a sustained analysis of two of Cherokee playwright Lynn Riggs's works—*Green Grow the Lilacs* (1930), which served as the basis for the musical *Oklahoma!*, and the lesser-known play *The Cream in the Well* (1940)—for how each depicts the often unrecognized Cherokee characters as they struggle make sense of their lives and their relationship to the land on the eve of Allotment.

The third chapter, "Disappearing Grounds and Backgrounds of the Gulf," uses the visual and narrative concepts of background and foreground to analyze texts that address the constant and ever-present land loss in the Gulf wetlands. In particular, I look to how this loss affects Indigenous communities in this area, building from ideas forwarded by Louisiana Indigenous writers/artists Rain Prud'homme-Cranford and Monique Verdin. I then examine erosion in John James Audubon's Louisiana writings and paintings of the early nineteenth century alongside George Washington Cable's representation of this period in the short story "Belles Demoiselles Plantation" in his collection *Old Creole Days* (1879). Each author relies on anxieties of loss as they attempt to examine what whiteness means in a plantation context for southern Louisiana. By examining the visual background to Audubon's work (focusing more on the land in the background than the birds of the foreground) and the narrative background of Cable's short story, I build the direct and present links to questions of erosion, Indigenous control of homelands, and the pathos of loss in the plantation economy. The majority of the chapter focuses on Verdin's 2019 photographic essay collection *Return to Yakni Chitto* to illustrate how an Indigenous understanding of the land and waterways along the bayous complicates popular narratives of the region that often focus on lost causes rather than long-held strategies for continued survival. Through an analysis of the way that Verdin uses the background of landscape within her portraits of foregrounded Indigenous community members, I build from my examination of Baltz's landscape work in chapter 1 and my interrogation of Lange's non-recognition of Indigeneity in her portrait work in chapter 2. Whereas Baltz's landscapes rarely include people, and Lange's portraits rarely include wide shots of the landscape, Verdin generates work that shows the profound connections between Indigenous homelands and Indigenous peoples. These connections challenge myths of Indigenous disappearance and ask audiences to re-see the foregone conclusions of land loss in the Gulf.

Chapter 4, "Gullies and Removals of the Plantation South," considers the stakes of erosion within a critical context that draws from both histories of southeastern Indigenous Removal and analyses of what Katherine McKittrick theorizes as "plantation futures."[74] In doing so, this chapter examines Margaret Mitchell's 1936 novel *Gone with the Wind* and the popular archive surrounding the geological site of Providence Canyon in the southwestern part of the state, including the understudied African American poet Thomas

Jefferson Flanagan's depiction of the eroded site. This analysis builds from the Removal history subtending the Allotment narratives of chapter 2 and the white pathos explicated via an analysis of the background and foreground in chapter 3. Mitchell's novel repeatedly refers to both erosion and Removal, often in deeply interrelated ways. Most tellingly, the title of the book, which was written during the Dust Bowl, is itself an erosion metaphor, signaling how prominent concerns over agricultural instability were during the period. At the same time, Providence Canyon—a massive eroded gully created following forced Muscogee Creek Removal from the area in the nineteenth century—was being touted by tourism boosters in the state as a new kind of "natural wonder" despite the well-known fact that it was the result of the unsustainable agricultural practices of the enslavement and sharecropping systems. I place each of these objects within the contexts of the geological and geoscience theories laid out by Charles Lyell, who traveled and recorded gullying in the US South during the 1830s on the heels of Indigenous Removal, and Hugh Hammond Bennett, who revolutionized soil science during the 1930s era of Mitchell's and Flanagan's writing and the "rebranding" of Providence Canyon. I close the chapter with a consideration of visual artist Elizabeth Webb's 2019 installation *For the Mud Holds What History Refuses (Providence in Four Parts)* and Flanagan's short collection *The Canyons at Providence (The Lay of the Clay Minstrel)* (1940). While both Webb's and Flanagan's works attempt to make sense of the canyon and the legacy of plantation enslavement for the erosion of the material earth, neither project seems entirely able to account for the legacy of Indigenous Removal and homelands theft that subtends the history of the US South. This limit point demonstrates the legacy of entanglements that make reading the work of settler memory alongside the projection of naturalized plantation futures difficult but nonetheless all the more necessary. Taken together, the texts examined here demonstrate the insidious ways the plantation attempts to replicate its form through the repackaging of its thefts and horrors as natural.

The final chapter, "Littoral Cells and Literal Sells of the Atlantic," focuses on the Eastern Seaboard, examining the clash of real estate development with erosion along barrier islands designed by nature to shift and adjust within the littoral cell system of the Atlantic currents. This tension between change and stasis informs nearly every erosion controversy along the Eastern Seaboard. Beginning with an analysis of Monacan poet Karenne Wood's work in *Markings on Earth* (2001), I take on subsequent stories of coastal erosion as bound up with the legacy of early English colonization along the Atlantic coast. As Wood's poetry illustrates, both Indigenous homelands and communities have long had to be dynamic to survive in this space. However, settler colonialism has consistently attempted to fix each in a static place or identification. To illustrate this point and how it works within an anxiety over white settler disappearance, I offer a short analysis of Ellyn Bache's 1993 short story "Shell Island" alongside the history of the island as a resort for Black beachgoers in the early twentieth century. I go on to examine the effort to save Cape Hatteras Lighthouse during the same period that Bache's story appeared, showing how the lighthouse works as a literal beacon of settler colonialism. The debates over how to best protect the enormous structure relied heavily on the visual rhetoric exemplified in well-known North Carolina photographer Hugh Morton's aerial photography. From there, I analyze Earl Swift's *Chesapeake Requiem* (2018), in which he recounts living on a quickly eroding Tangier Island where the population almost unanimously supported Donald Trump in 2016 and is skeptical of climate change. Instead, residents refer to their central problem as "erosion," a telling elision of the scale of perceptible local change versus unfathomable global change over time. I place the island in its historical context as recorded by John Smith in the sixteenth century, as a kind of tempestuous "Limbo," and I query what perpetual limbo defines the global condition of living under a regime of climate change and erosive sea-level rise. I close the chapter with a return to Wood's poetry, reminding the reader that in this particular landscape, Indigenous people know that change has long been the way to survive.

I conclude the book by considering the global scale of erosion. Rather than view this closing as a return to the universalizing global, I intend for it to be understood as a thought exercise in the multiplicity of the local, where erosion is never exceptional even as it may always be as particular as it is particulate. This multiplicity of the local mirrors Blaser and de la Cadena's discussion of the pluriverse, or the Zapatista theory of the one world where many worlds fit. I focus on international debates, including the history of apartheid and erosion in South Africa; the rising tides of Venice, Italy; Edward Said's gestures to erosive agricultural projects by the Israeli state in Palestine in *After the Last Sky* (1986); erosion in Hungary (the location of the first international conference on soil science, in 1909); and the soil conservation practices of pre-invasion Peru. These international examples demonstrate how anxieties about erosion engage narratives of the Global North and the Global South in locally specific ways. This concluding material also allows me to return to Baltz's European work alongside Bennett's clear 1930s summations regarding colonialism and erosion. This closing international scope reframes the relevance of literary regionalism when facing planetary problems and illustrates how visual and literary narratives can both limit and enliven possibilities for the future.

Across these chapters, I find that although I am examining the long narrative life of erosion in the Americas, this topic has consistent resonance with and relevance to many present-day concerns in areas across the globe. The way people understood erosion in the past matters for how they will handle the challenges of the future. These obstacles face humans in regions of the world as different as Alaska and Indonesia, and yet routinely the brunt of these uncertain futures crashes against the lives of Indigenous peoples fighting to hold on to homelands that are eroding away. While many readers might imagine that the traditional hard sciences will offer the singular path for humans' continued existence in the era of extreme climate change, all of my previous intellectual work leads me to believe that what we learn from narrative study might ultimately be the thing that carries humans forward on this planet and allows us to retain our humanity.

1

Landslides and Horizons of the West

IN THE SUMMER OF 2019, the *Los Angeles Times* released a feature article on coastal erosion in California. The title emblazoned across an ever-shifting loop of ocean tide on the paper's website reads, "The California Coast Is Disappearing under the Rising Sea. Our Choices Are Grim."[1] The article, by Rosanna Xia, highlights a number of coastal locations across the state, from Gleason Beach an hour north of San Francisco to Imperial Beach just above the US-Mexico border.[2] There are also interactive maps showing what seawater inundation will mean for various cities, as the user can slide a dot along a scale to flood Fisherman's Wharf and send storm surge up to the edge of San Francisco's financial district or completely level Imperial Beach to the south. The article also links to "The Ocean Game," where users have eight turns to save an imaginary beachfront town from sea-level rise. The introductory page of the game notes, "The sea is rising higher and faster—California could see a jump of more than 9 feet by the end of the century. Flooding and erosion threaten homes. Beaches could vanish. But everyone insists: This is a game that can be won."[3] Placing the user in the position of an elected official for a small beach town, the game offers three initial choices and proceeds on a choose-your-own-adventure model with a looping background of ocean sounds and animated cartoon citizens that resemble pegs from the board game Life. Notably, each choice is marked by a number of dollar signs, and in addition to eight turns, the user has ten available dollar signs to save the town. The first choice, "Build a rock wall to protect the homes," costs two dollar signs; "Add sand to widen the beach" costs three; and the cheapest option, with one dollar sign, is "Hire a consultant for more information." The most conservative and bureaucratic choice, "Hire a consultant," results in this advice: "Your consultant says: 'Adding sand will make beach-goers happy, but beware: The sand washes away over time. Seawalls can protect homes, but they aren't good for beaches. If you can afford it, *consider buying out beachfront homeowners to get them out of harm's way.*'"[4] The most expensive initial option, buying sand, yields good results for the summer, but the game tells users that the sand ultimately washes away after a year, leaving them in the same predicament as before, except now the townspeople are anxious. As the game explains, "Your town is thriving and the economy is good. But people around town are fretting about the rising ocean," and now the middle-of-the-road option, building a wall, gives the user this descriptive logic for their choice: "I'm worried too. It's time to build a rock wall." Thus, at every turn of the game, the user is confronted by two intertwined inevitables: erosion and anxiety.

Regardless of how one plays "The Ocean Game," the only way to save the town is what the accompanying article refers to as "managed retreat," a buzz phrase in the coastal erosion debate across the state. In the logic of the game, sand replenishment and rock wall construction both ultimately fail, and the game becomes less about geological processes, infrastructure, and innovation and more about real estate negotiations with beachfront homeowners—all of whom have different buyout prices and attachments to their homes, ranging from the emotional to the economic. As the water comes closer and closer, homeowners drop their selling prices. Depending on how aggressively one bids, variables emerge that affect whether or not the managed retreat can occur before the ocean washes away both public and private property.

The game, then, needs the anxiety of the ever-encroaching surf to make managed retreat (the only "winnable" path here) viable for the town's budget. This anxiety is akin to the ticking-clock temporality that Kyle Powys Whyte outlines as part of a settler-organized linear time, as I discuss in the introduction.[5] Moreover, the game's central logic is that of winning and losing, where the ocean always wins but the user can also win if they concede that the ocean will be the ultimate winner. In this simulation the path to winning against erosion is finding a way to convince the townspeople to retreat, through a combination of capital and the imminent threat of disappearance, both of which seem to suggest some form of paradoxical acceptance: accept that you cannot win against coastal erosion, or accept this amount of money to forget what you cannot accept. The overall framework of winning and losing calls up the idea of lost causes, which for many are more associated with neo-Confederate ideologies of the US South. However, the game's location in California perhaps allows it a more optimistic tone, assuring the user that battling erosion "is a game that can be won." Either way, whether the user focuses on winning or losing, the game's structure pits humans against the earth, competing with the planet for their own survival, a dubious strategy that seems akin to the bad decisions that produced the problem in the first place.

Taking this tension between erosion and anxiety as its cue, this opening chapter places all of the book's cards on the table at once in what might strike some readers as an unlikely beginning point for what is ultimately a book about settler colonialism in the literature of the United States: California. Many people imagine the history of colonialism as a wave of progress or terror that washes from east to west, beginning somewhere around present-day New England and ending on the Pacific Coast. This is obviously a myth. Settler colonialism arrived in what we currently call Florida, invaded north from what we know now as Mexico, skirted east from Russia, then south from the present-day Alaskan coast, and appeared in multiple places in multiple times in a way that leaves no neat narrative wave intact. In other words, California is as good a place to start as any. Moreover, the ongoing process of settler colonialism is not history, and it doesn't count for all—or even most—of it. Every place on the continent has a history that precedes, runs parallel to, entangles with, and will ultimately outlast the history of settler colonialism. California is not a frontier to be won or lost. As multiple Indigenous scholars from the region have shown, it is a place with a long history all its own, something that even progressively minded US citizens often seem to forget.[6]

This strategic forgetting frequently creeps into conversations about erosion in the state. For example, the *Los Angeles Times* article opens with this setup: "The California Coast grew and prospered during a remarkable moment in history when the sea was at its tamest. But the mighty Pacific, unbeknownst to all, was nearing its final years of a calm but unusual cycle that had lulled dreaming settlers into a false sense of endless summer."[7] The growth, prosperity, dreaming settlers, and endless summer mentioned here are straight out of a warped postcard, one that hides the violence of the mission system, the genocide of the Gold Rush, the land (and human) abuse of industrial agriculture, and any number of stark realities of the state.[8] Interestingly, the article attempts to place erosion within its own siloed vision of the state's contemporary problems, noting, "Wildfire and drought dominate the climate change debates in the state. Yet this less-talked-about reality has California cornered. The coastline is eroding with every tide and storm, but everything built before *we knew better*—Pacific Coast Highway, multimillion-dollar homes in Malibu, the rail line to San Diego—is fixed in place with nowhere to go."[9] Two curious things happen in this passage: first, it positions wildfire and drought as separate issues from erosion when they are not; and, second, it imagines some type of collective "we" that is presumably related to the possessive adjective *our* from the article's title. When one takes California's Indigenous history and contemporary Indigenous knowledges into consideration, this sentence is simply wrong in the inverse: California's myriad ecological issues are connected through the singular event of settler colonialism rather than forming many unrelated, separate events. Residents of California are not a collective "we" now facing these crises in the simultaneous possessive togetherness of "our." As Kathryn Yusoff explains, "The passage to universalism in ecological or planetary terms without a redress of how that humanity was borne as an exclusionary construct, coterminous with the enslavement of some humans and the genocide of others, remains a questionable traverse."[10] In other words, even if not everyone has previously recognized the erosion issues before them, Indigenous peoples in California have long *known better*.

This chapter, then, takes up a particular type of regional exceptionalism. The popular representation of California often renders it as the national space most associated with progressivism. However, when one focuses on erosion, the same declension narratives that inform representations of other regions also appear in narratives about the loss of California as an ever-disappearing prosperous future. In fact, the *Los Angeles Times* article on erosion sets up almost exactly this premise: "Elsewhere, Miami has been drowning, Louisiana shrinking, North Carolina's beaches disappearing like a time lapse with no ending. While other regions grappled with destructive waves and rising seas, the West Coast for decades was spared by a rare confluence of favorable winds and cooler water."[11] All of these locations—notably in the US Southeast, despite the fact rising tides will also affect New York and Massachusetts—are rendered as lost spaces somewhere else, places where settler Californians could seemingly export their anxieties as distant declensions rather than core tenets of their own progressive fantasies. In order to elucidate the connections among anxieties of loss, landscape, and settler fantasies, I examine Lewis Baltz's landscape photography and Octavia Butler's novels of the speculative near future alongside Indigenous studies sources about Indigenous ecological knowledge and land management in the region. This examination demonstrates how even progressively minded discourses of land loss and climate change often rely on the erasure of Indigenous claims to homelands and California exceptionalism.

These fantasies of California exceptionalism appear even in critically informed projects such as Rebecca Solnit's *Infinite City*, where she writes, "The place that is San Francisco has both a literal geography as the tip of a peninsula that juts upwards like a hitchhiking thumb and another, cultural, geography as the most left part of the left coast, the un-American place where America invents itself."[12] Echoing new media scholar Lev Manovich's "California ideology," which is described by Mark Tribe as "a deadly cocktail of naive optimism, techno-utopianism, and new-libertarian politics popularized by *Wired* magazine," Solnit posits that the "left coast" is simultaneously the heart and extremity of America.[13] This paradox of un-American American invention space suggests that California becomes a kind of separate geography, cordoned off both for the benefit and to the detriment of the rest of the United States. Richard Barbrook and Andy Cameron build on Manovich's eastern European skepticism of this idea, writing their own critique of what they call "Californ*ian* ideology," placing the burden on the residents rather than the topography itself by asserting:[14]

> The widespread appeal of these West Coast ideologues isn't simply the result of their infectious optimism. Above all, they are passionate advocates of what appears to be an impeccably libertarian form of politics—they want information technologies to be used to create a new "Jeffersonian democracy" where all individuals will be able to express themselves freely within cyberspace.... However, by championing this seemingly admirable ideal, these techno-boosters are at the same time reproducing some of the most atavistic features of American society, especially those derived from the bitter legacy of slavery. Their utopian vision of California depends upon a wilful blindness towards the other—much less positive—features of life on the West

Coast: racism, poverty and environmental degradation . . . Ironically, in the not too distant past, the intellectuals and artists of the Bay Area were passionately concerned about these issues.[15]

The leveling of this critique at California exceptionalism necessarily exposes how one region's narratives of exceptionalism are equally dangerous to another's, as each allows American exceptionalism to float free in myths of an ideal Jeffersonian democracy. Although Barbrook and Cameron almost ironically do not name it themselves, Thomas Jefferson's dream for America was always about romantic and removed Indigenous peoples and the landscape fantasies that support those concepts in policy, narrative, and practice.[16]

Almost two centuries after Jefferson, Baltz and other members of the New Topographics photography movement pushed against a "California ideology" rooted in sentimentality and nostalgia, by showing the physical topography of California as it was in the late 1970s: a banal capitalist development full of pavement and exhausted soil—the refuse of the California fantasy, from the Gold Rush to the postwar era. While such work is important, a lost-cause narrative of an earlier mythical West and California often undergirds these geologically stark photographs, and the New Topographics movement's disruption of the Western fantasy never fully comes to terms with the region as Indigenous homelands. In this way, their work—and Baltz's in particular—forms a compelling conjunction with Butler's novels in the Earthseed series, *Parable of the Sower* (1993) and *Parable of the Talents* (1998), where the protagonist draws from an alternate understanding of the California landscape, one based in Indigenous knowledges, without fully investigating the romance and pathos of Indigenous loss that has produced the contemporary ecological collapse depicted in the novels. Baltz's and Butler's work shares a central concern of representing a postindustrial California, and their personal generational and geographic affiliation (Baltz was born in 1945 in Newport Beach and Butler in 1947 in Pasadena) makes them a provocative, if not unexpected, pairing for analysis. Placing them together within a critical Indigenous studies framework of California allows the contours of California exceptionalism to come forward, where even the most progressive non-Native critiques of and from the state still struggle to understand that the disappearing land of California was, is, and will always be Indigenous homelands. Taking this into consideration, I close this chapter with a short meditation on Esselen/Chumash author Deborah Miranda's *Bad Indians* and attempt to think outside of the *Los Angeles Times*'s collective anxiety of loss over an eroding California future, focusing instead on a return of homelands to Native people in the state.

The history of colonialism along the West Coast is entirely bound up with the eroding coastline, and not in the way that the *Los Angeles Times* frames the issue as about a namesake *Pacific* Ocean lulling settlers into California dreams. As Cornelia Dean explains in her work on erosion, "The European settlement of California, especially the twentieth-century development boom that took off in the 'Southland' around Los Angeles, is a saga of increasingly elaborate efforts to bring nature to heel."[17] As she outlines, "For thousands of years, California's beaches were nourished in sporadic but enormous gulps when fire denuded the landscape, intense rain flooded it, and mudslides carried tons of sediment away. Hundreds of tons of dirt regularly thundered down the mountainsides, through the canyons and onto the beaches, where waves and currents sorted it and carried it along the coast. These floods of mud supplied the beaches with 75 to 95 percent of their sediment and created sandy beaches along much of the coast."[18] This process was largely interrupted by the damming and corralling of waterways following the large 1938 flood, a plot point in John Steinbeck's *The Grapes of Wrath* (1939), which I consider in the next chapter. These projects attempted to control rivers and create a water supply for the rapidly expanding populace. As a consequence, they "trap sand destined for the beach ... and reduce the kind of high-velocity water flow that picks up dirt and carries it along," while concrete-lined waterways such as the Los Angeles River have no available sediment to even acquire.[19] Dean goes on to offer that "by some estimates, more than one hundred million cubic yards of sand is locked away behind California's dams inland," and she quotes University of California, Santa Cruz, coastal scientist Gary Griggs, who notes that from 1979 to 1999, "this coastal region has been starved of enough sand to build a beach 300 feet wide, three feet deep and sixty miles long."[20] Compounding this beach starvation by dams, "sand and gravel operators have mined hundreds of thousands of tons of sand from California beaches, dunes, and river beds" for construction projects, while "people eager to live on the coast cut off its last desperate sand source, the seacliffs that erode and collapse when a beach does not have enough sand," which they have "armored ... with rip-rap, revetments, or seawalls."[21] All of these projects—damming, sand mining, and building of artificial hardscape sea barriers—constitute a material theft of land not only from the beach itself but from the Indigenous peoples who claim these homelands. Thus, settler colonialism has not merely taken land title—a rough bureaucratic metaphor for the space it represents—but stolen the very earth.

Interestingly, California is ground zero for the legal concept known as *sand rights*, as developed by lawyer and legal scholar Katherine Stone. Based on the jus publicum, or principle of public interest in land use, the concept of sand rights essentially states that as places of public interest, beaches have a right to their sand. When this sand is trapped by dams and jetties or affected by private construction, the littoral zone is deprived of its rightful sediment needed to maintain its equilibrium. This idea could extend to a scenario where if a cliff would need to erode to nourish a beach, and one's private construction has prevented that cliff from doing so, then the private property owner has stolen that sediment from the beach and, by extension, the public. This concept of sand rights also applies to dredging and sand-mining projects, which take sediment from one area for use in another or in construction projects.[22] Stone's work on sand rights has been used in cases in Florida and Maryland, and as Dean explains, even the Army Corps of Engineers acknowledges the principles set down by the theory in its calculations for how to fund beach nourishment projects. It has not, however, been heard in a case by the US Supreme Court. One wonders how soon, on a warming planet, the immediate concerns of coastal erosion will force the high court to determine what damages and recourse apply to disappearing earth along the shore or to sediment held hostage by those upstream.[23]

There is, however, an ironic twist to the recognition of erosion along California's beaches that reveals the tension between ideas of untouched nature and settlerism. The increased industrial, commercial, and infrastructure projects in California that were causing sand depletion also hid it as the state dredged offshore waterways and then deposited the resulting sand along various beaches as "landfill." Thus, in some ways the problem masked itself as a solution. However, in some places the deepening of offshore waterways actually increases the intensity of wave activity along the shore, causing the beach to erode even faster. In this way, the process of erosion in California is a prime example of how settler colonialism completely dismantled a carefully managed environment where California Native people had learned to work with the contingencies of their landscape over thousands of years. While popular settler sentiments toward California rendered it a wild, untouched Eden, it was anything but. As M. Kat Anderson explains, "The indigenous people of California had a profound influence on many diverse landscapes," and "without an Indian presence, the early European explorers would have encountered a land with less spectacular wildflower displays, fewer large trees, and fewer parklike forests, and the grassland habitats that today are disappearing in such places as Mount Tamalpais and Salt Point State Park might not have existed in the first place."[24] Moreover, she offers a compelling example of this nonrecognition of Indigenous land management: "John Muir, celebrated environmentalist and founder of the Sierra Club, was an early proponent of the view that the California landscape was a pristine wilderness before the arrival of Europeans."[25] And she points out specifically how in "staring in awe at the lengthy vistas of his beloved Yosemite Valley, or the extensive beds of golden and purple flowers in the Central Valley, Muir was eyeing what were really the fertile seed, bulb, and greens gathering grounds of the Miwok and Yokuts Indians, kept open and productive by centuries of carefully planned indigenous burning, harvesting, and seed scattering."[26] Put another way, California wasn't beautiful and desirable because humans were absent; it had been *made* beautiful and desirable through thousands of years of trial-and-error land management practices developed by active human communities.

This is not to romanticize Indigenous peoples as magically in tune with the earth—just the opposite. Indigenous peoples had millennia to develop scientific systems suited to their environment, which settlers first misidentified as untouched wilderness and later romanticized and mythologized as resulting from the unearned talent of purely spiritually ecological Natives who did not have science or strategy guiding their deliberate choices to survive in place. As Yurok/Hupa/Oneida scholar Kaitlin Reed argues via Hupa scholar Brittani Orona in their work on California, "This environmental ethos ['We are part of the land, and the land is us']—far from a New Age metaphor—guides Indigenous ecological management practices. We mean it literally. When a group of people live in the same place for thousands of years, our ancestors become the soil, they become the Earth."[27] Here Reed forwards a direct rebuttal to those who might attempt to undermine the connection to Indigenous homelands through their suggestions that it is either mystical affectation or merely a metaphorical political or rhetorical talking point. This ongoing misapprehension of dynamic land management practices as romantic fantasies about wilderness and Indigeneity informs my reading of the California texts in this chapter.

The two central artists I discuss in this chapter, Lewis Baltz and Octavia Butler, both query the relationship between a destructive humanity and the earth it inhabits, each seemingly suggesting that humanity is bound to be destructive. As Butler does in her speculative fiction, Baltz offers a deep investment in the question of Western landscape as a product of materialist historical conditions, specifically the colonial obsession with "exploration." In addition to being a prolific photographer, Baltz was a prolific writer, composing various reflections on his own work as well as critical commentary on other contemporaneous photographers' exhibits. In these written reflections on his earlier 1970s landscape photography, Baltz begins to think through his own place in the genealogy of artists who photographed the American West, and these reflections carry through to his work in the late 1980s. This body of prose work frames my analysis in this chapter of two of Baltz's most important California exhibitions—*The New Industrial Parks near Irvine, California* (1975) and *Candlestick Point* (1989)—which bookend his work in the genre of landscape photography. Even though he created series about places as far away as Maryland, the majority of his landscape period centered on the American West, and much of his accompanying writing draws from his background of having grown up in Newport Beach in the post–World War II era. Even from the materials that are not explicitly about California, there emerges a clear point of view in how Baltz framed questions of land use, photography, and the romantic (or, in his view, tragic) history of the American West.

Situated in the middle of this landscape photography period, his most concrete descriptions of his landscape work perhaps come in his notes for the Utah-based *Park City* (1981) series, where he writes, "When I first saw Park City it was the land, rather than the structures, that held my interest. Other than in films and photographs of natural catastrophes—or in disaster movies—I have never seen a landscape so bleak, blasted and chaotic. Much of the land was churned and pocked, littered with fragments of wood, twisted wire, rusted metal and broken glass, and supported only the sparsest vegetation.... Beneath the world of boom-time optimism, the condition of the land suggested a sense of ultimate futility."[28] When considering the question of land abuse, Baltz posits an ultimate futility that mirrors anxieties of a lost cause; the land has been too abused to ever return to something he imagines it was before. Moreover, he underscores his work as participating in a larger framework of how humans make meaning from their fragments of information and experiences under late capitalism. For instance, he continues, "Our responses to landscape are conditioned by our experience as inhabitants of an industrial society, insofar as since that society inculcates in us the values that we project upon nature."[29] Baltz seemingly does not trifle with abstracted emotional mystery but instead strives to lay bare the historical conditions that have created the scenes he photographs. As he explains of Park City, "It was no mystery why the land looked as it did. Many of the construction sites were on land previously used as a dump for mining wastes. The age of the dump sites did little to mitigate the appearance of destruction; with time, wind and water further pulverized the soil. The land held a visible record of its use, and the residue had its own order that was neither pictorial, nor non-pictorial."[30] In this way, Baltz attempts to set up a central "formal" tenet of his work, which is to simply point the camera at what is before it and trust that it exposes the historical layers of land abuse from extractive practices. While such a process may seem to be hyperrealistic, the photographs resulting from Baltz's nontechnique come across as almost alien. And Baltz himself notes, "If this sounds as though I'm suggesting that I decided to simply level the camera and point it at the horizon, that is substantially correct. . . . The formal qualities of these images echo an order implicit in the scene in front of the camera, an esthetic similar to the NASA photographs of the surface of Mars."[31] In this way, Baltz's work can be understood as placed between two romanticized pillars of exploration: the Euro-American invasion of the West and the "final frontier" of colonized space, a theme I return to later in the chapter when engaging Butler's *Parable* novels.[32]

Baltz clearly sees the legacy of American landscape photography in the West as defined by this anxiety over an industrialized future. Along with John Muir's writings, Ansel Adams's photography is central to the visual imaginary of California in the minds of many audiences. Though not named as such, the famous photographer of Western landscape seems at the front of Baltz's mind as he writes, "Only after a certain historical development took place could the land be thought of as a suitable object for esthetic contemplation. Among these developments, the most prominent was the evolution away from the land as the primary site of economic production. As the agrarian model was eroded by its industrial successor, it became possible to take the distance necessary to contemplate the land as 'landscape.'"[33] Baltz may not be entirely wrong in this assessment, but his analysis leaves out one central feature: this perspective is true only if one does not recognize the Western agrarian model as also a centrally destructive force of settler colonialism. Indeed, an agrarian model may have been both literally and figuratively "eroded" by industrialism, and this anxiety certainly subtends Adams's earlier landscape works on California. However, this completely misses the point that the "nature" Adams records is not a pristine wilderness but the traces of an ecology carefully cultivated and managed by California Native peoples for thousands of years.[34]

This mythology continues to pervade critical understandings of Adams and the Western landscape even in the early twenty-first century, particularly as Americans confront an ever-shifting relationship to climate change.[35] As Kenneth Brower writes for the *Atlantic*, "Evidence mounts that most of the world may indeed be beyond salvation. The notion that the American West is the last chance for a New Jerusalem was not just an Ansel Adams conceit; it was a conviction that pervaded America for generations ('Go west, young man'). The conviction was still fresh for Adams's generation, born just a decade after the official closing of the frontier. And the fact is that most of America's unspoiled landscape and big scenery still reside out west."[36] This romance of the frontier and the assumption of what constitutes an "unspoiled landscape" have their basis in a settler colonial myth of empty virgin land that began even before the advent of photography. As Wolfgang Scheppe explains, "American landscape painting rests on a myth of virginal non-historicity, which positions the genesis of the nation proximate to an act of creation."[37] In other words, it isn't incorrect for Baltz or other critics to diagnose the central tenet of Western landscape art as a psychological retreat from the conditions Western settler ideology created, but it also isn't the entire story.

A closer look at two of Baltz's series, *New Industrial Parks* and *Candlestick Point*, reveals something about a particular late twentieth-century moment of erosion anxiety for California. As Scheppe argues, "[Baltz's] singular oeuvre . . . put an end to the image regime of the landscape."[38] Despite this upending of a regime that often "excis[ed] the apparitions of force that were part of this land acquisition," Baltz's work is still inflected by his own kind of horizon line.[39] In other words, just as interesting as what Baltz sees in his own images is what he cannot see. While he unquestioningly challenges the mythology of California, from the Hollywood fantasies of Los Angeles to the techno-optimism of San Francisco, via his hyperrealist photographs, he also misidentifies the culprit for the waste he sees before him as an alienating industrialism rather than an invasive settler colonialism of which he himself is a continuous part. In this way, his work has embedded within it its own kind of "white sight" as articulated by Nicholas Mirzoeff.[40] Indeed, no single person can see beyond their own horizon line of understanding; however, Baltz's photographs set up a compelling metaphor for the horizon of possibility as settlers try to make sense of their own limitations of understanding the Indigenous homelands of their own constructed American landscapes.

Despite the limits of his horizon of understanding, Baltz remains one of the best recorders of what settler land abuse has done to the region. *New Industrial Parks* is a monographic series of fifty-one photographs, eight of which were included in the 1975 exhibition *The New Topographics* at the George Eastman House in Rochester, New York. The complete series depicts building facades and landscape features of a new office park outside of Irvine in Orange County. The images contain numerous linear elements, and despite the fact that the architecture is new, the entire series, together, feels vaguely decrepit. The gray tones and flat surfaces evoke a sterility and alienation from what one might imagine as landscape, and almost nothing truly "monumental" is contained in any of the photographs. As a result, this exhibition not only ushered in a renewed understanding of American landscape photography but also queried a particular moment of eminent California Reaganism writ large, where every place is seemingly no place interesting.

Paradoxically, it's hard to find much *land* in Baltz's early landscapes. However, even in his photographs from *New Industrial Parks* that seem to offer virtually no representation of the land itself, there remains everywhere the evidence of erosive practices. The pile of dirt hemmed in by an asphalt parking lot in *South Corner, Riccar America Company, 3184 Pullman, Costa Mesa, 1974* seemingly renders the earth "unnatural" to itself, cordoned off as if the soil will ruin the rest of the surrounding landscape. Likewise, the single spindly, nearly bare tree—planted, ironically, in an attempt at landscaping—next to a fresh parking lot in *South Wall, Unoccupied Industrial Structure, 16812 Milliken, Irvine* appears as an uncomfortable anomaly among the neat, gray-toned straight lines of the building and rain downspout. The two saplings in *South Wall, PlastX, 350 Lear, Costa Mesa, 1974* stand behind another asphalt foreground, where one can see the remainder of water running in a particular gullying pattern over the newly paved parking lot (figure 1.1). The company's name, PlastX, casts a shadow against the smooth-walled building behind the two trees, making every element appear almost ghostly. In nearly every photo in the series, the lack of healthy vegetation, layers of sod, expanses of asphalt, and struggling saplings lining newly painted parking lots all signal an imminent ecological disaster where water overuse followed by rain runoff will cause even more rapid land degradation. The concrete buildings and asphalt parking lots alongside the rain downspouts appear as a visual metaphor for the dams and rivers across the state that have had their courses manipulated and beds lined in an effort to control flooding and that as a result lack any sediment to nourish beaches. These aren't cliffs tumbling into the Pacific Ocean or dramatic dust storm clouds sweeping over the state's Central Valley. Instead, the industrial parks are banal and quotidian in their relentless destruction.

Every image in the series offers unnaturally flat and linear surfaces, causing any living object in the frame to appear disturbingly uncanny in its refusal to get in line. Nearly every component in the photographs refuses a sense of depth. As Chris Balaschak points out, "The most heated criticism of Baltz's work has pointed directly at the photographer's tendency to only engage the surface."[41] He goes on to quote Allan Sekula's critique to this effect: "[Baltz] suggests that the oxymoronic label, 'industrial park' is somehow natural, an unquestionable aspect of landscape that is both a source of pop disdain and mortuarial elegance of design. Baltz's photographs of enigmatic factories fail to tell us anything about them."[42] I concur with Balaschak that such a

1.1   Lewis Baltz, *South Wall, PlastX, 350 Lear, Costa Mesa*, 1974. From the exhibition *The New Industrial Parks near Irvine, California*, Leo Castelli Gallery, New York, 1975.

critique seems to miss the mark in that Baltz does anything but naturalize the industrial/park contradiction. I also wish to extend Balaschak's rebuttal of Sekula to argue that of course Baltz's work stays on the surface and fails to tell us anything about this sprawling suburban landscape. This surface of nothing strikes me as the very point Baltz's images make: the settler colonial, late-capitalist industrial park has no depth. Like haphazardly thrown sod and misplaced saplings surrounded by concrete, it has no effective roots. Even though Baltz may have been unable to link his own image-making to the settler colonial history of California, it does not mean that his images do not record it. He levels his camera at a horizon entirely predicated on a shallow history of place, and the resulting image offers just that. If the elegance of these images is "mortuarial," then it's likely because such land use practices do in fact represent death.[43] Instead of offering a romantic, wilderness California like earlier generations of landscape photography—a genre that certainly covers over the death and exploitation of both Indigenous peoples and homelands in the region—Baltz offers an assessment of the land that makes that destruction plain. His photographs stay on the surface because that is where settler colonialism lives and where erosion occurs.

Baltz himself acknowledges that the United States is a surface nation, in a 1975 review essay on photographer Robert Adams's work *The New West*. He writes, "Posing an ecological threat that is only now being taken into account, the sprawl-city is as hostile to genuine urbanism as it is the land it recklessly consumes. For as much as the landscape is deformed by the encroaching suburban juggernaut, traditional urban (*i.e. civilized*) values are equally under attack; American civil culture, never so very deep, disappears if spread too thin."[44] Again Baltz hits a mark that it's not clear he was aiming for. American civil culture *is* shallow, especially when compared with Indigenous civil cultures. His emphasis on "civilized" urban values gestures toward a particular kind of mainstream American life as a height of correct thinking that is under attack by suburban middle-class sprawl. The anxiety here, as Baltz makes plain, is not really about land damage at all. Rather, it's about a disappearance of something that is always on the verge of not existing because of its inherent unsustainability: the United States of America. The anxiety of erosion that the *Los Angeles Times* explores in 2019 is only a continuation of what Baltz attempts to diagnose in the 1970s. Neither project, however, seems able to take a step back and realize that the core issue underlying all of this is an American project built on stolen Indigenous homelands.[45]

Of course, it's up for debate whether or not the ecological underpinnings and social critiques that I offer as embedded in Baltz's topographic work are indeed part of the composition. There is some disagreement on whether the New Topographics of the 1970s intended their work to engage in social critique or whether they had less political, more artistic aims (as if those two are ever entirely separate). They did, however, alienate audiences, and this alienation could likely be born of a latent anxiety in the settler colonial project.[46] At the time, Baltz explains, American landscape photography seemed to have passed its moment as a useful medium, and "redeeming the American landscape for photography seemed a worthy task and more difficult than one might imagine."[47] He reasons that this is because "the present generation of Americans, for the most part, never experienced the landscape without experiencing its counterface, industrialism, and it seemed likely that without this dialectic the entire notion of 'pre' landscape might seem so estranged from ordinary reality as to appear escapist and sentimental."[48] This explanation reveals just how limited the historical aesthetic perspective of the earlier landscape photography is, as it also failed to offer some "pre" landscape or to make the viewer aware that there is no "pre" landscape. It only pretended to offer non-Native people some view of the West, as if there were an untouched there out there. The entire history of landscape photography is bound up with an artistic practice of not seeing Indigenous homelands as part of the work of settler memory. Baltz fails to realize in these remarks that rather than breaking with or updating this prior work, he merely extends it by imagining that settler colonialism ever had a "before" on this continent that did not depend on its romantic rendering of Indigenous homelands in its own psyche of imagined wilderness *or* industrial park. He is correct that the New Topographics resisted the romantic, escapist, and sentimental view of land, but he somehow still clings to the idea that there was an agrarian ideal eclipsed by industrialism. In truth, that very agrarian ideal also comprises a romantic, escapist, and sentimental view of the land.[49]

Despite his avowed resistance to romantic narratives, even Baltz's landscape work ends on an unexpected and confusing point of optimism that almost turns over on itself from today's present view. His 1989 *Candlestick Point* represents the end of an era begun largely with *New Industrial Parks*. The photographs of the late 1980s depict the landfill point (i.e., the land was built out of dredged earth from the San Francisco Bay and fill dirt from other construction and infrastructure projects) on the San Francisco Bay that would become the San Francisco Giants' and 49ers' Candlestick Park parking lot. The jutting piece of land "reclaimed" from San Francisco Bay appears as a site of refuse and desolation in the photographs, which are not individually titled in the series itself. Construction detritus, old tires, patches of tangled and dying brush all stand in the foreground of the initial black-and-white landscape images, while Hunters Point shipyard looms across the water in the background. As in *New Industrial Parks*, everything appears mortuarial, and the earth seemingly sustains almost no life. Nearly every tree is bare, the grass seems dead, and even the trees that appear to have leaves are difficult to distinguish from piles of debris across the panoramic landscape, which is dominated by horizon lines. In his last interview, Baltz mentions that shortly before this work at Candlestick Point, during his shooting of San Quentin Point just to the north, he had finished reading the nuclear apocalypse treatise *The Fate of the Earth* (1982). He remarks, "I was in dark mood myself. I'd just read Jonathan Schell's book . . . and [was] seeing the landscape as a dead man might see it. [San Quentin Point] had both wet and dry areas—not yet $1.5 million homes."[50] This wry nod to San Quentin's real estate development from what he photographed as an imagined postapocalyptic wasteland foreshadows the very problem that challenges the player of the *Los Angeles Times* game, who must buy out people's waterfront homes to avoid a coming apocalypse of erosive rising tides. Notably, according to land-loss-projection maps, most of these expensive homes on San Quentin Point will be underwater even with a sea-level rise of thirty-three inches, at the low end of predictions. The high-end projections, at sixty-six inches, drown nearly everything on the peninsula north of San Francisco, including the point's notorious prison.[51] Of course, Baltz may not have known these predictions at the time. Nonetheless, his anxiety about nuclear apocalypse in the series at San Quentin sets up his final work farther south on the bay, at Candlestick Point. He may not have been aware of what the coming apocalypse was, exactly, imagining the punctuated event of nuclear annihilation rather than the slow, quotidian creep of land loss from an ever-warming planet's rising tide combined with rivers that carry no sediment. What remains, however, is a clear anxiety of disappearance. The anxiety of a shallow American civil culture, the lurking threat of a nuclear catastrophe during Ronald Reagan's presidency, and the slowly emerging recognition of rising seas all inform what Baltz's work depicts and what he diagnoses in earlier landscape photography—that it covers over the tacit recognition that the United States' relationship to the land it occupies is never more than an affair with the surface.

Notably, something curious happens toward the end of *Candlestick Point*. Baltz suddenly switches into color. His final interviewer, Jeff Rian, calls it "the Wizard of Oz moment," to which Baltz explains, "I was winding down from something. This was the piece in which I took everything I knew and used it and didn't want to repeat it again, ever."[52] When Rian notes the unusual and uncomfortable scrub landscape of *Candlestick Point*, Baltz again loops back to his earlier work in Orange County, claiming, "When I think about the landscape in Orange County, when and where I grew up, and looking at it today, they took paradise and created New Jersey in a generation. William Burroughs said that America was always evil, waiting there lurking even when the first settlers walked across the Bering Strait."[53] And here we have the point where Baltz cannot see past his own constructed horizon line. Those weren't settlers that walked across the Bering Strait (a now largely disproven theory to explain the entire peopling of the American continent), and while the nation-state or idea we call America might be evil, the idea of it lurking behind Indigenous presence creates a tone of inevitability that makes settler colonialism a

1.2   Lewis Baltz, *Unnamed Image 23*, 1989. From *Candlestick Point*, n.p.

foregone conclusion.[54] The inevitable so-called evil of America did not happen as unimpeded destiny. It was made in the choice to believe in the myths of an empty wilderness, or settler agrarianism, or to believe that land destruction happened first in a place along the East Coast, like New Jersey, and had only lately arrived in California.

Still, despite this pessimistic tone, the switch to color at the close of *Candlestick Point* injects an uncharacteristic optimism at the end of the cycle of Baltz's work (see figures 1.2 and 1.3). In the 2011 edition produced by fine-art publisher Steidl, the change happens all at once and without warning as one flips the page. Suddenly, the hills in the background become subtly green with life. The sky is a pleasant pale shade of blue, and one can even make out tiny yellow flowers among the weedy green grasses. The desolation of the

1.3 Lewis Baltz, *Unnamed Image 24*, 1989. From *Candlestick Point*, n.p. In Baltz's original, *Unnamed Image 24* begins the color portion of the series, which cannot be replicated here. Subtle color makes the hills, tree, and wildflowers in the foreground come alive, marking a shift from the preceding black-and-white images. See Pier 24 Photography, "Lewis Baltz 'Candlestick Point' Flip Through" on Vimeo to experience this shift to color firsthand. https://vimeo.com/215743628.

location remains, but somehow even the orange rust of oil drums suggests a movement beyond death. Baltz, who spent his career forcing viewers to see beyond a California ideology, somehow manages to take a disturbingly resilient bow, leading one to wonder, Is this optimistic point of view his or the viewer's? The land and water suddenly seem to come alive as if to suggest in clichéd terms that "life finds a way" even in what Baltz has presented as a near-apocalyptic landscape. Suddenly, there seems to be land in this landscape. And after all, this land has been in color all this time. Just as Adams did a generation earlier, Baltz showed viewers a black-and-white place that existed more in the imagination than in reality.

In this *Wizard of Oz* moment, I argue that what viewers experience more than the switch into color is perhaps the futility of wizards behind curtains, who will not ever be able to help recover a "no place like home" for settler colonialism in the Americas. In other words, in Baltz's landscape work viewers see how both the invention of a California ideology and the rejection of it still struggle to come to terms with settler colonialism and the recognition of Indigenous homelands. Such a reading resonates with the reference to Baum's novel and the 1939 film adaptation. As Jerry Griswold explains, "A few intrepid scholars have suggested, however, that more than middle America, Oz is a portrait of California—or, at least, the dream of California, that verdant vision that haunted dirt-farming families in the Dust Bowl even before *The Grapes of Wrath* and the Great Migration of the thirties."[55] Although scholars more regularly attach political meaning to the novel's time of publication, in 1900, it's easy to imagine the Dust Bowl allegory that the 1939 film would have inspired. The sparse, gray-toned Midwest opens up to a fantasy of a lush, green California. However, this is a tale of regional exceptionalism. In this way, Baltz's images of the consequences of accumulation—sprawl, waste, development—lurk behind this ever-moving settler logic, demonstrating the way he works with a visual inversion of American desires and realities. In the words of K. Wayne Yang writing as La Paperson, even "'off the grid' does not describe a place, but a set of redemptive behaviors—it is a *terra nullius* imaginary of a somewhere, nowhere, neverplace where one is no longer a settler."[56] In the fantasies of Edens and Ozes, settlers project a kind of longing for a home that can never be returned to. No amount of wizardry or structural engineering (social or physical) can fulfill a settler colonial desire for homelands. It can merely project fantasy onto the realities of Indigenous homelands and then attempt to construct narratives that naturalize these desires into landscapes. In his own *Wizard of Oz* moment, Baltz forwards this ambivalence in living color.

But what of actual, material erosion? Projections suggest that all of this reclaimed land of Candlestick Point is going to be underwater with a worst-case sea-level rise of sixty-six inches by 2100. At the same time, San Francisco is currently turning the point into a multimillion-dollar mixed-use development community echoing the real estate development of San Quentin Point thirty years prior. Considering this, does California exceptionalism look like managed retreat or some fantasy of "realistic" real estate? According to a 2016 pro-development article by John King in the *San Francisco Chronicle* regarding the Candlestick Point real estate venture, "After the dirt is graded and compressed, the shops scheduled to open on the site in 2019 will be located as much as 10 feet above the former level of the entrance to the stadium used by the Giants and 49ers—a height intended as a safe perch above the rising sea for another century or more."[57] While the author covers some of the potential challenges that might face such a water's-edge development, he notes, "The fact is, this land exists. It would be overly cautious to rule these tracts off-limits because of environmental pressures that might not be felt in our lifetime. San Franciscans could benefit in decades to come if carefully planned projects create space for housing and jobs, as well as parks that bring people to the water."[58] The troubling logic of this argument implies that problems that extend past one's own lifetime are not problems to be considered at all. For King, the very existence of the land means that it must be developed, in real estate terms, without a totally clear picture of what will happen for following generations. In other words, it is a solution that refuses the responsibilities of kinship time. King goes on to paint a picture of developers who have thought of everything, and he describes their renderings of what will become of the point: "The plan is to regrade the 28-acre site ... so that it angles up from the water to building parcels that would be high enough to stay dry even if the sea level does rise 5½ feet." He asserts that this is part of a plan "involv[ing] a concept called 'managed retreat'" where "the slope would be terraced in a procession of paths and native plants, culminating in a landscaped strip located above the most extreme sea-rise projections for 2100."[59] The landscaped strip echoes the planning depicted in Baltz's *New Industrial Parks*, where every planted tree has an effect, as if simply adding the "natural" to the industrial development solves all ecological challenges. This return to planned development on Candlestick Point mirrors precisely what Balaschak argues was the point of departure for the New Topographics: "The critical aspects of the 1970s American landscape photography were not aimed at the built environment, but at the bureaucratic visual representation of these places."[60] And he explains that "we can understand American landscape photography of the 1970s as reacting to corporate boosterism with a formally stylized mode of documentary."[61] The fact that Baltz managed to photograph Candlestick Point forty years before it became a new booster-planned mixed-use development site, and not after, is equal parts accidental and prescient. It illustrates on some level a tacit awareness that while, as a citizen of the 1980s, Baltz may have had the anxiety of a sudden nuclear holocaust at the forefront of his mind, there was perhaps an awareness of another slower-moving apocalypse continually reimagined and rendered by real estate developers as settler progress.

The *Chronicle*'s King also praises San Francisco for being better, more cautious, than other places in the United States, noting that the development project "incorporates a variety of measures to cope with climate change that is expected to alter the bay. They also show that the science of sea level rise, still heresy in some parts of the nation, is taken seriously by local decision makers."[62] This idea that somewhere out there, in other parts of the nation, people regard rising waters as heresy does nothing if not continue a mythology that somehow California is different, better, more progressive than other communities facing the reality of sea-level rise. Even while the article notes that Orrin Pilkey, a leading expert on coastline adaptation and shore erosion, offers his positive impression of the plan, one still has to wonder how smart it is to build what the developer describes as "apartments, condos, townhomes and high density towers" that constitute "thousands of planned mixed-income dwellings" in a location that is certainly going to face at least several feet of rising water.[63] Currently, the proposed development is wading through a sea of bureaucratic delays, but in the meantime, the developer's projections for the area all promise a rosy future where people live and work in a tactical urbanist dreamscape. But the problem remains. It is unlikely the sea is going to rise for the next eight years and then simply stop in place. Other cities such as Galveston, Texas (a place one can imagine King thinking of as a part of the nation where climate change is "heresy"), tried a hundred years ago to simply elevate the town by several feet and hope that the ocean would never catch up.[64] On a warming planet, the ocean always catches up. Notably, the national news media offered coverage of the Isle de Jean Charles Indigenous community in Louisiana, designated as the first US "climate refugees," who initiated plans to move their community strategically inland to avoid the consequences of Gulf Coast land loss (an issue I explore in chapter 3).[65] However, I think one would be hard-pressed to find San Franciscans depicted as climate refugees, despite the fact they are facing the same issues and even repeating unsuccessful strategies tried by Gulf communities in the US South over a century ago to avoid the erosion associated with rising sea levels. If San Francisco, California, is, as Rebecca Solnit asserted, "the un-American place where America invents itself," then what does building thousands of people's homes on Candlestick Point say about the next reinvention of America?

This question of the next reinvention of American life permeates Octavia Butler's Earthseed novels. Beginning with *Parable of the Sower* and continuing with *Parable of the Talents*, Butler outlines a near future plagued by climate change and a resulting societal collapse. Each of the novels features clear scenes of erosion. Coastal erosion appears in the first novel when the citizens of Robledo are comparing their situation with that of the coastal town of Olivar. Additionally, the two fulcra of the second novel, *Talents*, are erosive mudslides. Moreover, both books are invested in landscape and Indigenous traditional ecological knowledge of the region. However, somewhat troublingly, Butler never confronts the Indigenous history of the state directly, and Indigenous peoples and communities are largely absent in her fictional world. Just as Baltz connected his investment in landscape to NASA photography, the explicit aim of Butler's protagonists in the Earthseed series is for their group to take "their place among the stars," and *Talents* ends with members of the Earthseed religion leaving for the moon.[66] The planned third book, "Parable of the Trickster," would have likely followed the group to space and outlined their struggles there, but sadly Butler did not have the opportunity to complete the novel before her untimely death in 2006. Just as Baltz cannot see beyond his own horizon line to understand how settler agrarianism was not some ideal "pre" landscape to industrialism, Butler, too, seems to struggle to break through an ideological ceiling to recognize the existence of Indigenous communities within the novels, which offer a near future of a wasted California. Thus, I want to consider these two problems together: What are readers to make of Butler's use of erosion events as almost a deus ex machina of her Earthseed world, and how are audiences to understand her works' nonengagement with actual Indigenous pasts, presents, and futures, despite the fact she is otherwise prescient in nearly every way? Simply put, the erasures and erosions of the Earthseed novels may constitute a horizon line of possibility even for Butler.

The titles of the books right away signal an investment in agriculture and soil science. The biblical parable of the sower, where Jesus explains how seed falls and can thrive or not, depending on the soil, serves as his metaphor for the ability of a faith to take hold and grow. As he explains, when spread on rocky soil, the seed cannot grow roots and survive hardship. While, as a biblical parable, this story is a metaphor and not actual soil science, it's worth pausing on Butler's use of it for her novel's title. Indeed, the protagonist, Lauren Olamina, is invested in the foundation of her own religion, called Earthseed, and in the idea that "God is change," and Butler also makes the question of growth, food supply, and rootedness central to the series.[67] Moreover, the second book, *Parable of the Talents*, also relies on a namesake biblical parable involving the dirt. In the biblical parable of the talents, a master gives each of his three servants an amount of currency (a talent) to manage, before he leaves them to go on a journey. On his return, he asks what each man has done with the money. The first two have increased its worth through various investments, while the third servant hid his talent by burying it in the ground, where it did not reproduce. For this, his talent is taken from him, and he is cast into a dark exile. To some, the lesson of this is to use rather than hide what one has been given, and one's rewards will increase. To other theologians, it seems to represent a lesson about a master reaping what his servants have sown through labor exploitation and usury.[68] In either case, however, the most striking feature for this analysis is that burying money in the earth yields nothing. Money does not grow in dirt; only seed can be sown for a return, as the previous parable indicates. Therefore, an examination of soil erosion in the two novels allows us to more fully understand how the series, seemingly preoccupied with getting to space, is actually invested in what to do with the material earth.

In the first novel, readers find the fifteen-year-old protagonist living in the year 2025 in a walled suburb in southern California while the larger Los Angeles metropolis seemingly crumbles around her and her family. The setting of her fictional Robledo seems to be near the real suburb of Mission Hills (the location of the Mission San Fernando Rey de España), as Butler tells us it is twenty miles outside of Los Angeles and near the 118 freeway. One of the first large conflicts of the novel comes when Lauren Olamina loans a book to her friend Joanne and begins to talk to her about preparing to survive what is obviously a set of events signaling the surrounding apocalypse. As they discuss what they will need to know to live through the chaos, Lauren narrates how she tells Joanne to, "'Read this,'" and she "hand[s] her one of the plant books. This one was about California Indians, the plants they used, and how they used them—an interesting, entertaining little book. She would be surprised. There was nothing in it to scare her or threaten her or push her."[69] When Joanne's parents learn that Lauren has been talking to their daughter about the world outside their community and a coming collapse, Lauren gets in trouble. She explains to her father, "I loaned Joanne a book about California plants and the ways Indians used them. It was one of your books. I'm sorry I loaned it to her. It's so neutral, I didn't think it could cause trouble. But I guess it has." He replies, "Yes, I will have to have that one back, all right. You wouldn't have the acorn bread you like so much without that one—not to mention a few other things we take for granted."[70] When Lauren expresses surprise that the acorn bread they eat comes from a recipe in this book, her father explains further, "Most of the people in this country don't eat acorns, you know. They have no tradition of eating them, they don't know how to prepare them, and for some reason, they find the idea of eating them disgusting. Some of our neighbors wanted to cut down all our big live oak trees and plant something useful. You wouldn't believe the time I had changing their minds."[71] This exchange leads Lauren to consider the deeper history of the state she calls home and the way it affects her life. However, this recognition stops incredibly far short of recognizing that there would likely be living Indigenous people in the surrounding southern California of the novel. The fact that all of this conversation places Indigenous people in the past and describes their knowledge as an "entertaining little book" signals to the reader that there is a disconnect between what the novel seemingly wants the reader to see as valid history and knowledge and what the novel sees as a valid history (and present) of its own.

Acorns are, of course, a central feature of a California Indigenous diet and agricultural system. The growth and treatment of the nuts to remove tannins involves a complex process that California Native people have developed and maintained over millennia.[72] Butler's use of this knowledge for her Earthseed novels does, on the one hand, connect the narrative to a recognition of and respect for this history. On the other hand, the lack of Indigenous characters, Indigenous place-names, or even a focus on how surely some Indigenous peoples and groups are also surviving yet another apocalypse in the region seems like an odd omission. California Native people and communities had already faced down several apocalyptic events, including the advent of the mission system stretching from southern California to Sacramento; the conversion of lands to Mexican rancheros, with a labor system similar to African American enslavement in the US South; the later Gold Rush and massive influx of settlers into northern parts of the state; and the ongoing theft of Indigenous homelands and waters for industrial agriculture and petroculture. Butler positions the near future of the apocalypse and collapse of the United States Lauren Olamina faces as an exceptional event, yet this event is exceptional only if we imagine non-Indigenous people as the central characters of the California story.[73]

Indeed, the descendants of enslaved African and African American people are not settlers in the sense that Europeans and, later, Euro-Americans are. They are the descendants of kidnapped Indigenous peoples from other places. Uncomfortably, however, Butler places her protagonist in an in-between of the littoral cell where she might run aground on what Tiffany King theorizes as the Black shoals, which "constitutes a moment of convergence, gathering, reassembling, and coming together (or apart)."[74] This dynamic space is not a neat or tidy area where Black and Indigenous futures and pasts simply join without friction. Rather, they are defined by that very friction and movement. As Mark Rifkin argues, "Even in good faith efforts toward meaningful engagement, the assumption of a shared set of terms, analyses, or horizons of political imagination between Black and Indigenous struggles may be premature or may obfuscate significant distinctions."[75] However, as he also explains, "Approaching Black and Indigenous political struggles and imaginaries as oriented in different ways—as following their own lines of development and contestation that are not equivalent to each other—does not mean understanding them as utterly dissimilar or as having no points of intersection or mutual imbrication."[76] These obfuscations and points of intersection inform my analysis of the *Parable* texts. For example, the novel's use of the acorn as a symbol throughout both novels suggests a California Indigenous past to these stories, but it ignores what would certainly be an Indigenous present, and it certainly does not offer an Indigenous future alongside what readers more readily can see as Butler's theory of Black futures. However, this does not mean that it fails to gesture toward a point of connection, however tenuous or complex. Rather, the novel opens up a space to consider what Yang, writing as La Paperson, considers via Mishuana Goeman as "storied land," where the "temporal analysis implied by Indigenous struggle and Black resistance" adds to the spatial analysis of critical cartography.[77] This reconsideration brings the "when of land" in conversation with the "where of place" in order to show the convergences and divergences of Indigenous memory with Black futurism.[78]

In this framework, the acorn is used as a telling device that calls up home for the characters, and this homing makes such elisions all the more pronounced. Lauren explains in one diary entry, dated Monday, August 2, 2027, "Tonight we cleared some ground, dug into a hillside, and made a small fire in the hollow. There we cooked some of my acorn meal with nuts and fruit. It was wonderful. Soon we'll run out of it and we'll have to survive on beans, cornmeal, oats—expensive stuff from stores. Acorns are home-food, and home is gone."[79] This risky campfire to cook acorn meal gives Lauren the double pang of nostalgia. The fact that she and the other travelers have stepped off the highway—almost surely the same route as the Camino Real that connected the twenty-one missions of what was known as Alta California—to have this nostalgic meal as they flee north, away from imminent danger, could make them akin to Indigenous peoples who had to move and survive as they lost their own homes to the settler invasions of the Spanish missionaries, or it possibly makes them like the Spanish missionaries traveling the colonizing road.[80] The reminder that "home is gone" signals both a sense of loss and a troubling question: Whose home? Is this the loss of a settler home and land? Or are audiences supposed to read Lauren and her growing community as a new group of survivors indigenized to this place through their diet of acorns? This gesture and the questions it calls up create an uncomfortable tension inside and outside of the novel over the stakes of whether readers should see Lauren Olamina and her compatriots as stand-ins for Indigenous survivors or—in the absence of any Indigenous characters or community—as settlers eventually displaced by their own unsustainable ways.

Even more telling is the penultimate scene of *Sower*, which again links the characters to an Indigenous symbol without an Indigenous presence. The group of refugees from the south include an older man named Bankole, who will become Lauren Olamina's love interest and life partner. His own family, he reveals, has a small plot of land in the northern part of the state, where the group can hopefully find some respite from the chaos. When they arrive, Bankole's family is found dead and the land deserted. Lauren Olamina explains to him how everyone in the group shares this experience: "We'll bury your dead tomorrow. I think you're right to want to do it. And I think we should bury our dead as well. Most of us have had to walk away—or run—away from our unburned, un-buried dead. Tomorrow, we should remember them all, and lay them to rest if we can."[81] When Bankole inquires about how they will conduct this ceremony, Lauren responds, "I have acorns enough for each of us to plant live oak trees to our dead."[82] This use of the acorns to plant new trees that will offer new sustaining food and life from death offers Sower a certain kind of optimism that is rare for Butler's novels. In this moment, the reader sees the glimmer of a chance that the community they begin on this land will survive for generations.

However, Bankole offers words that temper this optimism and reveal both his and Lauren's seeming attachment to the settler colonial nation of the United States:

> "You're so young," he said. "It seems almost criminal that you should be so young in these terrible times. I wish you could have known this country when it was still salvageable."
>
> "It might survive," I said, "changed, but still itself."
>
> "No." He drew me to his side and put one arm around me. "Human beings will survive of course. Some other countries will survive. Maybe they'll absorb what's left of us. Or maybe we'll just break up into a lot of little states quarreling and fighting with each other over whatever crumbs are left. That's almost happened now with states shutting themselves off from one another, treating state lines as national borders. As bright as you are, I don't think you understand—I don't think you can understand what we've lost. Perhaps that's a blessing."[83]

Bankole's anxiety over the loss of the United States as well as Lauren's prediction that it will change and remain itself showcases their anxiety surrounding an investment in what the United States was, even though each character must be aware that what the United States *was* has created their present condition of what the United States *is*. Bankole has indeed lost his optimism for the future, but he seems to maintain an optimism of the past—a California ideology that has produced for him in the year 2027 his own sense of a lost cause.

As he and Lauren Olamina continue this reflection on where they are going and how they will make a new start on this new plot of land, Bankole continues, "You know, as bad as things are, we haven't even hit bottom yet. Starvation, disease, drug damage, and mob rule have only begun. Federal, state, and local governments still exist—in name at least—and sometimes they manage to do something more than collect taxes and send in the military. And the money is still good. That amazes me. However much more you need of it to buy anything these days, it is still accepted. That may be a hopeful sign—or perhaps it's only more evidence of what I said: We haven't hit bottom yet."[84] When Lauren asserts that this place will offer them a buffer against sinking any lower, he responds, "I don't think we have a hope in hell of succeeding here."[85] Just before he asserts this, Lauren mentions that he looks a bit like an older Frederick Douglass, directly connecting the plot to a time in the nineteenth century marked both by the optimism of emancipation and by the genocide of California Native people.

This ambivalence between hope and loss carries across the novel's closing paragraph, where Lauren tells us, "So today we remembered the friends and the family members we've lost. We spoke our individual memories and quoted Bible passages, Earthseed verses, and bits of songs and poems that were favorites of the living or the dead. Then we buried our dead and we planted oak trees. Afterward, we sat together and talked and ate a meal and decided to call this place Acorn."[86] Butler then adds the verses from Luke that recount the titular biblical parable. This linking of a Christian tale and the community name Acorn creates a tension in the book over the religious and spiritual heritage of what will become the new religion of Earthseed that Lauren Olamina hopes to found, beginning with the newly established community at Acorn. In one way, the naming of the community and the cooperative whole-earth justice of their founding suggest a recognition of and return to Indigenous ways of knowing within the postapocalyptic California landscape. Conversely, the Christian overtones, the group's following of the Spanish missionaries' Camino Real north, and their kind of assumed possession of the purchased land provoke a small feeling of appropriation of Indigenous knowledges within a structure that seems vaguely reminiscent of settler colonial logic. This isn't to say that Butler herself is wrong but that in the character Lauren Olamina she presents an uneasy conflation of Indigenous knowledges and neo-settlerism in the postapocalyptic landscape of the imagined 2027. The fact that in the previous scene Lauren and Bankole each still cling alternately to optimism and nostalgia for the United States of America makes this friction all the more perceptible. This central tension marks a compelling cautionary point with regard to Butler's work: Black futurisms must work against inadvertently taking the form of settler colonial pasts. This challenge is all the more real for her characters who are attempting to survive a late-capitalist America and must almost literally walk the line (the Camino Real) between future liberation and historical repetition. When Butler's novel is plotted alongside the historical contingencies of this specific place, there emerges a much more entangled story than one that simply imagines private landownership and Indigenous knowledges gleaned from a book as the solutions to the earth's ecological crises and the collapse of the United States.

For example, the location of Acorn also adds a layer of tension to the question of Indigenous histories, homelands, and knowledges within the two novels. Lauren notes, "We've reached our new home—Bankole's land in the coastal hills of Humboldt County. The highway—U.S. 101—is to the east and north of us, and Cape Mendocino and the sea are to the west. A few miles south are state parks filled with huge redwood trees and hordes of squatters. The land surrounding us, however, is as empty and wild as any I've seen."[87] These descriptors place the fictional town of Acorn somewhere due west of present-day Redcrest, California, and due south of Stafford; this is one of the only places where US 101 could be both north and east of a location. The park to the south is almost certainly Humboldt Redwoods State Park. This places Acorn somewhere along the northern portion of Sinkyone homelands, with the neighboring homelands of the Mattole to the west, Nongatl homelands across the main Eel River to the east, and Wiyot homelands to the north. Notably, this also situates the Acorn camp within the areas affected by the multiple Eel River Basin massacres of the 1860s, a genocidal campaign against Native people waged by US troops not in the direct theater of the emergent and ongoing US Civil War.[88]

In addition to these geographic and historical factors, there is also a question of land speculation and the timber industry. Bankole tells Lauren of the land, "I own three hundred acres. . . . I bought the property years ago as an investment. There was going to be a big housing development up there, and speculators like me were going to make tons of money, selling our land to the developers. The project fell through for some reason, and I was stuck with land that I could either sell at a loss or keep. I kept it. Most of it is good for farming. It's got some trees on it, and some big tree stumps. My sister and her husband have built a house and a few outbuildings."[89] Given the history of land and labor dispossession of African American families in the United States, there is little room to critique any Black character for acquiring property. At the same time, Butler directly tells us that Bankole acquired this land originally as a speculator, calling up both the history of mining speculation in the region and colonial land acquisition that holds on to Indigenous homelands as mere real estate. In other words, Acorn is built on lands taken through genocidal campaigns against Indigenous people during the period of the Civil War, and these lands are in turn now owned by a Black family as a result of real estate speculation for land development. When these moments are taken together, the spinning wheel of settler capitalism in the novel seems all the more furious, endangering the characters and the lands they have settled.

While the condition of the soil factors more directly in the second novel, Butler previews the situation toward the end of *Sower*. Lauren tells the group, "Bankole owns this land, free and clear. . . . This area was logged sometime before Bankole bought it. Bankole says it was clear-cut back in the 1980s or 1990s, but we can make use of the trees that have grown since then, and we can plant more."[90] As Kaitlin Reed explains, the waves of timber extraction in California created a number of ecological problems, including stream and salmon egg suffocation, water temperature increases, and erosion.[91] Land along the Eel River is notoriously affected by soil erosion in landslides, and aside from the Mississippi River, if left to its own devices, the Eel River would carry more soil sediment than any other river watershed in the United States.[92] Active, living tree roots help mitigate erosive landslide activity, so when Butler reveals that this land, in her novel, has been clear-cut for lumber just a few decades earlier, that indicates that it stands a higher risk of landslides. Importantly, while landslides sound like a bad thing, the transportation of sediment is an important part of how California watersheds maintain ecological balance and replenish beaches along the coast, as I mentioned earlier.[93] The long view of land management practiced by Indigenous peoples in the region kept these factors in view. However, the land degradations of hydraulic mining (a process that blew away whole mountains with waterpower in order to extract gold), the timber industry, and settler damming projects have created excessively unstable conditions in a location already known for its shifting soils. That Lauren Olamina thinks that her group can reverse almost two centuries of damage by planting both figurative and literal acorns represents a fair amount of optimism and ignorance. It also signals to the reader that she might not completely grasp the history of the place where she has chosen to settle, nor how her plans for the future can fully take the past into consideration. It is unsurprising, then, that the two structuring events that fundamentally change the plot of the follow-up novel, *Parable of the Talents*, are both landslides.

The novel opens with a prologue from Lauren Olamina and Bankole's daughter, Larkin, as she describes the journal writings she has acquired from her late mother and father. Chapter 1 is framed as taking place in 2032 and opens with a set of journal entries—one from Bankole contextualizing the apocalypse they are living through and another from Lauren Olamina on the fifth anniversary of Acorn's founding. Lauren Olamina records her recurring nightmares that feature her father's retelling of the biblical parable of the talents; she notes, "My 'talent' . . . is Earthseed. And although I haven't buried it in the ground, I have buried it here in these coastal mountains, where it can grow at about the same speed as our redwood trees. But what else could I have done?"[94] Like many of Butler's works, the novel sets up an uncertain assessment of its protagonist and her mission. Lauren Olamina and Bankole's daughter offers a distanced reading of her parents, tinged with a resentment of her mother's primary attachment to the religion she founded. This forces the reader to begin the second novel wondering what happened to the fledgling eco-conscious, holistic, dynamic community of Acorn to result in what the daughter believes to be a dogmatic cult. Readers see this past through Lauren Olamina's writings as framed by her daughter Larkin's narrative voice.

The first major turning point of *Talents* comes when a neighboring town, Halstead, has an erosive landslide that kills the local doctor. In an entry dated December 18, 2032, Lauren Olamina explains:

> Halstead has a major problem. Halstead used to have a beach and above the beach was a palisade where the town began. Along the palisade, some of the biggest, nicest houses sat, overlooking the ocean. On one side of the peninsula were the old houses, large, well-built wood frame structures. On the other side were newer houses built on land that was once a seaside golf course. All of these are . . . were lined up along the palisade. I don't know why people would build their homes on the edge of a cliff like that, but they did. Now, whenever we have heavy rains, when there's an earthquake, or when the level of the sea rises enough to saturate more land, great blocks of the palisades drop into the sea, and the houses sitting on them break apart and fall. Sometimes half a house falls into the sea. Sometimes it's several houses. Last night it was three of them.[95]

This event causes the town of Halstead to need Bankole's medical services. However, Bankole expresses frustration at the town for not understanding managed retreat and chastises Halstead: "Why do you still have people living on the cliffs? . . . How many times does this kind of thing have to happen before you get the idea? . . . Move the damned houses inland, for heaven's sake. Make it a long-term community effort."[96] The messenger from Halstead who has been sent to retrieve him responds, "We're doing what we can. . . . We've moved some. Others refuse to have their houses moved. They think they'll be okay. We can't force them."[97] The town's inability to enact a policy of managed retreat to mitigate the effects of erosion causes its own doctor to die in the landslide. As a result, Halstead representatives ask Bankole and Lauren to leave Acorn and move to their town on the coast.

Lauren does not want to move and leave Acorn behind. Nonetheless, she humors Bankole with a visit to the neighboring community to see what their life might be like there. She notes in her journal, "[Bankole is] in love with the place. It's just what he wants: long established, yet modern, familiar, and isolated. There are comfortable big houses—three and four bedrooms. And, thanks to the wind turbines in the hills, along the ridges, there's plenty of electricity most of the time. And there's modern plumbing. We have a little of that now, but it's been a long struggle. Halstead, except for its crumbling coastline, is about as well protected as any town could be."[98] Lauren even enjoys the house they have been given to stay in, explaining, "The house rattled and creaked, but it was warmed both by electric heaters and by fires in the fireplaces, and it was set far enough back from the coastal bluffs to be in no danger for many years, if ever."[99] Even though Lauren Olamina finds the oceanfront town beautiful and relaxing, the anxiety of erosion is ever constant in the background, and she notes, "We can see the ocean every time we travel up the highway to the Eureka-Arcata area and farther north. Up there, it has washed away long stretches of sand dunes and done real damage along the Humboldt and Arcata Bay coastlines. This is all the fault of the steadily rising level of the sea and of occasional, severe storms."[100] It would be misleading to say that Lauren Olamina does not want to leave Acorn simply because she knows Halstead is crumbling into the sea, but the slow, background ticking clock of sea-level rise gives her an added justification for arguing to Bankole that they should remain with the religious community she has founded rather than take refuge in the amenity-laden Halstead. Additionally, what Lauren Olamina cannot see is that the literal collapse of Halstead is not simply because of rising tides. Sediment has been trapped by structures built on cliffs that should erode, crumble, and flow downstream to feed these beaches. In other words, Halstead is in danger because towns farther upland are also holding land back from where it wants to go.

This decision not to move to the oceanfront Halstead may have ultimately been a bad one. As Larkin writes in her own narration, "Should she have left Acorn and gone to live in Halstead as my father asked? Of course she should have! And if she had, would she, my father, and I have managed to have normal, comfortable lives through [the] upheavals? I believe we would have."[101] And later Larkin reiterates, "If only my mother had agreed to go with my father to live peacefully, normally in Halstead, it wouldn't have happened. Or at least, it wouldn't have happened to us."[102] Thus, from the perspective of a generation later, Lauren Olamina's refusal to leave Acorn seems what her daughter calls "shortsighted"—a diagnosis of a kind of horizon line of possibility that Lauren Olamina cannot see past. This opportunity to move arises in the novel only because of the eroding coastline. The refusal of this opportunity is also bolstered by a sense that Halstead itself is destined to disappear, as the sea will continue to claim the cliffs of the town. Lauren Olamina remains committed to what she has begun at Acorn, and even as the political climate of the country becomes even more dangerous for religious outsiders, she sees a future for her emerging Earthseed religious ideology.

Lauren Olamina's optimism for her religious movement proves to be misguided, however. Shortly after the birth of Larkin and her second refusal to relocate to the eroding Halstead, a group of Christian paramilitary missionary forces invades Acorn. They ultimately kill Bankole and enslave the rest of the town in a series of nearly unspeakably violent events that echo the Eel River Basin massacres against the area's Indigenous communities in the 1860s. Lauren Olamina records the events in a journal entry written on Monday, September 26, 2033, where she recounts that the invasion began the previous Tuesday.[103] The surviving community members are controlled via remote slavery collars that prevent them from retaliating, and the Christian zealot missionaries enact a program of religious reform and forced labor against the community. They also sexually abuse numerous members of the community and enact a community apparatus not unlike the Spanish missionary system of eighteenth-century California.

The violence continues for the next year and a half, until Lauren Olamina and the others begin to devise a plan to overthrow their captors. However, once again an erosion event changes the course of the plot. Writing in secret on Wednesday, February 28, 2035, Lauren Olamina recounts, "Day before yesterday, we had a terrible storm—truly terrible. And yet, it was a wonderful thing: wind and rain and cold . . . and a landslide. The hill where our cemetery once was with all its new and old trees, that hill has slumped down into our valley."[104] She explains that because the missionaries thought that the Acorn community worshipped the trees, they had made them cut down the founding oaks of Acorn where their families were buried. The lack of vegetation created an erosion threat in an area already known for unstable soils. Lauren Olamina explains that, as a result, "the hillside has broken away and come rumbling down to us."[105] This landslide destroyed the nicest cabins of the community, including the one originally used by Lauren Olamina and Bankole, which the missionaries had commandeered for themselves, and killed everyone inside. It also destroyed the central monitoring system for the slavery collars, freeing all of the community from control. As she explains, "Last Sunday, we resolved to free ourselves or die trying. Now, instead, the weather, and our 'teachers'' own stupidity has freed us."[106] The fortuitous erosion event gives the group their freedom and allows Lauren Olamina and the others to escape to whatever future awaits each of them. Sadly, the community at Acorn is gone forever, but from Lauren Olamina's perspective, the larger project of Earthseed is not dependent on the first founding community. These two events combined—the Halstead and Acorn landslides—reveal the novel to be concerned with erosion in California in a way that puts the process into a perspective that is not wholly negative or positive. Landslides may seem bad, but the beaches have a right to that sediment. Events that may on the surface seem either celebratory or tragic become more complicated when one considers the deeper temporality of the California Indigenous landscape.

This tension mirrors the novel's own seeming ambivalence toward Lauren Olamina—is she a hero invested in returning to Indigenous ways of knowing the land, or has she become her own kind of neo-settler? The close of the novel does little to answer this question. We know from her daughter Larkin's framing that this is Lauren Olamina's last journal entry, as she watches the Earthseed faithful depart to colonize the moon. As Lauren Olamina explains, "Traveling with the people are frozen human and animal embryos, plant seeds, tools, equipment, memories, dreams, and hopes. As big and as spaceworthy as they are, the shuttles should sag to the Earth under such a load. The memories alone should overload them. The libraries of the Earth go with them. All this is to be off-loaded on the Earth's first starship, the *Christopher Columbus*."[107] The name is a shocking return to exploitative colonial logic, but Lauren Olamina offers a justification and clarification: "I object to the name. This ship is not about a shortcut to riches and empire. It's not about snatching up slaves and gold and presenting them to some European monarch. But one can't win every battle. One must know which battles to fight. The name is nothing."[108] While Lauren Olamina claims the name Christopher Columbus is nothing, one has to wonder: Do the space colonizers on this ship see this name as nothing, or do they see it as indicative of their legacy and mission? Moreover, if this name is nothing, was Acorn's name also nothing? Uncomfortably, the novel maintains an ambivalent stance that leads the reader to question who exactly they have been pulling for across the two novels. *Sower* lulls the reader into seeing a justice community informed by Indigenous traditional ecological knowledges as a path to survival. However, given the novel's erasure of any living California Indigenous people, I argue that readers have to take a hard look at some facts of Acorn. It's founded on land acquired through admitted land speculation and deforested by the timber industry, and it was the site of a particularly destructive genocidal campaign against California Native people. Moreover, Acorn is founded by a religious group that has followed the Camino Real all the way into northern California. Together, these factors should give readers pause: Is Butler's hero actually among our villains?

This ambivalence reveals the two novels' ultimate paradox of anxiety. As Gerry Canavan offers:

> The Parables series is structured at its core by antinomy, exemplifying the philosophical tensions we see in Butler's work across her lifetime.... [T]he novel's utopian aspirations versus its anti-utopian presuppositions; its critique of neoliberalism coupled with its production and valorization of a kind of ultra-neoliberalism; the tension between helping *the species* survive versus abandoning many or most of *the individual members of the species* to misery; and its two overlapping but ultimately incompatible faces of our lawless, anarchic future, the space of the frontier versus the space of the slum, neither of which feels very much like "progress."[109]

This antinomy, according to Canavan, could have been resolved in the third, unrealized book, "Parable of the Trickster," which according to Butler's archival materials would have followed the space colonists as they tried to adapt to a new environment in space. Canavan argues, "Without it, *Sower* and *Talents* remain bound in a kind of antinomic relationship: *Sower* (utopian dreaming) and *Talents* (anti-utopian reality) are locked in a death struggle that cannot be overcome because the sequel that would have synthesized and moved beyond both can now never be written."[110] This schema of the novels isn't entirely off base. However, I think it overlooks how the very structure of anti-utopian reality already subtends *Sower*. It is not simply that readers encounter a reality in *Talents*; it's that perhaps Lauren Olamina's project has *always* been imbued with a settler impulse that her eventual disciples traveling to the moon name even more accurately than she would like to admit. Rather than place a Black-futures protagonist uncritically in the realm of an unquestioned good, Butler forces her readers to consider what five hundred years of America have done to make some people's vision of Black freedom look and act like settler colonialism. She seems to caution in the novel that Black futurism cannot be modeled on settler colonial pasts. The discomfort of this convergence between Black futurism and the work of settler memory causes readers to run aground on one of Tiffany King's conceptual shoals, where they must ask what tensions and solidarities emerge out of a Black liberation that does not recognize and honor the Indigenous claim to that very same soil. As with Baltz's closing gesture in his landscape works, with the *Wizard of Oz* move to color in *Candlestick Point*, there could be an impulse to see the fulfillment of Earthseed's mission to the stars as a hopeful ending. This is the problem with utopia, however. It is no place—οὐ τόπος—after all. And for settler colonialism, there is no place like home.

But what does the very real place called home look like for Indigenous people in present-day California? Both Baltz's and Butler's work make it difficult to see beyond a horizon line of settler colonialism in the state. However, if there is a way to live through, and with, landslides and eroding coastlines, then the best hope is with California Indigenous people, who have a much deeper historical engagement with their homelands. Deborah Miranda explores this connection in her 2013 multigenre memoir *Bad Indians*. She recounts her story as she works across generations, covering the history of the mission system, the decades-long state-sponsored school project devoted to that history, her own personal trauma, and her location of her family's own Ohlone/Costanoan-Esselen land grant, "El Potrero," which was taken by American gold speculator and all-around aggressively violent land thief Bradley Sargent. As Miranda describes, "The loss of land is a kind of soul-wound that the Ohlone/Costanoan-Esselen Nation still feels; a wound which we negotiate every day of our lives [that] clearly presaged intergenerational trauma with the accompanying loss of self-respect and self-esteem."[111] While Miranda is speaking most directly about the loss of legally recognized title to the land, I argue that the increasing complications from erosion created by settler colonialism place another, more material and literal (and littoral) land loss on the horizon. The geological degradation is, however, also on Miranda's mind as she combs through historical maps and deeds, looking for any evidence of her family's El Potrero: "Perhaps it seems strange to others, but so much of our history has literally disappeared that I simply assumed any land that was once El Potrero couldn't possibly still be there—it had to be bulldozed, dug up for minerals, stripped for timber, sucked dry for water, blacktopped, graded, and subdivided into oblivion. 'El Potrero,' the land itself, couldn't possibly still be standing."[112] This is the difference between a white American anxiety of property disappearance via erosion and an Indigenous concern over land losses associated with home. While many settlers have sublimated their concerns over a loss of US ascendancy and white supremacy into anxieties about fading beaches and lost condominiums, there is an Indigenous dimension to this story that underscores how the loss of the land does not constitute the same experience for everyone. For Indigenous people, erosion isn't just about multimillion-dollar homes on the water or a collapsing Pacific Coast Highway. It's about homelands that have been irreparably destroyed by hydraulic mining, industrial agriculture, and damming, all to support the acquisition of wealth via American development. When settlers see erosion as an advancing danger, are they seeing this loss in terms of the earth or in terms of the white ascendancy built into the project of capitalist "progress"? This conflict of vision determines how one meets the challenges posed by erosion.

Miraculously, Miranda locates the El Potrero land grant from among a palimpsest of colonial records and places it in the landscape of present-day California. As she describes, "I didn't know whether to laugh or cry when I found that El Potrero is part of the Santa Lucia Preserve: a combination private, corporate-owned real estate holding and nature conservancy consisting of Bradley Sargent's former ranch."[113] Miranda's own ambivalence in her reaction—"laugh or cry"—showcases a relief that the land is "preserved" but also illustrates a certain kind of ironic California ideology where those with means attempt to operate within a system of imagined conscious capitalism. Miranda explains her reaction to this realization: "After my immediate relief that the land had not been developed beyond all recognition, I was almost equally disturbed to discover that part of it had become a luxury housing development with a golf course designed by Tom Fazio, as well as a fully equipped sports center, 'Hacienda' country club, and equestrian center with one hundred miles of trail. I imagined the worst: El Potrero beneath an emerald green, spotlessly groomed golf course."[114] Miranda learns that the land is luckily not under the golf course, but the situation of the preserve in general evokes its own melancholy. Moreover, the description of the community now at the site of El Potrero has an eerie resonance with the walled communities meant to insulate those with enough money in the ecological and societal collapse outlined in Butler's Earthseed novels. As Miranda explains, "Although the preserve is secluded and not accessible to the general public, its survival and care are paid for by those wealthy enough to have purchased million-dollar (or more) homes nearby. Still, I know the truth: despite the miracle of this land's existence, El Potrero and the other Indian ranchos and rancherias have not been, and will never be, turned back over to the descendants of tribes from that area."[115] Miranda makes us all the more aware that conscious land-care capitalism is still settler colonialism. Land "preserved" is not and cannot be the same thing as "land back." Just as Eve Tuck and K. Wayne Yang criticize the metaphorical slide into "decolonization" as a troubling catchall for neoliberal multiculturalism efforts, Miranda highlights the sliver of distance between the parabolic limit of settler-controlled "preserved" lands and the axis point of Indigenous control over their homelands.[116]

Despite this rather bleak truth, Miranda closes her memoir with a compelling dream where she imagines a series of surreal events when she and her family are allowed to visit the homelands that had been stolen from, then granted to, then stolen again from her Esselen ancestors. The dream weaves together connections among the land, humans-animals, and animal kin, and despite the compounding traumas of homelands theft and unearthed remains of ancestors, Miranda wakes from the dream "full of tears and wonder and pure joy."[117] She challenges her readers to see a world where connection to homelands becomes possible through a re-storying of the world that addresses the wound of land loss: "Tom King throws down a challenge that every listener must heed: Indian or not, haven't we lived under the burden of California mission mythology and gold rush fantasy long enough? Isn't it time to pull off the blood-soaked bandages, look at the wound directly, let clean air and healing take hold?"[118]

This challenge is the only one to accept if there is a future where California wants to grapple with an erosion problem without defaulting to the *Los Angeles Times* images of a "calm but unusual cycle that had lulled dreaming settlers into a false sense of endless summer."[119] The truth is that the past five hundred years have been anything but a calm, endless summer for California Indigenous people. Handwringing and myths of a progressive California that accepts the "heresy" of climate change better than other places in the nation are only more bandaging over an infected wound of Indigenous homelands' theft. So while the *San Francisco Chronicle* promotes real estate renderings of a Candlestick Point development and the *Los Angeles Times* insists on a zero-sum game where humans will "win" against coastal erosion, California writers and artists such as Baltz, Butler, and Miranda have consistently attempted to show California as it *is* and as it *will be*, stripped from the watercolor fantasies of developers and insulated gated communities that perhaps imagine only their own survival on a changing planet. Indigenous peoples in the state have outlined time and again how their own land management strategies—undamming sediment-carrying (and salmon-carrying) rivers, scaling back real estate development, conducting carefully controlled burning, restricting logging, reintroducing native plant life, and regulating agricultural runoff, to name but a few—all offer viable options to survive the erosion crises of the future. This is the ultimate glitch in "The Ocean Game" the *Los Angeles Times* proposes. In their interactive eight moves to save a coastal town in California from erosion, they have neglected the most crucial *first* move: return the land.

2

Surfaces and Allotments of the Heartland

WHEN MOST PEOPLE IMAGINE EROSION in the context of US history, almost no event springs to mind more frequently than the Dust Bowl of the 1930s. The appearance of impoverished farmworkers and great billowing clouds of airborne dirt remains lodged in the modern psyche. Among those often black-and-white images that crowd the imagination, Dorothea Lange's work stands apart. Even if viewers are not sure who Lange was or what her photographic work entailed, they often recognize and can call to mind her photography for the Farm Security Administration without its attendant specifics. Among these familiar images, Lange's photograph titled *Migrant Mother* is surely the image most associated with this period of economic and environmental insecurity (figure 2.1). However, like every other example of erosion in the popular imagination that I examine in this book, the image,

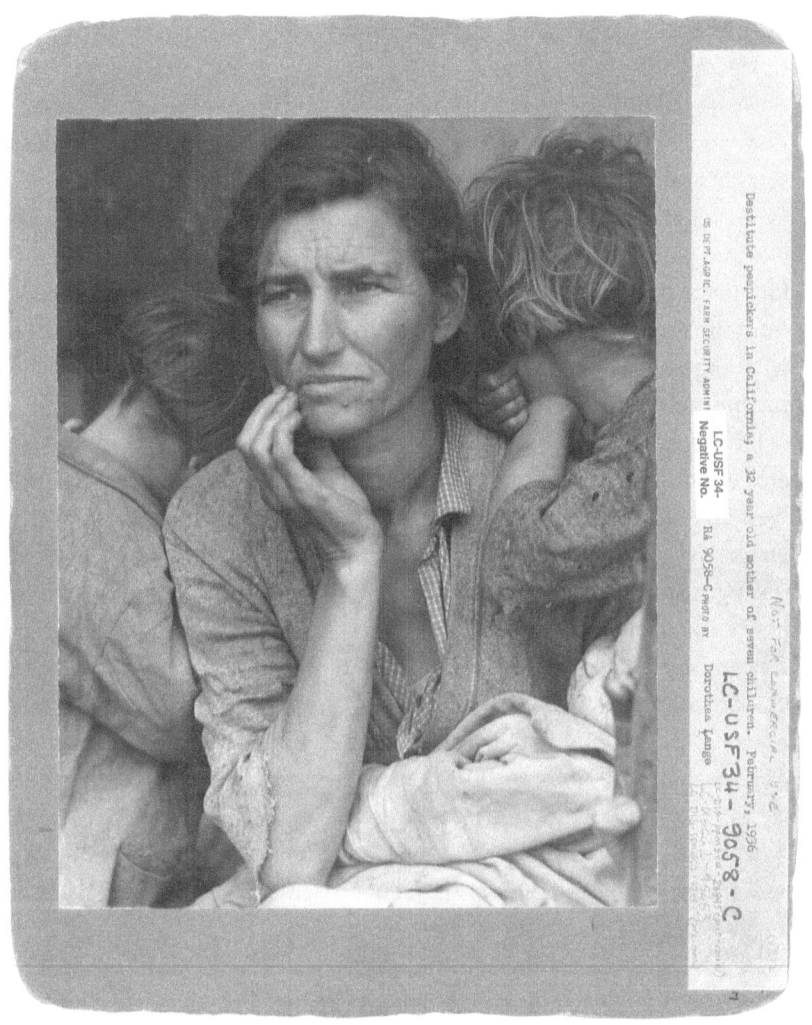

2.1 Dorothea Lange, *Destitute Pea Pickers in California. Mother of Seven Children. Age Thirty-Two. Nipomo, California* (unretouched original), February 1936. Portrait popularly known as *Migrant Mother*, published in the *San Francisco News*, March 11, 1936. Library of Congress Prints and Photographs Division, https://www.loc.gov/resource/ppmsca.23845/.

the place, and the narrative that has come to coalesce around the Dust Bowl conceal as much as they reveal.

To begin, this photograph most associated with so-called Okie migrant farmwork was taken in California.[1] It is no stretch to say that the human story of the Dust Bowl is as much about California as it is about Oklahoma.[2] Even though Cimarron County and Boise City, Oklahoma, located in the far-western panhandle of the state, were considered the epicenter of the environmental phenomenon, the mounting human consequences of the exodus from the southern Great Plains coalesced in the migrant camps associated with large-scale industrial agriculture on the West Coast. As I examined in the previous chapter, far from being an uncomplicated location of dreams, California in particular has its own multifaceted relationship to erosion and erosive anxiety. Furthermore, while it may be tempting to imagine the Midwest (and the resulting chapter of this book) as mere continually unfolding prequels to the first chapter, which focuses on the West Coast, I want to resist the urge to follow such a teleology—even in reverse—and instead think about how erosion and the anxiety of disappearance recur in pulsating flashes of concern across regions and across time.[3] Oklahoma, then, is not a prelude to California. Rather, Oklahoma is linked to California as it is linked to Georgia as it is linked to North Carolina and Louisiana and many places in between. As each continually informs the other, the images and narratives that audiences associate with discrete places exist in a continually recursive present. All of these places are Indigenous homelands, and all have been changed by settler colonialism. Lange's work is but one place where audiences can catch a glimpse of this interrelation if they know *how* to look.

Importantly, this photograph most associated with the impoverished condition of white migrant farmers during the Great Depression is not of a white Euro-American woman at all. The subject of the photograph, known only as the "Migrant Mother" for several decades, is a Cherokee woman named Florence Thompson (née Christie), who was born near Tahlequah, the capital of the Cherokee Nation, in the eastern part of the state.[4] As several critics have outlined, later in life when she was identified as the famous "mother," Thompson expressed feeling exploited and misrepresented by Lange's photograph.[5] Although Lange did not receive any specific monetary commission associated with the image's vast popularity, the photograph largely cemented her place as one of the most important photographers of the era. Moreover, Lange herself did not know Thompson was a Cherokee woman and did not record her name at the time.[6] Thus, this powerful image that evoked so much anxiety and response about what was happening to impoverished white farmers never depicted whiteness at all. It showed instead a Cherokee mother who by 1936—just under a hundred years after Cherokee Removal from the Southeast—had been pushed by settler colonialism, US policy, and its attendant industrial agriculture to the brink of existence at the other end of the continent. The white American audience of this image looked at Thompson's face and saw their own supposedly victimized whiteness of the 1930s, not the actual Indigenous victim of their own settler ways.

While some journalists and critics have discussed Thompson's identity as a Native American woman, they have often done so in generic ways that collapse Indigenous identity into a singular racial or ethnic category that fails to account for specific tribal histories and nationhood.[7] Sally Stein comes the closest to attempting to place Thompson in a specific critical context, but even her analysis rests on how nonwhite and racialized subjects are seen or not seen within American photography, without giving much attention to Cherokee history, lifeways, or worldviews. Even without this specificity, however, Stein works to correct recent analyses of *Migrant Mother* that still depend on white supremacy and white pathos even as they purport to critique it. As she explains, Gary Gerstle, writing as recently as 2001, said of the famous photograph, "Part of the photograph's appeal lay in the sheer brilliance of its composition, but part depended, too, on its choice of a 'Nordic' woman. Her suffering could be thought to represent the nation in ways that the distress of a black, Hispanic, Italian, or Jewish woman never could."[8] Indeed, Gerstle misses the point of what he cannot see, what Lange did not see, and what many people still do not see: how much the anxiety of white pathos has always depended on making Indigenous people invisible even when they are hyperpresent. These misreadings of the surface have consequences for narratives of the Dust Bowl even today. In many ways, they mirror the very phenomenon of surface erosion that created the conditions of the Dust Bowl, in that the particles of meaning have been dispersed widely in a way that clouds understanding of narratives of America's so-called heartland. Imagining Thompson as simply Native American and not specifically Cherokee also creates a type of invisibility, as it unwittingly subsumes all violence against Indigenous people as part of a grand, impersonal wave of colonialism rather than the result of concrete actions and policies enacted by specific individuals in specific places at specific times. In this version of the narrative, the consequences of these actions become nobody's and everybody's fault, allowing whiteness again to float free in its own occasional passing guilt. Reframing Thompson as a Cherokee woman illuminates the history of the Dust Bowl within a particular story of settler colonialism and Indigenous nationalism.

For my part, I remain most interested in how the story of the Dust Bowl has largely ignored the specific ways that Removal, Allotment policy, and Oklahoma statehood created the conditions for one of the largest erosion events in global history and the ways these histories intersect with specific Indigenous individuals and nations of the region.[9] The story of the "Okie" Dust Bowl exodus has largely only been thought about as the story of white people and the anxiety of white impoverishment. When this anxiety is attached to the loss of the land, it becomes difficult to distinguish how much a white audience is concerned about the earth from how much they are concerned about their own investment in settler colonialism. For while we know that Oklahoman artist Woody Guthrie often peppered his lyrics with ironic jabs at the elite and powerful, there remains a haunting refrain in the pronouns of his own Dust Bowl national anthem:

> When the sun come shining, then I was strolling
> And the wheat fields waving and the dust clouds rolling
> A voice was chanting as the fog was lifting
> This land was made for you and me.[10]

Guthrie's point is in the pronouns, and though there might be an ironic ambivalence given the larger context, the song is nonetheless steeped in the anxiety of claiming an earth that is blowing away. This chapter takes up these narratives of the Dust Bowl and the Midwest to demonstrate how erosion is always about who claims the material earth to what ends and whether or not those claims are recognized.

Much as Thompson's identity as a Cherokee woman has remained underrecognized by the larger public, so too have the roots of the most famous dramatic text set in Oklahoma: *Oklahoma!* (1943). While Richard Rodgers and Oscar Hammerstein II staged what is rightly critiqued as a settler colonial fantasy set in Indian Territory ahead of statehood, the script on which *Oklahoma!* is based is Cherokee author Lynn Riggs's *Green Grow the Lilacs* (1930). Riggs was a highly respected and prolific playwright before the adaptation of his original play, and no consideration of the region from the early twentieth century is complete without a thorough examination of his work. In many ways, he is the ur-chronicler of Oklahoma. As Riggs wrote in a 1928 letter about the people of his home, "And I know that what makes them a little special, a little distinct in the Middle West is the quality of their taciturnity.... [T]he people who settled Oklahoma were a suspect fraternity, as fearful of being recognized by others as they were by themselves. Gamblers, traders, vagabonds, adventurers, daredevils, fools. Men with a sickness, men with a distemper. Men disdainful of the settled, the admired, the regular ways of life. Men on the move. Men fleeing from a critical world and their own eyes. Pioneers, eaten people."[11] Across his works, he demonstrates how these eaten people eventually eat the earth, signaling the clear link between erosion (a word that, as I note in the introduction, had a cankerous sense of "to eat" in its earliest geological usages) and the conditions of mind that the settlers carried with them into the region. Riggs can be understood, then, as both an Indigenous playwright and a careful ethnographer of those who came to "settle" Oklahoma. His understudied body of work demonstrates not only his acumen in tracing the lives of the people who were both perpetrators and victims of environmental crises but also the way the erosive practices that shaped the earth also shaped an anxiety of disappearance.

Examining Riggs's work alongside that of his non-Native contemporaries, including John Steinbeck, who also attempted to catalog the people of the vanishing earth, allows this chapter to explore foundational questions about the interrelationship of settler colonial anxiety, literary regionalism, and Indigenous knowledges that forms a geographic and geological space. I begin with a brief overview of how specific policies related to Indigenous oppression and settler expansion created the environmental conditions of the Dust Bowl. I then trace how these concerns resonate across the popular texts regarding the event: Steinbeck's *The Grapes of Wrath* (1939) and Lange's (with husband and intellectual partner Paul Taylor) *An American Exodus* (1939). I demonstrate just how pervasively Indigeneity and white settler anxiety appear across these texts that might otherwise seem not to index that piece of the Dust Bowl story. From there, I spend the majority of the chapter outlining how Riggs as a Cherokee writer offered a prismatic view of Oklahoma that held Indigenous and settler lives together in view, revealing how anxieties of continued survival shaped narratives of the lower Midwest's history across the 1930s, leading up to and following the Dust Bowl. In particular, I demonstrate the importance of questions of a gendered distribution of power within Cherokee and settler contexts for narratives of this time and place. Ultimately, this chapter reorients considerations of the Dust Bowl to show how the erosion of the presumed heartland has long been a story of the (non)recognition of Indigenous women and their nations' sovereignty.

The events that culminated in the Dust Bowl are complex and multifaceted, but they are necessary to understand in their relationship to settler colonialism, and so I offer an admittedly broad outline here to help contextualize the authors and works that later sought to make sense of the unfolding tragedy of the region. As a landscape, the southern Great Plains, stretching from Kansas to the panhandle of Texas and west to the northern corner of New Mexico, had evolved over millennia as a place where mineral-rich water had run down from the Rocky Mountains and covered a vast area with thick loam (a fertile soil containing a balance of sand, silt, and clay). On this loam grew buffalo grass, a drought-tolerant indigenous grass that has a thick root structure held together by stolons (horizontal root runners) and rhizomes, which held the surface soil in place even in the violent windstorms common in the area. It also served as forage for the bison that populated the plains before European and Euro-American invasion. The grass functioned as a protective blanket for this earth, and it maintained balance in an otherwise volatile and dynamic ecosystem characterized by long periods of drought and regular strong winds.[12]

Settlers changed this landscape rapidly in the late nineteenth and early twentieth centuries, however. The Dawes Act, or General Allotment Act, of 1887 began breaking up and subdividing tribal lands in the West into individual 160-acre allotments, "giving" designated heads of household lands to control as their own homestead rather than having lands be held in common by the tribal nations.[13] Ideologically, the Dawes Act purported to encourage tribal citizens to become Euro-American-style farmers and give up what the United States considered an "uncivilized" collective lifestyle. Financially, Allotment policy worked conveniently for the United States, as it "freed up" tribal lands for westward settlement by Euro-Americans once all 160-acre allotments had been distributed to Native "heads of households." There is no way to understand Allotment policy without accepting the central fact that it was aimed to destroy the strength of Indigenous nations via land theft and territory fragmentation and to forcefully assimilate Indigenous individuals into Western forms of agriculture, land title, and taxation. It was attempted genocide from the land up.[14]

Originally, the Dawes Act did not apply to the tribes who had experienced Removal from the Southeast and were now relocated in Indian Territory, presently known as eastern Oklahoma. The Curtis Act of 1898 amended the Dawes Act to include the Five Southeastern Tribes (Choctaw, Cherokee, Chickasaw, Creek, and Seminole) as well as to dissolve their tribal courts and governments.[15] Thus, the promise that, in exchange for their homelands in the Southeast, the Five Southeastern Tribes would hold these lands in Indian Territory in perpetuity lasted just over sixty years.[16] The breakup of tribal lands via Allotment paved the way for the conversion of Indian Territory and Oklahoma Territory to the state of Oklahoma in 1907. Leading up to this moment, there had been a number of "land runs," where white settlers flooded into the state seeking to steal land and set up homesteads, the most famous taking place in 1889 on April 22, a day that would later become Earth Day, of all things. The Enlarged Homestead Act of 1909 increased the typical settler allotment under the Homestead Act of 1862 from 160 acres to 320 acres on lands considered marginal and in need of more sustained irrigation in order to produce crops.[17] So while Native nations across Oklahoma saw their lands divvied up into 160-acre allotments for themselves, their "conveniently" remaining or "surplus" lands were parceled out to white settlers, who, seeking to set up their own farms, were potentially eligible to receive double the Native land allotment.

Even though the lines between Allotment policy and the Dust Bowl are almost direct and unbroken, few scholars aside from Hannah Holleman in *Dust Bowls of Empire* have made these connections explicit. And when they do, some scholars misalign key details and speak in rhetorical claims drawn from settler colonialism. David Montgomery goes so far as to write, "The Territory of Oklahoma (Indian territory, in Choctaw) was set aside as a reservation for the Cherokee, Chickasaw, Choctaw, Creek, and Seminole nations in 1854. It did not take long before the Indians' practice of maintaining open prairie seemed a waste to land-hungry settlers. . . . Between 1870 and 1900, American farmers brought as much virgin land into cultivation as they had in the previous two centuries."[18] There are a number of problems with this assessment. For starters, Indian Territory and Oklahoma Territory were two separate federal territories with separate histories. Initially, Indian Territory, as people then understood it, was created in *1834*, by the Indian Intercourse Act; Oklahoma Territory was incorporated in 1890 and comprised the western part of the later state. Furthermore, the Five Southeastern Tribes, which mainly held lands in eastern Oklahoma, had been place-based agricultural-cultivation nations since time immemorial, and several of the Five Southeastern Tribes included citizens who had been plantation owners and enslavers. Montgomery's assessment of their supposed "open prairie" lifestyle ignores this well-documented historical reality, and it fails to account for how the invasion of settler agriculture via the plantation economy even changed the gender balance of land management in these nations, where women traditionally controlled agricultural processes.[19] Moreover, while the word *oklahoma* indeed means "land of the red people" in the Choctaw language, it doesn't directly translate to a concrete term—Indian Territory—created by the US government. Montgomery's use of *Indian territory* is too close to the place-name Indian Territory, creating confusion between those meanings and suggesting that the Choctaw word directly translates from US policy. And last, there is no such thing as virgin land. Anywhere. While these misapplied details from Montgomery might seem trivial, I contend that they demonstrate a profound disregard for the Indigenous stakes of this erosion event, making it a mere prelude to settler colonialism. It flattens and erases the attempted genocide through continued land theft via erosion, and it converts Cherokee women like Florence Thompson into nameless white migrant mothers.

To mitigate these mischaracterizations of the Dust Bowl, it is necessary to do more than see Oklahoma and by extension the Midwest as a vast, uncomplicated, homogeneous prairie. There were Indigenous people (Caddo, Osage, Wichita, and Kichai, among others) in present-day Oklahoma before Removal forced the southeastern nations to set up home in the region. The original Indigenous nations of the region were themselves then removed, pushed, and crowded into new and reduced territories by Removal policy, both directly and indirectly. The agricultural situations in eastern Oklahoma, populated mainly by members of the Five Southeastern Tribes by the late nineteenth century, differed substantially from those in western Oklahoma, where members of western tribes such as the Comanche, Cheyenne, Arapaho, Kiowa, and Apache from the southern Great Plains had been crowded into reservations.[20] Additionally, even during the Removal scheme of the 1820s and 1830s, the United States certainly saw the Five Southeastern Tribes as a kind of potential "buffer" in Indian Territory against these western tribes, who seemed even more unfamiliar in their customs or even "dangerous" to white settler ways.[21] As Alaina Roberts demonstrates, the Five Southeastern Tribes became then both agents and victims of settler colonial practices, as some individuals thought they might differentiate themselves from their western Indigenous neighbors.[22] Despite these differences in climate, lifestyle, and history, Allotment policy affected *all* of Indian Territory and later Oklahoma Territory, and it set the stage for one of the greatest ecological disasters North America had ever seen.

The ecology of western Oklahoma was not conducive to large-scale crop-based agriculture. Indigenous nations of the southern Great Plains knew this and had long used the area for grazing while maintaining more permanent settlements in other places.[23] The plains sustained the bison; the bison sustained the nations of the plains. However, as the United States slaughtered bison to the brink of extinction (itself an attempt to break the Indigenous nations of the region), the plains appeared to be an empty surface to those who wanted them to be. Even though practically everyone knew that there was not enough rainfall in the area to support any attempt at monoculture, developers and investors were fond of repeating the idiotic slogan "Rain follows the plow."²⁴ Developers also touted "scientific" claims suggesting that the climate was changing: that the long period of plains drought was ending and the earth was entering a new, "wetter" period. Both of these claims are absurd. However, they induced white settlers, who were eager to have their own homesteads, to believe that the Great Plains was the next great piece of the American story and that they were the protagonists. As a result, the settlers plowed up the southern Great Plains, destroying the buffalo grass that had held the earth down for millennia and becoming known colloquially as *sodbusters*, a pejorative term initially used by ranchers, who, due to their own investment in cattle grazing, did not think the grasses should be plowed up. These sodbusters (who later adopted the term as a source of pride) relied on one crop more than any other: wheat.²⁵

As World War I approached, the US government, on the advice of then head of the US Food Administration Herbert Hoover, buttressed the price of wheat, further accelerating what many historians have called "the great plowup." As Timothy Egan explains:

> No group of people took a more dramatic leap in lifestyle or prosperity, in such a short time, than wheat farmers on the Great Plains. In less than ten years, they went from subsistence living to small business-class wealth.... In 1910, the price of wheat stood at eighty cents a bushel.... Five years later, with world grain supplies pinched by the Great War, the price had more than doubled. Farmers increased production by 50 percent.... With Russian [wheat] shipments [to Europe] blocked, the United States stepped in, and issued a proclamation to the plains: plant more wheat to win the war. And for the first time, the government guaranteed the price, at two dollars a bushel, through the war.... Wheat was no longer a staple of a small family farmer but a commodity with a price guarantee and a global market.²⁶

This emergence of a government-backed global commodity signaled to famers in the southern Great Plains that they should plow up even more land, destroy even more buffalo grass, and plant even more wheat. Indeed, Hoover's plan likely saved many from starvation around the world.²⁷ However, on the southern Great Plains, it set into motion events that would later starve the very people who had sown the seeds of both the Allied victory and, eventually, their own destruction.

The wheat-growing craze continued beyond the war, even through the beginning of the financial collapse of the Great Depression. By the end of

1931, thirty-three million acres of the southern Great Plains had been effectively destroyed due to surface erosion, and in 1932 the land began to blow away in increasingly cataclysmic dust storms that traveled all the way to Washington, DC, and even affected navigation in New York Harbor.[28] Eventually, one report suggested that Oklahoma alone had lost 440 million tons of topsoil.[29] Soil scientist and chair of the Soil Erosion Service under the US Department of the Interior Hugh Hammond Bennett stated unequivocally, "Of all the countries in the world, we Americans have been the greatest destroyers of land of any race of people barbaric or civilized."[30] In his Great Plains Drought Area Committee of 1936, Bennett stated, "Mistaken public policies have been largely responsible for the situation ... a mistaken homesteading policy, the stimulation of war time demands which led to over cropping and over grazing, and encouragement of a system of agriculture which could not be both permanent and prosperous." Moreover, he continued, "The settlers lacked both the knowledge and incentive necessary to avoid these mistakes. They were misled by those who should have been their natural guides."[31] Simply put, even in the 1930s, even if they wouldn't have used these exact terms, experts knew that the Dust Bowl was caused by settler colonialism and white settlers directly encouraged by US policy. Although Bennett suggests that government agencies should have been "natural" guides to prevent such a crisis, such an assessment is dubious given that the US government also had a vested interest in settler colonialism and quick capital. In these remarks, then, we glimpse a tension between ideological beliefs about the function of government: to prevent and mitigate ecological crises or to prop up an economy. In many ways, this tension continues to inform the present-day United States, where those who are led (either actively or passively) to cause ecological damage also personally suffer the eventual environmental and economic consequences.

The most famous fictional assessment of these settlers who both created and suffered from the Dust Bowl is John Steinbeck's *The Grapes of Wrath*. However, many readers often forget that in Steinbeck's novel the final blow to the family whose story it tells comes not from surface erosion in Oklahoma but from erosive gullying from a flash flood that washes out their camp and possessions in California. Thus, as oxymoronic ever-moving settlers, the family cannot escape the phenomenon of vanishing soil, which follows them across the continent. As with Lange and Thompson, the story of the Dust Bowl often ties Oklahoma and California together in an ever-shifting play of historical memory.[32] Another often overlooked detail is that Steinbeck's migrant family, the Joads, aren't from the western Dust Bowl region of the state at all. Rather, they hail from Sallisaw, Oklahoma, which is in the Cherokee Nation, near the border with Arkansas. Some critics, such as Donald Worster, have criticized this oversight, observing that "Steinbeck, like most other Americans, assumed too simply that people like the Joads were victims of a natural disaster that gave the banks and the landlords an excuse to put them off the land, but in truth, their somber story was only peripherally connected with the drought on the plains."[33] Others grant the author a bit more grace, such as Joel Hedgpeth, who notes, "Essentially a sage [saga] of people driven from their land by forces beyond their control (overlooking the fact that their own misuse of the land had set the process in motion), of confrontation in an environment where there was no new land for the taking, and no interest in the dispossessed on the part of the entrenched establishment except as seasonal field workers, *The Grapes of Wrath* was in no sense an environmental story. The Dust Bowl is a fait accompli as the novel begins, and the book is about what happens after environmental disaster."[34] However, I don't think it is necessary to get bogged down in questions of Steinbeck's intention or ignorance; instead, I want to pause on the fact that Steinbeck's novel *does* begin in the Cherokee Nation, outlining the choices and challenges of a family of white cotton sharecroppers who have managed to destroy the land in one generation.

Few critics have considered the Indigenous undertones of *The Grapes of Wrath*, with the notable exception being Choctaw- and Cherokee-descended scholar Louis Owens. Owens discusses the Indigenous figure within Steinbeck's novel at the level of the symbolic, and while I mostly agree with him that for Steinbeck, "the Indian is of significance only as a symbol of the destructive consciousness underlying American settlement and the westerning pattern," I also interrogate how this symbolism connects with the material reality of Indigenous land dispossession, erosion, and white pathos.[35] Steinbeck alludes to settler colonialism on the first page, beginning the novel with the lines "To the red country and part of the gray country of Oklahoma, the last rains came gently, and they did not cut the scarred earth" and ending the second paragraph by noting, "The air was thin and the sky more pale; and every day the earth paled."[36] This paling earth serves as the opening metaphor for a land dying from industrial settler agriculture—or put differently, whiteness—while faraway banks are "monsters." Speaking in the collective, disembodied third person, Steinbeck outlines the "owner men's" position, with the "squatting" tenant farmers speaking in response: "And at last the owner men came to the point. The tenant system won't work anymore. One man on a tractor can take the place of twelve or fourteen families.

Pay him a wage and take all the crop. We have to do it. We don't like to do it. But the monster's sick. Something's happened to the monster," to which the tenants reply, "But you'll kill the land with cotton."[37] The owner men respond, "We know. We've got to take the cotton quick before the land dies. Then we'll sell the land. Lots of families in the East would like to own a piece of land."[38] When the owner men state clearly, "You'll have to get off the land," the tenants respond tellingly, "Grampa took up the land, and he had to kill the Indians and drive them away. And Pa was born here, and he killed weeds and snakes. Then a bad year came and he had to borrow a little money. An' we was born here. There in the door—our children born here."[39] They continue to resist as the owner men insist the bank monster will have its way, saying, "But it's our land. We measured it and broke it up. We were born on it, and we got killed on it, died on it. Even if it's no good, it's still ours. That's what makes it ours—being born on it, working it, dying on it. That makes ownership, not a paper with numbers on it."[40] And then ultimately, "The tenants cried, Grampa killed Indians, Pa killed snakes for the land. Maybe we can kill banks—they're worse than Indians and snakes. Maybe we got to fight to keep our land, like Pa and Grampa did."[41]

This white agrarian fantasy of land justice not only reveals the tenants' understanding of where their land comes from but also demonstrates that their roots are as shallow as the monoculture cash crops they have depended on. They are people of the surface, and the surface is blowing (and washing) away. Their anxiety for the vanishing earth is not about the land itself but about their own loss of dominion over it. In this way, in their articulation of their own dispossession, the farmers illustrate the logic Robert Nichols describes in his *Theft Is Property!* whereby "new proprietary relations are generated but under structural conditions that demand their simultaneous negation."[42] While the white settler farmers have most certainly not been dispossessed in the same way as the Indigenous people they killed for the land, Steinbeck uses them to articulate how "theft is the mechanism and means by which property is generated."[43] This is not to say that Steinbeck undercuts the farmers' position as victims of capitalism, but any initial sympathy for these white settler farmers is undermined by their self-admitted roles as agents of both genocide and ecocide who used theft to create the very property whose loss they now mourn.

Steinbeck's novel is divided between these interstitial vignettes and the plot centered on the Joads. Many contemporaneous critics expressed their disregard for the intercalary chapters that waxed philosophical or imagistic on the surrounding context.[44] I, however, agree with eventual readers and scholars who came to appreciate Steinbeck's use of the form to weave a particular story of one family into a larger context of the economic and ecological collapse of exploitative agriculture.[45] The interstitial chapters are like the plow ruts themselves, turning up histories that make the very question of settler land attachment uncomfortable. In these instances, Steinbeck erodes the reader's sympathy for the white families that they might otherwise be supposed to see as uncomplicated victims in a great anticapitalist novel. In one interlude chapter, Steinbeck presents a character explaining how he was "a recruit against Geronimo" and giving a graphic account of killing a young Native man; he then converts this to a metaphor about hunting pheasants, to explain his own feelings after seeing the murdered Native man's body up close: "An' bang! You pick him up—bloody an' twisted, an' you spoiled somepin better'n you; an eatin' him don't never make it up to you, 'cause you spoiled somepin in yaself, an' you can't never fix it up."[46] This tragic regret requires parsing. For one, the old man storyteller dehumanizes the Native man into the figure of a pheasant (a fact that Louis Owens notes is offensive in and of itself). And two, the tragedy seems not to be in taking a life but in spoiling *himself*, as he attempts to sublimate his murderous actions into a pathos of his own loss of character. As Owens puts it, the characters ultimately only "mourn for their own loss or diminishment, not for the actual people killed or driven from the land."[47] This vignette, then, offers a striking metaphor for the entire narrative of the Dust Bowl, where destruction of Indigenous peoples and lands can only be understood insofar as the white settlers stay on the surface of their own consciousness rather than imagine the deeper history of how their actions have affected others.

In addition to these moments, Steinbeck peppers the main Joad plot and the alternating expositional chapters with Cherokee characters. Besides a Cherokee girl dancing with a Texas boy in chapter 23, just after the old man tells his "Geronimo story," there is Jule Vitela, who is described as "half Cherokee."[48] When Tom Joad says to him, "They says you're half Injun. You look all Injun to me," Jule responds, "No . . . Jes' half. Wisht I was a full-blood. I'd have my lan' on the reservation. Them full-bloods got it pretty nice, some of 'em."[49] This exchange makes almost no historical sense, given that, one, Allotment policy allowed mixedblood Cherokee people to register for allotments; two, dissolution of reservations had been attempted throughout this period, rendering the political Cherokee reservation as technically "nonexistent"; and, three, Cherokee identity worked within the community in far more nuanced ways. Steinbeck's ignorance of these details speaks to the larger ignorance of Native histories across American narratives—about erosion or otherwise. Owens notes that this grossly inaccurate detail is somewhat odd for Steinbeck, who was known to have conducted a lot of research for his novel, but like Owens, I contend that what Steinbeck knew or didn't know about Cherokee reality and identity is not the point.[50] Rather, the important fact is that Steinbeck very well understood that *there was* a Cherokee diaspora tied up within the Dust Bowl migrations. Reading for the Indigenous presence within the narrative of Dust Bowl erosion is not an exercise in retrospective imagination. Indigenous lands and Indigenous peoples have always been a part of this story, regardless of any one author's monocrop understanding of history.

Like Steinbeck, Dorothea Lange and Paul Taylor did not necessarily understand the Indigenous contours of the erosion story they were telling. For her part, Lange was certainly aware of Indigenous history in the Southwest, as some of her earliest nonstudio portrait photography had been of Native people in the region when she had traveled and lived there with her first husband, Maynard Dixon.[51] However, her images of Indigenous individuals never move beyond the more clichéd contemporaneous representations of southwestern Native peoples.[52] Thus, the fact that the most iconic photograph of her career was of an Indigenous woman whom she didn't recognize as Indigenous illustrates how a portrait of a person can be determined as much by the person behind the camera as the one in front of it. Ironically, as James Swensen quotes in his study *Picturing Migrants*, Lange *thought* she saw what was before her, explaining in a later interview, "I saw these people. And I couldn't wait, I photographed it. . . . Luckily my eyes were open to it. I could have been like all the other people on that highway and not seen it. As we don't see what's right before us. We don't see it till someone tells us."[53] Given her historically revealed oversight, this assertion reads all the more like an instance of Nicholas Mirzoeff's theory of "white sight," where "what is visible . . . is not simply the result of vision" but rather "an event that takes place between more than one person." As he explains further, "Vision is personal, always interfaced with other forms of perception. . . . [W]hat white sight sees is made, not found."[54] Lange did see a poverty that others refused to see, but she could not see the story of Indigenous Removal before her—despite her quest for insight, she misunderstands and misaligns key details of the surface in front of her and her camera. Notably, however, Lange and Taylor did not think the story of the Great Depression or Dust Bowl was exclusively about whiteness or the southern Great Plains, even while Roy Stryker (Lange's boss) and the Farm Security Administration believed, perhaps correctly, that documentary work featuring impoverished white people would render the most sympathetic response from the larger white public.⁵⁵

As Lange and Taylor conducted their work for the Farm Security Administration, they frequently trafficked in two central themes even as they pushed against them: the depleted soil and the loss of status among white people. As Lange biographer Linda Gordon explains, "Taylor and Lange liked the metaphors *social erosion* and *human erosion* to describe the correlate and consequence of soil erosion."⁵⁶ Taylor had leaned on geological metaphors since at least 1933 when he wrote, "As the faulting of the earth exposes its strata and reveals its structure, so a social disturbance throws into bold relief the structure of society."⁵⁷ Settler colonialism creeps in everywhere around the erosion metaphors both in their own work and in later retellings. In 1935 Taylor began a project to count the migrants coming into California across the Yuma bridge, and as Gordon explains, "He realized that it was a mass migration ... that would equal the gold rush," tacitly linking this environmental and labor crisis to a previous crisis of invasion, violence, and land theft for Indigenous people on the West Coast.⁵⁸ Gordon adds, "Lange, thrilled to be a *pioneer* investigator, tended in her later years to magnify what they had witnessed at the time," and she quotes Lange's Archives of American Art interview with Richard Doud, where the artist remarked, "That was the beginning of the first day of the landslide that cut this continent. . . . [T]his shaking off of people from their own roots started with those big storms, and it was like a movement of the earth."⁵⁹ Gordon continues by asserting that "[Lange] was right these people were refugees, uprooted from generations-old farms and dispossessed of all but what they could wear or stuff into jerry-rigged jalopies."⁶⁰ However, for these white migrants from the lower Midwest, this assessment simply cannot be true. As of 1936, white settlers had only been able to set up farms in the region and "put down roots" for roughly (and generously) forty-five—maybe fifty—years. These are not deep-rooted people leaving behind ancestral lands. These remarks from Lange and Taylor reveal the extent to which they saw the migration both in geological terms and in direct relation to previous waves of settler colonialism. When Gordon picks up the language of Lange as a "pioneer" and of migrants as from "generations-old farms," one has to stop and consider to what extent contemporary critics continue unwittingly to create a nexus among lamenting erosion, valorizing colonialist endeavors, and trafficking in an anxiety of lost white supremacy.

Despite the fact that the Dust Bowl and larger Great Depression affected people from every racial and ethnic background, Taylor and Lange's Farm Security Administration work came to focus exclusively on whiteness. As Gordon explains, "Taylor and Lange did not explicitly call for the aid *because* these migrants were white. But the photographs in their context did. Their conscious use of white images to win support for their recommendations reflected an unconscious shock that even whites lived in such terrible conditions. And, of course, the Okies and Arkies sang that tune to Lange and Taylor: 'We *did* live like white people.' 'We ain't no paupers. We hold ourselves to be white folks.'"[61] And ultimately, according to Gordon, even though "Okies were rarely treated as badly as people of color," nonetheless "New Deal progressives like Lange and Taylor believed they could not afford *not* to exploit racialized sympathy for the Okies as a means of mobilizing support for better treatment for farmworkers in general."[62] Despite any intention they may have had, Lange and Taylor became key players in linking erosion to the anxiety and pathos of white loss.

However, once Lange and Taylor embarked on their own publishing venture in *An American Exodus*, they took the opportunity to break from the constraints of simply marketing poverty as a concern of white migrants from the Dust Bowl. In contrast to contemporaneous photographic books featuring rural poverty, such as Erskine Caldwell and Margaret Bourke-White's *You Have Seen Their Faces* (1937), Archibald MacLeish's *Land of the Free* (1938), and James Agee and Walker Evans's *Let Us Now Praise Famous Men* (1941), all of which primarily or exclusively featured white people as their photographic subjects and protagonists, *American Exodus* begins with two chapters focused on African American rural life and poverty. Across the rest of the book, Lange and Taylor frequently center the words and images of Mexican and Filipinx subjects. *American Exodus* also makes clear that erosion had not appeared as an agricultural problem only in the lower Midwest of the 1930s. It includes a large portion of Lange's documentary photography from the US South, and it deals extensively with cotton as the centrally destructive crop. For one caption Taylor and Lange quote southern sociologist Arthur Raper, who argues that "the collapse of the plantation system, rendered inevitable by its exploitation of land and labor, leaves in its wake depleted soil, shoddy livestock, inadequate farm equipment, crude agricultural practices, crippled institutions, a defeated and impoverished people."[63] Raper's assessment is largely correct: the Dust Bowl, with its impoverished conditions, was not the first ecological crisis of US agriculture, nor was it exceptional in its results. The event had only come to seem this way in the larger popular imagination because the depicted victims had been white people of the imagined heartland. The exceptional perception of the Dust Bowl depended on certain fantasies of regionalism, which psychologically quarantines some crises to certain parts of the country while seeing other regions and peoples as metonymic for the nation as a whole. Despite Lange and Taylor's attempt to provide a more holistic account of erosion among all US peoples, *American Exodus* was largely a flop. As Gordon observes, "The book might have won greater popularity had it featured only 'Okies.'"[64] Ironically, even though Lange and Taylor sought to link the ecological and human destruction of the US South to the crisis in the lower Midwest and the resulting migrant labor emergency on the West Coast, they decided not to include perhaps the one photograph that could have told that story better than any other (had they known what it actually depicted): *Migrant Mother*.

By all ethical and moral standards, Florence Thompson should have been born in the Cherokee Nation of the southern Appalachians, in lands currently occupied by the states of Alabama, Georgia, North Carolina, and Tennessee. Had she grown up among Cherokee relatives and homelands, she would have experienced a matrilineal clan system where women largely controlled both the land and the agricultural policies and practices of the Cherokee Nation.[65] These practices were informed by the respect attached to a central figure in the Cherokee worldview: Selu, the Corn Mother. As Theda Perdue explains, "The connection between women and corn gave women considerable status and economic power because the Cherokees depended heavily on that crop for subsistence."[66] The world where Cherokee women controlled the agricultural decisions of their nation (and by extension many other facets, as every society knows that the people who control the food supply have a say in multiple arenas) was damaged largely by the invasion of the plantation economy of the US South. As white settlers began to establish plantations based on their kidnapping and enslavement of African peoples, they influenced many affluent Cherokee people to do the same if they wanted to "compete" in the new imposed economy.[67] As a result, across the eighteenth century, many prominent Cherokee families adopted at the very least the outward appearance of the white settler family, with a man running the use of plantation lands, enslaving African people and their descendants, and sending their children to white missionary schools.[68] The Cherokee Nation also set up a new government structure based on representatives who were men, although some scholars point out how the specific contours of this new constitutional government managed to fold in older, more traditional forms of leadership.[69] Nevertheless, this quick transition fundamentally destabilized the long-practiced gender balance of Cherokee society. It disrupted not only women's power but also the stewardship of the land as it related to the traditional ecological knowledge women had honed in the region for thousands of years.

Eventually, white settler aggression culminated in the Indian Removal Act of 1830. The state of Georgia, supported by the federal government, forced the majority of Cherokee Nation citizens from their homelands in the late 1830s.[70] As Perdue explains, there was a gendered component to Removal itself:

> Opposition to removal found strength and sustenance in the traditional culture of Cherokee women. Their association with the land and common title to it as well as their centrality to family and community quietly challenged the men who accumulated vast wealth, acted out of self-interest, and asserted political authority over other Cherokees. But the institutional changes implemented by these men had largely eliminated women from public action. They seemed to retreat into the shadows, and while they exerted considerable moral force during the removal crisis, they had even less power to resist removal than the male leadership whose authority the federal government subverted.[71]

This gendered angle of Removal and later Cherokee history in Indian Territory adds an important layer to how one might understand the texts surrounding Allotment and the eventual Dust Bowl.

Removal to Indian Territory devastated the Cherokee Nation, but despite this, the Nation survived and rebuilt, with their emergent center at Tahlequah, just outside of which Florence Leona Christie (Owens Thompson) was born in 1903.[72] By 1927 she and her first husband, Cleo Owens of Mississippi, had already migrated to Oroville, California, where Owens died of tuberculosis. When she became pregnant by a "well-to-do Oroville business owner" in 1933, she returned to Oklahoma in fear that the man, his family, or the authorities would intervene, and that she would lose the infant or even all of her children.[73] In 1934 she and her children returned to California with her parents and other family members, becoming at that time a part of what many would recognize as the Dust Bowl migration. However, the circumstances that led Florence Thompson to be in California are better understood as multiple removals—of the status of Cherokee women leaders under the pressures of a patriarchal plantation economy, of Cherokee people from their homelands in the Southeast, of Cherokee nationhood in their new adopted lands, and of Cherokee territory following Allotment and Oklahoma statehood—not migrations. Because of these removals, Thompson and her family had been forced to live close to the surface of the earth rather than grow from the roots that should have been theirs in the Southeast. Lange and Taylor's decision not to include this portrait in *American Exodus*, where they explicitly try to connect the disasters of the US South to the lower Midwest to the West Coast, is almost certainly due to the photograph's by-then ubiquity. However, it's hard not to see their ultimate unwitting ignorance in this decision as yet another removal of an Indigenous story from the narrative of American erosion.

Like Florence Christie Thompson, Rollie Lynn Riggs was born in the Cherokee Nation in Indian Territory during the time of Allotment. His own Cherokee mother, Rose Ella Duncan Gillis, passed away as a result of typhoid fever in 1901, just two years after his birth, and his white settler father married another Cherokee woman, Juliette Scrimsher Chambers, six months later. Before her death, Riggs's mother had enrolled her children under the Dawes and Curtis Acts, securing herself a 160-acre allotment that was then willed to her children following her passing. Riggs never developed a good relationship with his stepmother, and by his own accounts, his father also struggled to see the value in his son's gifts and interests as a writer and artist.[74] Riggs's identity as a gay man also likely complicated his relationship with his father, but many descriptions of their supposedly fraught relationship lean on euphemistic descriptions of Riggs's sexuality as the cause for their disconnect. Thus, this understanding of their relationship might best be understood as more nuanced than some critics have allowed.[75] While some scholars seem to downplay Riggs's association with and attachment to his Cherokee family, instead emphasizing his father's white settler identity, I don't follow this thinking.[76] There is enough evidence in Riggs's work, correspondence, and personal journals to support the claim that he very well understood his own Cherokee identity.[77] Furthermore, because Cherokee kinship traditionally follows the matrilineal line, it's likely that Riggs *always* understood his belonging in the Cherokee Nation. Growing up in the Nation outside of Claremore, Riggs lived in a world not unlike Thompson's and even that of Steinbeck's fictional Joads: a place where Cherokee citizens, Cherokee Freedmen, and white settlers negotiated lives that intertwined by both necessity and choice. This is a life that many outsiders are not predisposed to see, just as Lange and so many others could not see Thompson as a Cherokee woman.

Because of the time and place where he came of age, perhaps no person other than Lynn Riggs could have written the play that would become *Oklahoma!* While the 1940s musical became something of a redemptive ode to the place that had suffered immeasurably during the 1930s, Riggs's original play, *Green Grow the Lilacs*, was composed ahead of the Great Depression and Dust Bowl. Despite this, it's impossible not to read his work as an uncanny foreshadowing of the tragedies to come. When the play is read in conjunction with his tragedy *The Cream in the Well* (1940), written after the Dust Bowl, a prismatic view of Oklahoma emerges that takes into account the complicated destruction of the region's land via settler colonialism. For their part, Rodgers and Hammerstein converted Riggs's uncertain ending for Curly and Laurey into a happy marriage plot for their musical. However, Riggs's play—especially when placed within the scope of his other works—reveals a more nuanced and destructive image of Indian Territory, where individuals negotiate hard realities that may or may not have happy endings and that may or may not warrant enthusiastic exclamation marks setting off a place as an unmitigated triumph.

Riggs himself thought about his plays as working in tandem to paint a portrait of his home in all its complications. He writes, "People are always asking me about Oklahoma. Sometimes they say: 'I know a lot about Oklahoma, from your plays.' This always makes me ill at ease. The range of life there is not to be indicated, much less its meaning laid bare, by a few people in a few plays. Some day, perhaps, all the plays I will have written, taken together, may constitute a *study* from which certain things may emerge and be formulated into *a kind of truth* about people who happen to be living in Oklahoma instead of South Dakota."[78] Here Riggs rejects the exceptional and the totalizing, instead recognizing the particular and diverse. His comment on building "a kind of truth" demonstrates his understanding of what literature of and about a particular region can do in terms of representing and shaping perception of a place. Rather than appealing to a place-based determinism, Riggs forwards "a kind of truth" akin to what Scott Romine articulates in relation to constructing the US South, where he sees a region (in his case the South) as "situated" and "always implicitly narrative: a way of mobilizing space in efforts of immense variety and scope."[79] Additionally, it seems Riggs invites his audience to read his Oklahoma works as episodes of a larger dramatic representation of a place during a time of transition and conflict. In the remainder of the chapter, I attempt to do just that by thinking of *Green Grow the Lilacs* and *The Cream in the Well* as pieces of a larger study in how Riggs wrote through the drama of his own home in his own lifetime: the creation of the state of Oklahoma out of Indian Territory, and its rapidly resulting ecological and economic crises. If Steinbeck uses the Dust Bowl as a fait accompli, then Riggs renders it as dramatic irony.

Many critics have asserted the importance of Riggs's work for Cherokee literary traditions and for its representation of a specific moment of uncertainty during the turbulent time of Oklahoma statehood and the attempted dissolution of Indigenous nations in the region. Craig Womack, Jace Weaver, Daniel Heath Justice, Chadwick Allen, and James Cox and Alexander Pettit have all created rich analyses of Riggs's work, and in many ways, my readings merely extend their superlative scholarship to query specifically how Riggs managed to render the causes of the Dust Bowl onstage via his portraits of settler colonialism in process.[80] Most recently, Kirby Brown places Riggs's work squarely within the context of the challenges to Cherokee nationhood in the early twentieth century. He writes:

> Only a single published book and a few other chapters, essays, and dissertations touch on the period between Oklahoma statehood in 1907 and the political reorganization of the Nation later in the century. Of these, most gloss the period as an uneventful, if not regrettable, way station on the way to the contemporary moment of sovereignty and self-determination. In these and other studies, allotment is so devastating, statehood so catastrophic that even the possibility of Cherokees continuing to imagine themselves as a national community becomes unthinkable. The years that follow are thus left as an unqualified (and unexplored) rupture, while Cherokee nationhood languishes in the lamented "Chief for a Day" era.[81]

However, among his study of Cherokee writing from this period, Brown offers an alternative way to understand how neither Cherokee authors nor the Cherokee Nation simply withered into nonexistence. Instead, Brown outlines "two paths that Cherokees might have followed in the wake of statehood." He explains how

> one [is] *rooted* in commitments to political reorganization at home, the other [is] *routed* through the emergent mobilities, technologies, forms, and expanding social relations attending modernity, and both *grounded* in the people, places, and histories that continued to define contemporary Cherokee life. Thus, while it might be accurate to say that the Cherokee *Nation* entered a roughly sixty-year period of political dormancy following Oklahoma statehood in 1907, Cherokee *nationhood* remained very much a part of how Cherokees from this period continued to understand themselves and the multiple worlds in and across which they moved.[82]

Working with these earth metaphors from Brown—rooted, routed, and grounded—I come to my own reading of Riggs as a playwright pushing against a white pathos imposed onto the eroding earth and onto Cherokee power, even within his plays that seemingly have little to do with Cherokee identity. In this way, I follow Brown's reading of Riggs's Cherokee-focused play *The Cherokee Night* (1936), which in "its self-conscious disruption of linear, national time and its devastating interrogation of postallotment blood politics . . . refuses teleologies of settler progress and Native declension, turning instead toward alternative modes of recognition and belonging that are anchored to flexible understandings of kinship and shared historical experience."[83] I extend Brown's argument to show how Riggs also accomplishes these same outcomes in one play seemingly focused on settlers, *Green Grow the Lilacs*, and another where the characters' Cherokee identity does not initially seem to be of much significance to the plot, *The Cream in the Well*. Together, these plays illustrate how the damaged earth intertwines with the damages wrought by settler colonialism on the minds, bodies, and souls of both perpetrators and victims—and victims who might become perpetrators themselves.

For starters, it is entirely possible to see Riggs's *Green Grow the Lilacs* as a typical marriage comedy depicting a happier time (Riggs himself sometimes alluded to this), and this is certainly what the adaptation *Oklahoma!* sells its audience. However, I argue that *Green Grow the Lilacs* offers a much more ambivalent rendering of Indian Territory, one whose ending borders more on tragedy than comedy. As Womack argues and I agree, "In *Green Grow the Lilacs* a number of really weird tensions exist that might challenge the innocence critics, and even Lynn Riggs himself, ascribe to it."[84] For example, whereas in the later musical adaptation Curly is freed from the law in a quick, virtually sham trial for his killing of Jud (called Jeeter in the original play) in self-defense, in Riggs's original play the curtain closes before the trial, leaving the audience uncertain about what the future holds for Laurey and Curly. Moreover, if one follows Weaver's lead that the play provides ample evidence to suggest that Curly is indeed Cherokee and not a white settler, the air of uncertainty grows thicker.[85] When this uncertainty is combined with the dramatic irony for some audiences of knowing that Oklahoma statehood hangs on the horizon and with it the attempted dissolution of the Cherokee Nation, where these characters live, the cloud on the horizon grows darker. And most important for my purposes, when one adds all of these factors to the repeated references to settler agriculture that pepper the play, *Green Grow the Lilacs* becomes a veritable prophecy of the twinned catastrophes of Allotment and ecological destruction via erosion.

One of the most important underpinnings to my reading is Weaver's proposal that Curly is a Cherokee character and that Rodgers and Hammerstein either missed this coding or believed it necessary to whitewash Riggs's play for their celebratory musical. As he argues:

> Though it is easy to talk about a Cherokee mixed-blood like Riggs being so culturally alienated that he wrote a play about Oklahoma with no Indian characters, a close reading of his work leads to a suggestion that something far more subtle is at work in the ethnic cleansing of *Oklahoma!* . . . [B]oth the play and the musical take place in Indian Territory—not Oklahoma Territory. Claremore is in the heart of the Cherokee Nation. What I am driving at is the suggestion that *Green Grow the Lilacs* is not devoid of Indian characters at all but is in some sense a play *about* them.[86]

Weaver continues this argument with a quick smattering of contextual evidence, including that Curly's last name, McClain (or McLain) "is a fairly prominent Indian surname" and that "the nickname 'Curly' could have come about because Curly, as a mixed-blood like Riggs himself, had curly hair, an uncommon trait among Natives."[87] He adds that Curly's profession as a cowhand would have been marginally more likely for a Native person at this time, and he outlines all the ways that staging such a romance between the directly described white woman Laurey and the Cherokee man Curly would have been more than a 1940s white audience would have been willing to accept. In other words, like Florence Thompson, Curly became a white hero of Oklahoma rather than a Cherokee person negotiating yet another attempted destruction of his nation by the US government.

In fact, there is much more textual evidence for Weaver's reading of Curly than even he catalogs. In the opening scene, as Curly comes to the house singing, he jokes with Aunt Eller, "Must think I'm a medicine man a-singin' and passin' the hat around," essentially beginning his presence in the play with a direct and self-deprecatingly joking reference to himself being Native.[88] And later in the play, when he breaks out of jail to see Laurey, Aunt Eller asks him, "Who's after you, the old Booger Man?" again referencing a specific Cherokee tradition, the Booger Dance, where Cherokee performers donned terrifying masks and represented intruders who were coming to terrorize the community.[89] Notably, the Booger Dance is thought to have arisen from specific historical fears of Euro-Americans bringing death and disease to Cherokee settlements, and some say it was developed to teach Cherokee children to be skeptical of white settlers, who could kidnap and enslave them.[90] These quick asides, therefore, do more than offer comic relief. For audiences in the know, even in the passing teases between them, Aunt Eller and Curly clearly share a mutual understanding of a Native context where these jokes would make sense for a Cherokee speaker and, more troublingly, where this humor is paired with a suggestion of historical trauma.

However, as Womack argues, "In *Green Grow the Lilacs*, the Cherokee sense of place and Cherokee nationhood are evident in the concrete naming of recognizable places within the nation and stage directions such as '*Indian Territory, 1900*' more than in some kind of ethnographic scrutiny of Cherokees."[91] I agree with Womack's assertion and offer the above evidence more to convince the skeptical reader that one of the largely accepted heroes of the American stage has been Native all this time.[92] Following from Womack, then, more important for my investigation is what Curly offers about *this* place and *this* time in the *Cherokee Nation*—not as a person with some simple romantic land attachment through his Indigeneity but as a witness to an ever-encroaching US settler colonialism that will change his material way of life and the earth itself. For instance, Curly makes repeated references to Allotment and the world shifting to even more white settler agriculture, exposing his own anxieties about what this means for his future. He laments to Laurey, "If I'd ever a-thought—! Oh, I'd orta been a farmer, and worked hard at it, and saved, and kep' buyin' more land, and plowed and planted, like somebody—'stid of doin' the way I've done! Now the cattle business'll soon be over with. The ranches are breakin' up fast. They're puttin' in barbed w'ar, and plowin' up the sod fer wheat and corn. Purty soon they won't be no grazin'—thousands of acres—no place fer the cowboy to lay his head."[93] Curly thinks maybe this "new" path would have been a better one to follow, where he, too, could have been a sodbuster, and he directly laments the demise of the grasses of "thousands of acres." His crisis is virtually existential as he catalogs the quick changes being wrought on the landscape. Remarkably, in the 1920s, ahead of the destruction to come, Riggs writes into the mouth of what would become his best-known protagonist the exact conditions that would cause the Dust Bowl.

Later, on his wedding night, before the chaos of the shivaree and the eventual tragedy ensues, Curly again expresses deep attachment to the land, accounting in optimistic terms for both the changes and the continuities:

> Look at the way the hay field lays out purty in the moonlight. Next it's the pasture, and over yander's the wheat and corn, and the cane patch next, nen the truck garden and the timber. Everything laid out fine and jim dandy! The country all around it—all Indian Territory—plum to the

Rio Grande, and north to Kansas, and 'way over east to Arkansaw, the same way, with the moon onto it. Trees ain't hardly a-movin'. Branch bubbles over them limestone rocks, and you c'n hear it. Wild flower pe'fume smellin' up the air, sweet as anything![94]

Again at the end of the play, as he spends time with Laurey after he has broken out of jail while awaiting trial for the death of Jeeter, he ponders what all these changes on the land mean for his future with Laurey on the brink of the dissolution of Indian Territory:

> It come to me settin' in that cell of mine. Oh, I got to learn to be a farmer, I see that! Quit a-thinkin' about dehornin' and brandin' and th'owin' the rope, and start in to git my hands blistered in a new way! Oh, things is changin' right and left! Buy up mowin' machines, cut down the prairies! Shoe yer horses, drag them plows under the sod! They gonna make a state outa this, they gonna put it in the Union! Country a-changin', got to change with it! Bring up a pair of boys, new stock, to keep up with the way things is goin' in this here crazy country! Life is just startin' in fer me now. Work to do! Now I got you to he'p me—I'll 'mount to sump'n yit! Come here, Laurey. Come here, and tell me 'Good bye' 'fore they come fer me and take me away.[95]

In these lines there is an air of hope, but audiences know it is shadowed by uncertainty. Moreover, he assents to the necessity of "cut[ting] down the prairies" and "plow[ing] under the sod" with the help of machines that ecological historians blame directly for massive and unprecedented land degradation in the region.[96] In this moment we see Curly thinking through how to adjust his worldview to the conditions he sees facing Indian Territory in 1900.

The tragedy that hovers above all of this action is that Curly has killed Jeeter Fry in a scuffle, after Jeeter attempted to murder him and Laurey on their wedding night. Throughout the play Jeeter possesses a disturbing vendetta against the world, consistently beleaguered by what he perceives as people "no better than him" constantly looking down on him. He loves Laurey and is violently angry that she has married Curly. If we follow Weaver's premise that Curly is Cherokee, then Jeeter's anger and attempted murder of Curly and Laurey starts to read a whole lot less like a good-hearted shivaree gone wrong and a lot more like an attempted lynching of an interracial couple. As he tries to ignite the haystack where Laurey and Curly are standing, Jeeter yells, "Yanh, you thought you had it over me so big, didn't you? And you, too, Missy! Wanted sump'n purtier to sleep with. Yanh, you won't be a-havin' it long. Burn you to cracklin's!"[97] Given the contemporaneous hangings and burnings of Black men during this time for perceived, manufactured, or actual sexual contact with white women, these lines from Jeeter become a far more sinister and pointed attack.[98] He cannot accept that a white woman would choose a Native man over himself, and his retribution is their death by fire.[99] When Curly jumps from the haystack to stop Jeeter from killing him and Laurey, the two men tussle, and Jeeter falls on his own knife. However, Curly is still arrested for murder and taken into custody by the federal marshals of the US territories.

Recently, however, there has been a reevaluation of the character of Jeeter/Jud. Director Daniel Fish and actor Patrick Vaill staged a 2015 performance of the Rodgers and Hammerstein musical where Jud was rendered more sympathetic. Fish remarked that for him the play signaled "a community's need to create an outsider to sustain itself" and encouraged Vaill to play the character as more sympathetic. Vaill connected to the character as an outsider via his own identity as a gay man, and he drew from thinking about Riggs's life and sexuality in terms of difference.[100] On the one hand, this reconsideration of Jeeter via Jud has compelling elements, and indeed, Riggs does create the character as a complicated villain. However, I argue that it's also possible that it is so difficult for non-Native audiences to imagine a Cherokee protagonist that they might also inadvertently misread an antagonist's motivations. It is possible, then, that like the contemporary media obsession with understanding the beleaguered whiteness of a so-called real America (often associated with middle-American heartland politics), audiences are also drawn into trying to find a justifying motivation for Jud/Jeeter's murderous actions. The character is certainly affected by class and community politics. He, however, also spies on Laurey, makes repeated unwanted sexual advances toward her, and attempts to murder her and Curly when he doesn't get the romantic or sexual attention he desires. So while class prejudices might contribute to his sense of oppression in the community, he is far from an innocent victim of a generic American classism. Instead, Riggs's original character is entangled in racial, class, and gendered politics that are far more complicated than Rodgers and Hammerstein allow.

Rather than the ending of the original performance and film of *Oklahoma!*, which brushes all of these complexities under the rug just in time for the entirely white cast to sing an anthem to their forthcoming statehood with the lines "We know we belong to the land, and the land we belong to is grand!" in *Green Grow the Lilacs* the curtain closes on an entirely different scenario. The Cherokee Curly has been arrested by US federal marshals for killing a white man. While awaiting trial, he breaks out of jail to see his white wife for what he thinks may be the last time. The only historical tidbit that might have been going in Curly's favor is that the notorious "hanging judge" Isaac Parker was no longer presiding over Indian Territory, as the US Congress had created new federal jurisdictions in 1895.[101] The ending of *Green Grow the Lilacs* does not show the audience what happens next. It offers no evidence as to whether Curly will be found guilty of murdering Jeeter or whether he and Laurey will get to begin a new life together. Instead, Riggs has Curly singing from offstage, in the stage directions indicating, "From the bedroom has come the sound of Curly beginning to sing softly, 'Green Grow the Lilacs.'"[102]

This titular and closing song might be the best evidence that the play represents an erosion tragedy of Indian Territory and not a marriage comedy of Oklahoma. Importantly, "Green Grow the Lilacs" is a traditional song where the narrator sings about his love leaving him and his joining the army. The chorus goes:

> Green grow the lilacs, all sparkling with dew,
> I'm lonely, my darling, since parting with you,
> And by the next meeting I hope to prove true
> To change the green lilacs to the red, white, and blue.[103]

Riggs has Curly first sing this song to Aunt Eller "half-sarcastically" in scene 1, but Riggs notes in the stage directions, "The song with its absurd yet plaintive charm has absorbed him. And he sings the rest of its sentimental periods, his head back, his eyes focused beyond the room, beyond himself—."[104] This song clearly talks about the conversion of a green plant to the "red, white, and blue," signaling that something green and alive will change to the colors of the US flag. In some sense, the song is an assimilation story. One does not have to "read into" Riggs's play to see that it offers an *at best* ambivalent tragedy of what is to come in Indian Territory as it becomes incorporated into the United States. Riggs makes his meaning plain in the title. The "wild flower pe'fume smellin' up the air" that Curly earlier praises across Indian Territory on his wedding night is likely the green lilacs (a notably strongly scented flower) that are about to turn red, white, and blue. When Riggs closes the curtain on Curly singing this song to Laurey ahead of his certain rearrest and trial for the murder of Jeeter, it's hard to hold on to any hope for what the future holds for this couple, this land, or Indigenous sovereignty in the ever-expanding and invading United States. Instead of the Dust Bowl redemptive *Oklahoma!*, Riggs's *Green Grow the Lilacs* stands as a prophecy of the crisis to come.

At the other end of the crises of the Dust Bowl and Great Depression lies *The Cream in the Well*, written in 1940. This play follows a Cherokee family in 1906, the year before Oklahoma statehood, as they negotiate a number of melodramatic fractures in their family, including murder, suicide, gay sex work, and sibling incest. A number of critics, including Womack and later Cox, have taken up this play as a more politically direct piece of Riggs's work. They place it in the context of his repeated travels to Mexico and resulting plays set there, as these events might have politicized Riggs, leading him to return to thinking about Indian Territory in a more concrete way. I, however, would like to consider what effect the ten years between *Green Grow the Lilacs* and *The Cream in the Well* had on Riggs as he—like every person in the United States—had watched Oklahoma be brought to its knees by the ecological disaster of erosion wrought by settler agriculture and statehood. We know that Riggs understood these factors, as he explicitly outlined them *before* the Dust Bowl in *Green Grow the Lilacs*. In his preface to the script for publication, Riggs states directly that *Green Grow the Lilacs* is about how the characters "relate themselves to the earth and to other people," making it clear that the relationship between humans and the earth was ever on his mind.[105]

In *The Cream in the Well*, he stages a Cherokee family negotiating a number of traumas, but these traumas are rarely if ever about their Cherokee identity as such. However, Allotment, statehood, and characters' own futures seem intimately tied up with the earth via the land, farming, and the role women will have in the family amid this confusion. Only Cox directly points out how Riggs "dramatizes the lives of impoverished Indian Territory families and communities in the midst of struggles over the relative merits of ranching, farming, or emigrating to more populous regions. Indeed, Oklahoma statehood for Riggs stands as not only a disruption of Cherokee sovereignty but also an attendant disruption of traditional land use in the late nineteenth century in Indian Territory."[106] Even though Cox labels the family of *The Cream in the Well* as impoverished when they clearly are not, I want to take up this quick mention from Cox to consider how the question of agriculture here cannot be divorced from the question of women's power and agency as the traditional agricultural stewards of the Cherokee Nation. In this way, I follow Womack's wish to see how "feminist readings of [*The Cream in the Well*] might be rich in terms of considering why these women are cast as witches and the reasons they have to be killed."[107] Simply put, it matters for the play that the protagonist, Julie, is a Cherokee woman born into a relatively well-to-do Cherokee family. However, she, like her mother, seems to have witnessed her real, material power—specifically as it relates to the land—shrink to almost nothing, with yet another catastrophe, Oklahoma statehood, imminent.

Without a doubt, Julie is a woman with gifts and challenges: she is beautiful and comes from a clearly prominent family, as evidenced by how they've managed and expanded their allotments, and she was educated at the prestigious Cherokee Female Seminary. But she is also stuck on her family's farm and tragically in love with her brother, Clabe. Notably, in his working drafts of the script and in his journals, Riggs calls the play "Julie Sawters," signaling that the play revolves around her psyche and struggles.[108] Before the action begins, we learn she insisted that her brother leave their farm and join the navy, presumably to avoid the temptations that their far-too-close bond has generated. Before she essentially sends him away, she also convinces him not to marry a white woman, Opal, because in addition to being in love with him herself, she thinks Opal is beneath their family. After receiving a letter where Clabe states directly that he will *never* come home and that his parents should "ask Julie" if they want to know why *and* learning that in the meantime Clabe has been writing Opal love letters, Julie drives Opal to commit suicide by drowning herself in the lake. Over the course of this action in act 1, there are also several scenes where Julie's mother reiterates to her how she and Julie are "alike," with a poisoning darkness in their souls that must be cured through love. However, Julie cannot accept this. She seems embittered at the limited scope of her own existence, empowered and trapped by her education and with nothing more to do than stay at home and pine.

By act 2, Julie has married Gard, Opal's widower, after he essentially blackmails her into it by revealing he knows that she emotionally tortured Opal to death. At the opening of the second act, Julie and Gard are visiting her parents' house on Thanksgiving, and it's revealed that Clabe will also be coming home despite his stated previous intentions. After his arrival the audience learns that Clabe has been dishonorably discharged, and he has more than a few harsh words for the US military. In a climactic confrontation, he also admits to Julie that he earned the money for the jeweled brooch he sent her from China by selling himself to other men on the docks. This horrifies her but not quite as much as when Clabe tries to convince her they should just throw off convention and admit their mutual desire. Instead, she decides to also drown herself in the lake by walking out onto the fragile ice, and she asks Clabe to make sure nobody misses her in the house until she can accomplish her aim. He agrees, and the play ends with him listening to their slightly less intellectually gifted younger sister, Bina, playing a cheery piano tune and enthusiastically musing on her own likely engagement to an equally simple man named Blocky.

The play offers no future for Julie and possibly, by extension, for any accomplished Cherokee women outside of marriage. As Womack states, "I might describe *The Cream in the Well* as Oklahoma gothic. More crudely, I call this one of Riggs's 'Oklahoma Sucks' plays."[109] He isn't wrong. However, I think there's something in *why it sucks* that Womack, Cox, and others have left on the cutting-room floor, and that's how shifting agricultural practices since the beginning of colonialism have damaged not only the land but also Cherokee women's role within the Cherokee Nation. Indeed, Womack sees that he has cut this analysis away in favor of his critique of how the play deals with queerness, writing, "In order to go home, the play seems to say, one must kill off one's queer self. *The Cream in the Well*, like many a Riggs play, shows the wear and tear this erasure has on people's bodies and souls.... A fascinating aspect of... *The Cream in the Well*... is that instead of killing off the gay characters in the play[s], Riggs gets rid of the women. Needless to say, this is not an improvement."[110] I want to pick up these remnants of argument that Womack offers. Instead of the term *erasure* for the damage to bodies and souls, I think that Womack could have just as easily used *erosion*, as not just bodies and souls but the land itself is damaged as the women become increasingly damaged and, eventually, are killed. In other words, Riggs links the erosion of women's power to the erosion of the land. As Mishuana Goeman argues regarding Chickasaw author Linda Hogan's work, "Native women's bodies, as markers against territorial appropriation, Indigenous futurities and contestations of colonial politics, are a locus of gendered colonial meanings and a site of contest," and "the bodies of Native women are dangerous because they produce knowledge and demand accountability, whether at the scale of their individual bodily integrity, of their communities' ability to remain on their bodies of land and water, or as citizens of their nations."[111] Riggs's play explores this connection, but notably he does not offer a redemptive outcome for the women or the land.

Although it is largely accepted that Riggs based the character Julie on his stepmother (they even share a name), the motivations of the women in the play go beyond simply the biographical. In one telling journal entry, Riggs writes of his work on the play, "I evidently have not laid the ghost of my stepmother. I evidently still hate her." He continues, "Now in *Cream*—if successful (even though the reasons for Jul.'s behavior are not the same) she does terrible things—but understandably—and stands as a poignant and tragic figure in the end." Whereas he acknowledges his own hatred toward the person who initially inspired the character, he also ascribes to her different and seemingly understandable motivations. It seems that this working out of the reasons behind Julie's actions has some attachment to Riggs's own understanding of himself, as he writes under this paragraph reflecting on Julie: "<u>This ought to solve my life.</u>"[112] It's unclear just how working out Julie's character will solve Riggs's life, but it seems that in the play he attempts to get at the root of something about the young Cherokee woman's predicament in this specific context on the eve of Oklahoma statehood.

The play begins with Mrs. Sawters and her daughter Bina discussing the merits of intellect and education, given that Bina has chosen not to go to school whereas her sister Julie went to the Cherokee Female Seminary, called in the play simply "Seminary."[113] Riggs writes:

> BINA [Still cheerful.] It wouldn't a-worked, nohow. Julie's got the brains around here. Julie—and you.
>
> MRS. SAWTERS Oh, don't count on me. I'm pretty rusty, I'm afraid. Way off here in the wilds. Brains rust, you know that?
>
> BINA You mean like a plough-share?
>
> MRS. SAWTERS Like anything else you don't use. Use 'em and use 'em. Brains, I'm talking about. Till they sparkle. Till they shine. Yes, siree.
>
> BINA Uh-huh. [Thoughtfully.] I'd rather be a rusty old plough-share for me.
>
> MRS. SAWTERS I think you would.
>
> BINA Settin' out in the rain. *Or* the *sun*. A plough-share don't worry— but takes the weather like it comes.[114]

This opening exchange reveals Mrs. Sawters's awareness that her mind has not been put to its best uses. The fact that it goes unused like a rusty plowshare makes explicit the relationship between her intellectual work and an instrument of agriculture. The reassertion that the youngest daughter, Bina, doesn't mind getting rusty signals a possible future where Cherokee women become even further detached from their roles as traditional agricultural leaders. Moreover, when the elder daughter, Julie, appears, Riggs offers these stage directions: "Striking and dark, she has inherited something from her mother. There is power here, a nervous but self-assured power and arrogance. But something is gnawing at her, something darkly troubling and dangerous. She, too, however, like her mother, is full of a deep, controlled cynicism; she usually means much more than she says."[115] This gnawed power seems to emerge from the same disuse that Mrs. Sawters describes earlier. I contend that their shared "deep, controlled cynicism" results from their awareness that as women they are the traditional agricultural leaders of their nation, now living on a farm where there is little left for them to do or lead outside of simple domestic arts. As I outlined in the introduction, the verb *gnaw* is directly related to the earliest use of the verb *erode*, as it comes from the Latin *rodere* ("to gnaw"), and the *Oxford English Dictionary* notes that the earliest transitive form of *erode* indicates "to destroy by slow consumption."[116] Thus, one can conclude that Julie is experiencing the effects of erosion, as have her mother and even several generations of women before them. Riggs's play links the erosion of the earth to the erosion of women's power within their nation; over the preceding two hundred years, their scope of power has been completely usurped by colonial patriarchy.

This slow consumption of women's power continues throughout the action of the play. Multiple family conversations revolve around the imminent statehood of Oklahoma. Riggs assigns the most critical words about this development to Mrs. Sawters as she and her husband discuss the issue:

> MR. SAWTERS Country's openin' up fast. Gonna join the Union, they say. Paper says.
>
> MRS. SAWTERS That'll just ruin everything.
>
> MR. SAWTERS No, I don't reckon it'll do that. It's bound to come. Lot's of people rootin' for it. Jine up the two territories together is the idy. Only trouble is what to call it—Indianokla or Indiahoma. Reg'lar knock-down drag-out about it.
>
> MRS. SAWTERS Well, there's lots of land out here.
>
> MR. SAWTERS And lots of people back east 'thout dirt enough to spit on.
>
> MRS. SAWTERS The march of progress, I guess you call it.[117]

Mrs. Sawters's feelings about statehood as "ruin[ing] everything" indicate a pretty clear condemnation. She sees her husband try to keep up with the farming and taxes of their own allotted land, and I argue that there is more than a touch of recognition here of what statehood and an increased number of white settlers might mean for their own holdings. Additionally, while Mr. Sawters worries about how to keep up with all his farmwork with only his "two hands," Mrs. Sawters has already arranged a solution that she tries to press subtly on her husband by inviting their neighbors Gard and Opal to dinner so that Mr. Sawters can try to hire Gard to help. Here we see that Mrs. Sawters has the intelligence, desire, and forethought to make these decisions about the farm, but her work has to appear as a casual, slightly manipulated suggestion rather than open leadership. Her premonition that statehood will ruin everything, then, lands as a recognition that she is already tenuously holding on to what power she has left. Her sarcastic and cynical remark about the "march of progress" seals her critique of settler colonialism as the white American hunger for more land will again crash over her only recently and imperfectly rebuilt world in Indian Territory.[118]

Even the titular metaphor, "cream in the well," signals a backdrop of women suffering, further tying the problems of the play to the erosion of women's power. Mr. Sawters recounts that when they first set up farming in the town Verdigris (named for a nearby river but also notably a word that indicates corrosion), they wanted to buy the Lowry farm. However, as he tells it, the farm had a strange feature they could not get past:

> They had a well there—a big old stone well with the clearest coldest water in the section. It was shore a treat to dip 'er up on a hot day and guzzle 'er down. But we'd hang the milk and butter and things down in it—the way you do to keep it cool—and ever' time the blame stuff 'd spoil. I couldn't figger it out. Why, the butter 'd get so rancid you couldn't stay in the same room with it. Good fresh cream and eggs we'd put down there—and ever'time the same dadburned thing. We never knowed what could be the cause. A funny thing. We shore let that place go like a hot potato, and found us another'n. [*Gravely.*] Yeah, more things in this life you have to watch out for. So many things—*and most of 'em you cain't even see.* It's too much for me.[119]

This passage calls on knowledge that people who have country backgrounds and families might recognize but that likely goes uninterpreted by those without. In rural places one would sometimes hear stories about women who committed suicide by throwing themselves down wells. As *The Cream in the Well* features not one but two women's suicides by drowning, it's not a stretch to say that Mr. Sawters's comment about the well suggests a long history of women driven beyond what they can take, such that they can only die in an act of defiance by contaminating the water itself. This energy permeates the play, and it spoils everything it touches. The symbols of fertility that go rancid—eggs and milk—offer an association with women's reproductive power and futurity.

And indeed, the women in this scenario find it hard to imagine their own futures. Julie laments to Bina, "I hate cooking and the little drudgeries you're so fond of."[120] When she speaks about Clabe leaving for the navy, she sounds less as if she misses him and more as though she wants to *be* him, exclaiming, "Here was his chance—forced on him, it's true—but he was smart enough to take it! All his life he'd wanted to see the world, to see what went on in towns and foreign lands, the way the world was made," and when her father insists that isn't true, she continues, "He did! You don't know him the way I do. He didn't have any secrets from me. He was sick to death of the smallness and lonesomeness, the killing work that never let up, year in, year out, the hum-drum grind, slaving away in the backwoods, while he was young and adventurous and all life beckoned to him and—" before her father interrupts her again.[121] There is more than a tone of projection in Julie's description of Clabe's feelings, suggesting that she is the one who feels circumscribed by her life. In one handwritten note on a typescript draft, Riggs writes of the two characters, "Isolation of the 2 together in childhood[;] Felt they were different[;] Nobody could separate them."[122] This note almost evokes the two as twinned in their own existence, two sides of the same coin—a metaphor that Julie offers to Clabe about love and hate. Even though the play makes explicit the presence of an incestuous attraction between Julie and Clabe, I think this can also be read as a play about Cherokee siblings still struggling to adapt to a world where a traditional Cherokee gender balance has been completely upended. This balance is intimately tied to the earth itself. Clabe pushes and pulls against the pressure to take up agricultural duties that have only in the past 150 years even been within a Cherokee man's designated purview. For her part, Julie does not want a conscripted domestic existence devoid of power, and that is all she feels she has despite her exemplary education. The care of the land should have been her birthright. In the absence of this, her power has nowhere to go other than to turn inward and gnaw on her own existence.

The attachment among men, women, and agriculture is made explicit in several other places in the play. Across the action, Mr. Sawters, who is certainly a kind and sympathetic character, nonetheless discusses little more than plowing up more and more sod. He is almost constantly talking about putting in another eighty acres of oats or plowing Clabe's untouched allotment to put in corn. Mrs. Sawters expresses skepticism about these ventures, but Mr. Sawters is adamant about expanding the crops, seeming to become an enthusiastic agent of the settler colonial agriculture that has damaged his own nation and family. Most significantly, however, Julie directly relates her very soul to the damaged land. She says:

> Oh, Clabe...
> > [*She bows her head, blind with tears. After a moment, she can speak again.*]
>
> When we were young, everything was so clear and bright.
> We were happy.
> > [*Simple and moving, without self-pity.*]
>
> If you could see my real self now—
> My soul
> If there is such a thing.
> It's a field that wagons have been driven over, over and over again in the rains.
> The wheels have cut the juicy earth to pieces.
> It's packed solid underneath the ruts—
> Solid—like rock.
> And no seed will ever grow there any more.
> Never.
> It's me that drove those wagons up and down,
> Me that wanted the field to be different,
> The crop that grew to be another kind of grain.
>
> I can't lick what I am.
> I see it now.[123]

Here Julie imagines her soul as a spent, soil-exhausted field that cannot grow a different type of grain. Julie cannot see a future where any seed will ever grow again; moreover, she blames herself for destroying the land because she wanted "the field to be different." While her soul is the field, she is the one who has "cut the juicy earth to pieces." In this way, she mirrors her own father as both agent and victim of the violence wrought by exploitative and exhaustive settler agriculture. In Riggs's play, there is no future for a young woman blocked at every path like Julie. While Mrs. Sawters managed to find an outlet in marriage, her own family, and her subtle agricultural management by suggestion, Julie can't seem to find this for herself. The world has changed so rapidly that her power can only destroy her and what she loves. Her demise is the demise of the soil.

Like Florence Thompson and Steinbeck's Joads, Lynn Riggs's plays come out of eastern Oklahoma in the Cherokee Nation, not the western counties of the Dust Bowl. This is not to say that all of these texts and images fail to demonstrate an "authenticity" to the event itself but rather to query to what extent the popular narrative of the ecological crisis simultaneously drew from *and* erased the Indigenous dimensions of the story across the region. The theft of Indigenous land was not merely a prequel to the Dust Bowl. Indigenous peoples across Oklahoma lived through the Dust Bowl just as surely as they had lived through numerous other crises. Riggs, in particular, shows Cherokee characters who are bound up with and even participate in the very system that is attacking their traditional relationships to the land and to one another. They have to make choices to survive the turn of the twentieth century. These choices are not easy ones; they are prescribed at every possible turn by US policies of land theft via title and taxation. Land theft via title eventually becomes the theft of the *material earth* as it blows and washes away before their eyes due to settler agriculture. Certainly Julie's suicide is not redemptive, and Riggs's later play is not a treatise of hope for the Cherokee Nation or for women. *The Cream in the Well* is a tragedy marked by poisoned relationships, land, and water. In his work Riggs attempts to answer a question of the heartland, which might be as deceptively simple as "How did we get here?" His artistic preference for setting his plays around the time of transition from Indian Territory to Oklahoma statehood shows us that for Riggs this was the moment he needed to examine in order to answer this question. For Riggs, the answer almost always seems to be found in settler colonialism and the exploitation of the earth. Of course, "How did we get here?" might be an easier question than "Where do we go from here?"

For Riggs, the theater was the way to get somewhere better. In 1939 (the same year as both Steinbeck's novel and Lange and Taylor's *American Exodus*), he begins work on a new theater project he calls "The Vine," with his creative and romantic partner Ramon Naya (born Enrique Gasque-Molina). Indeed, the title serves as an uncanny metaphor for something that holds the physical soil together with its roots as it grows. In other words, vines (a symbol I return to in the following chapter) often prevent erosion. The name of their theater is taken from the Alexander Pushkin poem "The Prophet." The inspirational line that Riggs highlights, "The green vine in the valley climbing," signals how "now we knew what the name of our theatre should be, 'The Vine Theatre'; for in the poet's concept it seemed as if he stood on a great height out in space viewing the world and seeing the strong, yielding, but always upward, always affirmative, and always green impulse toward the sun. Our theatre, green and resilient as a vine, must also thrust toward the sun."[124] This is not the green lilacs that will turn "red, white, and blue." This is an ever-green, ever-growing, living plant. Riggs and his unacknowledged but likely coauthor Naya see this life spirit as essential for what their theater will do and will require of the people who work in it: "We shall have only people who can not only endorse the essential nature of the Vine, but pour into it a fertile and continuously growing life spirit."[125] Their vision for what the theater can accomplish rests precisely on the ability to live and grow rather than destroy. Riggs writes, "This place is to be a place of creation, not of destruction. The way to combat destructive forces is not by destruction. Rather our accent will be always on creation. In the world today, forces in opposition to the triumphant, arrogant state are demolished by pogrom, by discriminatory laws—and the other tools of inhumanity and cruelty. We do not believe that those forces really achieve their ends. We believe that the way to destroy is not to destroy."[126] Significantly, Clabe speaks in near-identical terms in *Cream*, telling Julie, "The way to fight destruction is not with destruction—don't you see it yet?"[127] This correlation links the concerns of the play with the larger ideological vision of the Vine, suggesting that the existential questions explored between Julie and Clabe inform the very way that Riggs views the future.

Having just witnessed one of the most viscerally ecologically traumatic decades in history, Riggs places his faith in the power of a growing plant that starts to hold the earth together again. In combination with this, Riggs adds another rooted metaphor to the power of this project, stating that the Vine "will be aggressive in the way an oak is aggressive; not as the cannon is."[128] The oak, like the vine, is aggressive because it grows, because it has roots, and because in those roots it holds the ground together. Vines, for their part, create a breadth and strength as their tendrils map routes into new spaces, pulling together the metaphors Brown articulates as central to Cherokee literature of this period. Riggs and Naya together call for a radical reevaluation of the nature of power and the way it manifests. Such a call resonates with Holleman's recent assessment that "the case of the Dust Bowl of the 1930s ... was one dramatic regional manifestation of a global social and ecological crisis generated by settler colonialism and imperialism, illustrat[ing] the enormous consequences of relying on imperial 'politics as usual' to attempt a change in 'business as usual.'"[129] The literature of this period bears out her argument that "the ultimate source of the crisis was social, not technological, thus requiring massive social change to address."[130] In their call for a movement that is aggressive like an oak and ever growing like a vine, Riggs and Naya attempt to show how literary work is a viable venue for addressing the social crises of their time. They demonstrate how rootedness is key for holding this earth together. In a somewhat uncanny corollary, Holleman closes her own work on the Dust Bowl with a call to action for facing present ecological crises, asserting, "Our approach should be unapologetically *radical*—which simply means, from the Latin, that we must get *to the root* of things."[131] Indeed, the soil science story of the Dust Bowl *is* one of surface erosion, but the interpretation of its attendant literary and visual narratives must get beyond this surface, following the deep roots that stretch across the continent, aggressive in their radical commitment to an Indigenous story of future growth.

3

Disappearing Grounds
and Backgrounds of the Gulf

OF ALL THE CHAPTERS OF THIS BOOK, this one perhaps covers the most familiar present-day example of erosion for many readers. Many people in the United States are aware that the coast of Louisiana is eroding into the Gulf of Mexico. The current estimates suggest that on average Louisiana loses approximately 5,300 square yards, or 4,430 square meters, of land an hour. If this number is difficult to imagine, the colloquial saying in the United States is that every hour the state loses a "football field" of land.[1] For those that find yards, meters, or football fields obtuse, one could also call up the old denomination of the acre: Louisiana loses 1.32 acres per hour. Now, assuming that the average reader of English reads 200–250 words per minute and assuming that the reader of this chapter is at the high end of average, this means that by the time one finishes the approximately 14,000 words of this chapter, Louisiana will have lost 4,947 square yards *or* 4,136 square meters *or* 1 acre *or* an entire American football field except for one end zone. This present geological fact exists in the background of every task we complete every day. For example, in the skewed temporality that is a game of American football, where four fifteen-minute quarters take nearly four hours of clock time to complete, four playing fields have drowned. However, somehow, despite the urgency, this rather quick geological process clicks along behind us as if it is a slow, inevitable constant. But here's the thing: the land loss along the Gulf was never and does not have to be inevitable. It is not a narrative foretold. Yet.

In the highly praised and popular work *Unfathomable City*, editors Rebecca Solnit and Rebecca Snedeker's chapter introduction for United Houma Nation citizen and activist Monique Verdin's contribution ask, "What does it mean to be nearly native to someplace that's heading toward nonexistence?"[2] Similarly, in her work on Louisiana Indigenous music, Creole-Indigenous scholar Rain Prud'homme-Cranford asks "What happens when the land that supports the music's cultural meaning and reciprocal relationship ceases to exist?"[3] Together, these two questions shape this chapter's examination of erosion in the wetlands of the Gulf South. While much of the rhetoric around Louisiana land loss and the attendant cultural losses that accompany it focuses on the loss of Cajun (and occasionally white French Creole) cultures, much less attention is paid to the Indigenous communities that have survived on this land for generations.[4] Certainly, as French refugees from Acadia, many Cajun people have experienced a fraught relationship with larger US ideologies of whiteness and European settler colonialism, where they were always viewed as adjacent to—not entirely a part of—mainstream white US culture or opportunity.[5] In several instances they have been made the visible foreground of Louisiana cultural relevance even when that visibility and fascination from the larger public did not extend to providing them with more money, individual prestige, or power.[6] This foregrounding of Cajun culture has, in many ways, left Indigenous peoples even farther in the background, where their existence and attachment to their homelands are obfuscated despite their communities' efforts to save this land for themselves *and* their settler-descended neighbors. As the provocative questions posed by Verdin's and Prud'homme-Cranford's work suggest, the erosion crisis facing Louisiana has different resonances for different communities. Certainly, these communities overlap in interests and members, and their shared histories in the same place produce relationships that are not for outside scholars to pry apart. There are, however, benefits in foregrounding the Indigenous stakes of land loss and erosion, particularly in Louisiana, where recognition and visibility of Indigenous peoples and communities has been profoundly lacking for so long.

In fact, the entire cultural representation of Louisiana might be told in a consideration of the foreground and background. In many American imaginations, the Louisiana foreground looms large, distinct, and bright: New Orleans, Mardi Gras, Cajun food, Creole histories, voodoo, streetcars, and desire.[7] Even if the imagination offers a more somber portrait, the foreground is still made up of at least one large, dominating event: Hurricane Katrina. These are important, necessary components of Louisiana's story. However, their presence, always at the foreground of Louisiana representations, distorts the picture and makes it difficult for the viewer to see the temporal and spatial components of the background. For example, how many people realize that one can drive an hour and a half *south* of New Orleans, where whole communities support and sustain much of the state's economy? How many people can imagine Louisiana coastal devastation only via the singular, temporally bound event of Hurricane Katrina rather than the slow creep of the seemingly mundane geology of constant tidal erosion? How many people can name a single tribal nation in the state or say exactly who this ever-disappearing land belongs to? The story of erosion in Louisiana is one of the background, where quotidian spatial and temporal factors merge to create the present ongoing crisis.

This chapter attempts to read this background, not simply to invert the logic of background and foreground but to perhaps adjust the focus on how one might engage with literature and cultural productions beyond what catches the attention in the foreground. To do so, I constellate an analysis of erosion in John James Audubon's Louisiana writings and paintings of the early nineteenth century alongside George Washington Cable's late nineteenth-century representation of this period in "Belles Demoiselles Plantation," a short story in his collection *Old Creole Days* (1879). Both of these authors situate the loss of land in the southern part of the state as attached to concerns over a disappearing whiteness and white supremacy via the plantation. In a productive counter to these historical narratives, I then spend the second half of the chapter considering Monique Verdin's narrative and photographic work in *Return to Yakni Chitto* (2019). Verdin's writing and photography align with the work of Lewis Baltz and Dorothea Lange that I examine in chapters 1 and 2, as she composes human portraits against landscapes in such a way as to draw attention to the quotidian background that creates the present conditions of the foreground. Together, these texts allow an understanding of erosion along the Louisiana coast that disrupts popular narratives of the state where discussions of land loss are often tied to anxieties about the loss of whiteness and white supremacy.

Before launching into the particulars of this analysis, however, perhaps a bit of geological background is in order. The most basic question undergirding this chapter is, Why is Louisiana losing so much land so quickly? The simple answer is human interference. This interference traces back to settler colonialism's massive extractive projects to control the geological features of the region for profit. The more detailed and complex answer involves three core factors: a lack of sediment deposits from the hemmed-in Mississippi River; subsidence, or overall land sinking; and rising sea levels due to global climate change. Before massive levee projects, the Mississippi River shifted about the southern Gulf region, depositing rich alluvial soils from the Upper Midwest every year across a vast delta landscape. A giant roving river, however, is not always conducive to settler agricultural plans and plantations or massive human settlements along a river's banks. Thus, in an attempt to control these seasonal and shifting floods, settlers began erecting more and more constrictive levees to keep the river on course. This means that the incredible land-building power of the river along the coast was effectively cut off, causing it to dump all of its sediment into the Gulf of Mexico in a way that overwhelms the Gulf itself and denies the rest of the delta its regular soil deposits.[8] As for subsidence, many trace this to the oil and gas industry boom that followed the decline of the plantation agricultural economy in the region. Essentially, the oil and gas industry cut up the delta wetlands into a patchwork to lay miles and miles of pipeline for the ease of shipping oil to the mainland for further transport. The dredged canals allowed for increased tidal activity into the core of the dense vegetation of the existing wetland.[9] An apt illustrative metaphor might be "Break a twenty and it's gone." As these lands were fragmented via canals, and oil was pumped out from under the ground, the existing wetlands began to sink farther into the water. The third factor—rising sea levels—is depressingly ironically attached to the second factor of the petroculture economy. The very product being pumped out from under Louisiana's coastal wetlands and shipped off through the destructive canals compounds the problems for coastal communities everywhere.[10] Therefore, the geological factors affecting the Gulf Coast of Louisiana are highly localized *and* globalized, requiring us to think simultaneously in terms of geographic and temporal scale.[11]

Similarly, *background* and *foreground* are terms that require a consideration of the relationship between space and time. The two terms signal concerns over the spatial and temporal relationships among events, conditions, and objects. In referencing time, *background* and *foreground* appear in discussions over narrative construction. Common knowledge has it that writers foreground action or the immediate plot points of their story, and they intersperse background that might explain a character's larger motivations. The background often provides the space for historical consideration or the exploration of memory, while the foreground moves the narrative through the present and into the future. Similarly to how it works in the spatial composition of a photograph, background provides depth. Like the strata of the earth, the background creates layers of meanings that the foreground exists in relationship to, creating a connection between a diachronic deep time and the synchronic layer of the surface. Here I return to the temporal concept I outline in the introduction. Rob Nixon discusses this concept in terms of what Ulrich Beck calls a "shadow kingdom," and he extends the spatial metaphor to questions about temporality.[12] Given this idea's centrality to my analysis of erosion along the Gulf, I quote an excerpt from this passage from Nixon again here: "What forces—imaginative, scientific, and activist—can help extend the temporal horizons of our gaze not just retrospectively but prospectively as well? How, in other words, do we subject that shadow kingdom to a temporal optic that might allow us to see—and foresee—the lineaments of slow terror behind the façade of sudden spectacle?"[13] These questions and this framing from Nixon shape my interrogation of foreground and background. Together, *background* and *foreground* link spatiality and temporality, offering one way to think about erosion within the written and visual archives of Louisiana.

The language of temporal depth in narrative layers frequently appears in considerations of southern Louisiana. Histories of this place often call on the layers of colonial history: French, Spanish, French (again), and eventually US settler control. For example, Rebecca Solnit and Rebecca Snedeker describe New Orleans in terms of the "fathom" in their introduction to *Unfathomable City*: "'Fathom' is an Old English word that meant outstretched arms and an embrace by those arms. It came to mean a measurement of about 6 feet, the width a man's arms could reach, as well as the embrace of an idea. To fathom is to understand. Sailors kept the word in circulation as a measurement of depth, and it survives in the present day mostly as a negative, as unfathomable, the water so deep its depths cannot be plumbed, the phenomenon that cannot be fully grasped."[14] They continue, "New Orleans is all kinds of unfathomable, a city of amorphous boundaries, where land is forever turning into water, water devours land, and a thousand degrees of marshy, muddy, oozing in-between exist."[15] Despite their inclusion of Verdin's chapter in the collection, which clearly outlines just how deep and "unfathomable" this history is, Solnit and Snedeker still mistake the foreground of New Orleans for its own deep background, cutting off the Indigenous depth of place and connection in what is present-day Louisiana. They write, "New Orleans, founded in 1718 and drenched in the past, might be one of the oldest places in the United States in terms of culture and memory, but geologically it is part of the youngest, a region of soft alluvial soil that turns to mud, melts away, and erodes into the surrounding waters."[16] They are more or less correct to point out the relatively recent—in geological time—development of the region, as the Mississippi River Delta built the background of the place over time via soil deposits. However, the colonial perspective that renders New Orleans as "one of the oldest places" erases thousands of years in Indigenous knowledge about multiple places in the present-day United States and in the region currently known as Louisiana. This erasure persists even as the earthworks just to the north in Baton Rouge, on Louisiana State University's campus, were recently found to be the oldest extant human-made structures in the Americas.[17] Given this long history of forgetting an even longer history in southern Louisiana, throughout my analysis I pause to consider the difficulty of thinking about temporal background in a space with such a thick colonial residue.

The challenge of thinking about background and foreground in the terms of deep temporality also presents itself when thinking about the two terms spatially, as my invocation of Nixon's work suggests above. In visual works, background and foreground work together to guide the viewer's perspective. They orient the viewer in space and in relation to the items viewed. As with narrative, the background of an image asks the viewer to pause and consider context, setting, and motivation. One might think of narrative as fundamentally temporal, in that it moves an audience from moment to moment, and of visual texts as lacking this same temporal development, as they are locked into the singular moment depicting the spatial. However, the slippage of possible meanings between the concepts of background and foreground as both temporal and spatial phenomena allow a consideration of how image and text act together to create meaning for geographic and geological sites. It allows for the "temporal optic that might allow us to see—and foresee—the lineaments of slow terror behind the façade of sudden spectacle," as Nixon calls for. In this attempt to read for the temporal optic, one can begin to discern the relationship between the punctual event of the foreground and the quotidian struggle of the background.

One such place where the background and foreground seem in constant tension is Audubon's well-known portraits of North American birds. Notably, the French Creole Audubon invented his own Louisiana background, claiming at various times to have been born in New Orleans or on a Mandeville plantation, when in reality he was born in then Saint-Domingue, now Haiti. This false detail of Audubon's birth was initially generated by his father in acquiring a forged passport for his son when he was eighteen and newly arrived from France. It's unclear, however, why Audubon adopted this lie as truth, even among his own immediate family, long past his confirmation as a citizen of the United States.[18] As Ben Forkner speculates in his edition *Audubon on Louisiana*, "Louisiana stands far above every other region as the place that meant the most in his life and work. It was in Louisiana that he gave himself up totally to the project that finally became *Birds of America*."[19] Thus, it behooves the viewer of Audubon's birds to hold Louisiana always in the narrative background of the artist's work. Moreover, in many cases, Louisiana serves as the actual visual background in his paintings.

Obviously, Audubon's birds almost always stand at the forefront of the imagination. He quite literally foregrounds them. He positions many of his terrestrial birds against a seemingly neutral white space. However, as Nicholas Mirzoeff argues, this "blank" space actually signals a backdrop of Audubon's investment in the backdrop of a kind of white supremacy enabled and enforced by the emergent developments in so-called natural history.[20] Similarly, the backgrounds of Audubon's waterbirds are telling for how they depict a landscape that has changed drastically since his time, just two hundred years ago. The bayous of Louisiana are often rendered lush, dense with cypress trees and palms. The shores of the Carolinas and Florida are crowded with marsh grasses and watery coves. Notably, Audubon relied on the artist Joseph Mason to compose many of his backgrounds, with at least fifty documented instances. However, some speculate that Mason may have completed more uncredited work for the artist.[21] In addition to Mason, the Swiss-born George Lehman worked as a background artist for Audubon from 1831 to 1832.[22] Not only does Audubon record birds that would face enormous extinction dangers over the next century until the passage of the Migratory Bird Treaty Act of 1918, but, with Mason and Lehman, he also depicts landscapes that would face their own existential threats as plantations, logging, river diversion, and eventually the oil and gas industry overtook the scene.[23] In looking carefully at Audubon's backgrounds alongside his written narrative descriptions of place, one finds that he makes a perhaps unwitting account of erosion across the Atlantic coast and southern Gulf, tying these problems to a pathos of loss of the settler plantation economy.

Driven by the rarity and exclusivity of the quickly vanishing living avian archive, viewers are often seduced by Audubon's birds as the evidence of the thing not seen—a preplantation tropical paradise just at the foggy edges of their own agricultural moment. This fantasy sustains much of the continued fascination with his work. As Gregory Nobles explains, "*The Birds of America* was a tremendous artistic and ornithological achievement, a product of personal passion and sacrifice. . . . Audubon's avian images can seem more real than reality itself, allowing the viewer to study each bird closer and longer than would ever be possible in the field. The visual impact proved stunning at the time, and it continues to be so today."[24] But as Nobles goes on to explain, Audubon's work has a more complicated background than its visually impressive achievement: "Although never fully acknowledged, people of color—African Americans and Native Americans—had a part in making that massive project possible. Audubon occasionally relied on these local observers for assistance in collecting specimens, and he sometimes accepted their information about birds and incorporated it into his writings. But even though Audubon found Black and Indigenous people scientifically useful, he never accepted them as socially or racially equal. He took pains to distinguish himself from them."[25] Moreover, Audubon himself enslaved and sold human beings, and as Mirzoeff outlines, "[He] only turned to writing about birds after his debt-funded purchase of slaves to work at his Kentucky mill ended in bankruptcy in 1819."[26] Additionally, he later praised many of the wealthy plantation owners who financially supported his artistic work. Having crafted a story of his own "Spanish" mother as having been killed by Black revolutionaries in Haiti—another fabricated background, like his supposed Louisiana birth—Audubon sets up an extratextual "sympathetic" white plantation landscape behind his entire oeuvre.

For an initial example of Audubon's background, I turn briefly to plate 242 of his *Birds*: *Snowy Heron, or White Egret*, taken from his time near Charleston, South Carolina, in March 1832 (figure 3.1). Here Audubon encounters thousands of snowy egrets, "snowies," in the marshes. Despite depicting a scene on the Atlantic coast, Audubon opens the description by saying, "This beautiful species is a constant resident in Florida and Louisiana, where thousands are seen during winter, and where many remain during the breeding season."[27] As in many of Audubon's paintings, the bird is framed in the foreground, and it stares intently at the viewer. Lehman painted the background for this image, complete with a working rice plantation and a stalking, gun-toting human—a rare sight in any of Audubon's works, which often seemingly divorce the anthropo from the avian. The hunter, who many have speculated is a representation of Audubon himself, creeps up behind the foregrounded animal and in front of the plantation of the farther background. In this way,

3.1   John James Audubon, *Snowy Heron, or White Egret*, 1835. Plate 242 from *The Birds of America*, https://www.audubon.org/birds-of-america/snowy-heron-or-white-egret.

Audubon and Lehman create a suspenseful scene. While the bird stares at us, we see the future. This snowy egret will likely die at the hands of the famed naturalist, who indeed killed many, many birds in his pursuit of creating the perfect visual archive. In the framing of this scene, however, how much weight should the interaction between the bird and the human receive? As Mirzoeff argues, "The painting metonymically represented supremacy as the intersection of whiteness, settler colonialism, the Second Amendment, and the invisibility of enslaved African labor."[28] Farther back in the scene sits the rice plantation, rendered as a near-neutral backdrop. An elevated house sits near the water, surrounded by palms, with an expansive clear-cut lawn radiating out from there. A kind of fence and levee separate the paddy and trenches from the main area of the house. The space appears as ordered, the house as potentially inviting (depending on who the visitor is). This seemingly innocuous far background poses the multiple scales of death and destruction of the painting. This use of land—the plantation born of a European settler colonial system—*is* the harbinger of erosion and death.[29] This is the "early" America that Audubon only rarely records. And indeed, this was not recorded by Audubon at all. This plantation backdrop was rendered by the Swiss Lehman. In addition to the hunter approaching the egret, there is a sense of the clear-cut plantation landscape stalking the foregrounded marshy habitat of the bird, suggesting that the land itself is in danger.

We know from Audubon's writings that these southern landscapes were rapidly changing due to increasing settlement following the Louisiana Purchase in 1803. His descriptions of this process appear in his short pieces, or "episodes," that he published across his *Ornithological Biography* from 1831 to 1835 and were intended as a type of textual counterpart to his paintings in *Birds of America*. Just as Louisiana birds occupy considerable space in the early volumes of *Birds*, his self-titled "delineations of American scenery and character" focused almost exclusively on Louisiana during roughly the first twenty-five years of the century.[30] In these episodes, Audubon records the geological changes and challenges of the region.

For example, in "The Squatters of the Mississippi," Audubon outlines exactly how soil exhaustion induces the migration of settlers, similarly to how John Steinbeck outlines the Joads' journey, which I examined in the previous chapter. Audubon writes, "The individuals who become squatters, choose that sort of life on their own free will. They mostly remove from other parts of the United States after finding that land has become too high in price, and they are persons who, having a family of strong and hardy children, are anxious to enable them to provide for themselves. They have heard from good authorities that the country extending along the great streams of the West, is of all parts of the Union, the richest in soil, the growth of its timber, and the abundance of its game."[31] Despite the fact that that he asserts that these people come of their own free will, he continues by describing a hypothetical family from Virginia and what induces their move: "The land which they and their ancestors have possessed for a hundred years, having been constantly forced to produce crops of one kind or another, is now completely worn out. It exhibits only a superficial layer of red clay, cut up by deep ravines, through which much of the soil has been conveyed to some more fortunate neighbor, residing in a yet rich and beautiful valley. Their strenuous efforts to render it productive have failed."[32] In this description we see that Audubon recognizes the explicit connection between erosion and the migration of squatters. These squatters have suffered gullying via their poor agricultural practices in the East, a phenomenon I explore in more detail in the following chapter. Audubon paints the family's journey in tragi-romantic terms: "They dispose of everything too cumbrous or expensive for them to remove, retaining only a few horses, a servant or two, and such implements of husbandry and other articles as may be necessary on their journey, or useful when they arrive at the spot of choice."[33] However, this detail undercuts his lament for the supposedly poor squatter, for who are these "servants" induced to move with this family? Given that this family is from Virginia and that they have managed to exhaust their previous land in a hundred years, it seems likely that while *they* may have migrated of their own free will, these so-called servants they've brought along are likely enslaved. These "squatters" are not people moving to find a place to belong so much as they are active exporters of the US plantation economy eager to join with the Spanish, French, and German iterations of this exploitative practice already well underway in Louisiana.

Notably, Audubon's Virginia family doesn't even try to reform the habits that have caused the land exhaustion and erosion they fled. Although they initially "gaze in amazement on the deep dark woods around them," that doesn't stop them from starting to ruin it by cutting as many trees as possible, which is a sure way to induce massive erosion, as root density is one of the key mechanisms by which soil manages to stay in place.[34] The family begins by clearing only "a small patch of ground," but by October "the largest ash trees are felled; their trunks are cut, split, and corded" for sale to a steamer that comes to purchase the wood.[35] By the following fall, "the sons have by this time discovered a swamp covered with excellent timber, and as they have seen many great rafts of saw logs bound for the mills of New Orleans, floating past their dwelling, they resolve to try the success of a little enterprise. . . . Log after log, is hauled to the bank of the river, and in a short time their first raft is made on the shore, and loaded with cord-wood."[36] Audubon's remarkable "uplift story" of massive land degradation continues as "every successive year has increased their savings," and twenty years later, "the government secures to the family the lands on which . . . they settled in poverty and sickness."[37] It isn't enough that this family exhausted their land in Virginia; they have come to Louisiana and stolen even more land, both through their government deed and in the very fact that their wood business—and hence the clear-cutting of cypress swamp forests—is one of the earliest ways that erosion in the coastal wetlands of Louisiana began.[38] In other words, they didn't just steal the land "on paper"; they caused the material earth to disappear, again.

As if the singular horror story of this family (though Audubon does not seem to recognize it as horrific) isn't enough, he explains how they are not alone. He describes how "where a single cabin once stood, a neat village is now to be seen; warehouses, stores, and workshops increase the importance of the place.... Thus are the vast frontiers of our country peopled, and thus does cultivation year after year, extend over the western wilds."[39] This is not one family finding fortune. The effects of their settlement are not so isolated. He writes, "Time will no doubt be, when the great valley of the Mississippi, still covered with primeval forests interspersed with swamps, will smile with corn fields and orchards, while crowded cities will rise at intervals along its banks, and enlightened nations will rejoice in the bounties of Providence."[40] In his launching into the future, Audubon foregrounds the degradation of the very thing most popularly considered as part of his legacy: habitats for birds. I am not the first to point out the slippage between Audubon the man and Audubon the organization, and I will not dwell on this irony here.[41] My interest in his works from the early nineteenth century lies in the changes of the land he continually records in the background through his paintings and accompanying words. Only twice does Audubon record humans in his paintings (including the earlier-discussed inclusion painted by Lehman), and each time they are so tiny as to remain practically unnoticed. However, even though he does not place humans in the frame, his written works depict them flooding over the land, felling trees, and ruining the homes of the birds he seeks to record. He foregrounds this one Virginia family almost as if they are a family of birds, migrating from somewhere else to survive, but just as with the stalking hunter, his larger descriptions leave the reader with an uneasy background image of a providential future to come.

In addition to his direct descriptions of the destruction of cypress swamp forests, Audubon makes plain the uneasiness of those who have decided to settle (i.e., squat) and grow towns along the Mississippi River, a theme I examine in George Washington Cable's work shortly. In his "Improvements in the Navigation of the Mississippi," Audubon places the reader in 1808, when there was little settlement along the shores of the river, presumably because the Indigenous peoples of the area understood that the southern Mississippi was necessarily fickle in its course. This roving river regularly replenished the region with rich soil. Audubon describes "the thousands of sand-banks, as liable to changes and shiftings as the alluvial shores themselves, which at every deep curve or bend were seen giving way, as if crushed down by the weight of the great forests that everywhere reached to the very edge of the water, and falling and sinking in the muddy stream by acres at a time."[42] This is certainly erosion, but like a California landslide replenishing the coastline, this erosion is supposed to occur. The soil and land taken from the banks as the river shifts with the increased flow from melting snows farther north is the very thing that nourishes and builds the Louisiana delta farther south. Thus, while it might look like loss and destruction from Audubon's settler colonial perspective, it's important to recognize that this loss creates life and land farther south.

These natural erosive processes wreak havoc, however, for plantations and their owners. Audubon continues to describe these processes and their relationship to a pathos of loss in the plantation economy in his tract "A Flood." He notes that despite earthen levees created by riverbank communities, "the water bursts impetuously over the plantations, and lays waste the crops which so lately were blooming in the luxuriance of spring."[43] In addition to this destruction, he notes further losses: "Large sand-banks have been completely removed by the impetuous whirls of the waters, and have been deposited in other places.... The trees on the margins of the banks have in many parts given way.... Everywhere are heard the lamentations of the farmer and the planter, whilst their servants and themselves are busily employed in repairing the damages occasioned by the floods."[44] He states directly that via this process, "the rich prospects of the planter are blasted."[45] These losses are quite notably losses of a white plantation economy struggling to withstand the natural erosive patterns and fluctuations of the river. In describing the shifting, flooding river, Audubon attempts to create a sympathy for those who have seemingly lost everything to a capricious Mississippi. However, these farmers and planters have not lost everything. Their so-called servants remain, and in Audubon's descriptions they seemingly have the means to rebuild despite the loss of their "rich prospects."

Despite his portrait of plantation loss, Audubon makes it clear that in its fluctuations, the Mississippi River actually produces and sustains lands. He exhorts, "But now, kind reader, observe this great flood gradually subsiding, and again see the mighty changes it has effected. The waters have now been carried into the distant ocean. The earth is everywhere covered by a deep deposit of muddy loam."[46] In other words, this process creates the soil conditions that have allowed this plantation agriculture to thrive. These sediment deposits were creating and renewing the Louisiana wetlands that are now disappearing. While Audubon shows readers the small, ineffectual levees, his descriptions of the losses from flooding illustrate the logic that will eventually cause the US Army Corp of Engineers to institute larger and larger levee projects in order to prevent future floods.[47] This process will in turn starve the wetlands of the deposits they need to grow or, at the very least, sustain themselves. Ultimately, what Audubon paints in the background of his birds is the plantation economy of the Lower Mississippi that wanted the profits of its environment (the rich alluvial soil) without the payments (living with a capricious river).

With this, combined with the cypress-cutting squatters, Audubon outlines the exact conditions that will produce the present crisis while they were first in process. Indeed, the levee systems that will eventually cut off the river fluctuations altogether will not come along for another century, but the pathos over the loss of plantation wealth is already in process as Audubon records the landscape. He is not able to predict this future, describing it instead as "the Mississippi, with its ever-shifting sand-banks, its crumbling shores, its enormous masses of drift timber, the source of future beds of coal, its extensive and varied alluvial deposits, and its mighty mass of waters rolling sullenly along, like the flood of eternity."[48] Here he imagines that this process of cataclysm will go on forever, and he sees the losses and gains, the erosions and soil deposits, as part of a great background that will shape the life of this specific place. And who could blame him for not thinking that something that had happened for thousands of years would change so dramatically, so quickly? Who could blame Audubon for believing in the slow steadiness of change in the dynamic landscape? In asking these questions, however, it is all the more imperative to pay attention to the background conditions he describes. Audubon sees the changes as they occur in conjunction with settler colonialism, and yet he cannot bring those changes to the foreground of his imagination, believing instead that the river and its attendant earth will roll on forever despite settler colonial invasion.

Although later accounts of Louisiana will participate in an erasure of Indigenous communities and individuals, Audubon remains aware of their presence across his works. Of course, this awareness is due both to his presence in this place prior to southeastern Indigenous Removal and to his own work, which often involves relying on Indigenous knowledge of avian life and habitats. While audiences rarely see Audubon connect the dots between land degradation and colonialism, he makes his awareness of Removal explicit. In a letter to John Bachman dated February 24, 1837, he writes of the Removal he witnesses in Georgia: "We breakfasted at the Village of —— where 100 Creek Warriors were confined in Irons, preparatory to leaving forever the Land of their births! — Some miles onward we overtook about two thousands of these once free owners of the Forest, marching towards this place under an escort of Rangers, and militia mounted Men, destined for distant lands, unknown to them, and where alas, their future and latter days must be spent in the deepest of Sorrows, affliction, and perhaps even phisical want."[49] While I explore the particulars of Muscogee Creek Removal in more detail in the following chapter, it's important here to think about just how directly connected Audubon's record of the Southeast is to the precise moment of southeastern Indigenous genocide and dispossession of homelands. He is witness to the processes that will forever alter the landscape he records, and if audiences look beyond the birds in his foreground, this record comes into more distinct relief.

Indigenous people in the state currently known as Louisiana have long held a unique and complicated position in relation to the United States. As in every corner of the continent, there have always been Indigenous peoples who claim and know these lands and waters as home. When the United States completed the Louisiana Purchase in 1803 (land, I might add, that they bought from the French, who had previously stolen it from Indigenous people, making it one of the largest purchases of known stolen property in history), it was part of a larger design in Thomas Jefferson's initial architecture for what would eventually become the Indian Removal Act of 1830.[50] The act forced southeastern Indigenous nations to move to land west of the Mississippi River, specifically to what was then known as Indian Territory.[51] Indeed, the vast majority of Louisiana is west of the Mississippi River, and in many ways, the Indigenous nations there in total or in part—the Atakapa, Coushatta Caddo, Chitimacha, Houma, Tunica, and portions of the Choctaw—were not exactly included in "official" Removal designs. However, Indigenous communities in the state were constantly fleeing settlers. Different groups coalesced in out-of-reach bayous and swamps, and eventually, particularly in the case of the Houma, were pushed to take refuge along the Gulf Coast. These repeated "informal" removals, rather than a Trail of Tears journey that Audubon witnesses, created two uniquely bureaucratic problems for many of these tribal nations. One, it meant that they did not have the same "paper" relationship to the US government, as they had not engaged in the massive treaties under the Removal Act. And, two, because they had to relocate every few generations away from the ever-increasing squatters, the United States could effectively claim (as they do in some cases to this very day) that the archaeological record is not "deep enough" to tie them to place, one interpretation of part of the criteria for federal recognition under the US Bureau of Indian Affairs.[52]

Despite white Americans having been more than happy to engage in a largely out-of-sight, out-of-mind relationship to Indigenous peoples in Louisiana, writers across the nineteenth century record their continued presence in the Gulf wetlands. Audubon himself regularly notes interactions with Choctaw hunters, and at the other end of the century, George Washington Cable includes Native characters or characters with Indigenous backgrounds in several of his stories set across the 1800s. In his 1879 short story "Belles Demoiselles Plantation," Cable directly links questions of erosion, Indigenous land control, and the pathos of loss in the white plantation economy. The story begins with the background detail that a French count, whom Cable assigns the protective pseudonym De Charleu, had married a Choctaw woman before he had to return to France to account for a fire that had destroyed his colonial records. In France the count is granted a tract of land and "in a fit of forgetfulness" marries a second wife, who is a French noblewoman. As Cable sardonically sets up, "However, 'All's well that ends well;' a famine had been in the colony, and the Choctaw Comptesse had starved, leaving nought but a half-caste orphan family lurking on the edge of the settlement, bearing our French gentlewoman's own new name, and being mentioned in Monsieur's will."[53] Whether Cable intends it or not, this Choctaw orphan family, existing on the margin of the settlement, serves as a powerful narrative metaphor for how Indigenous people have been considered across the colonial history of Louisiana.

The count's marriage to this Choctaw woman, in addition to serving as its own background inside the story, has an even more telling background outside of it. Traditionally, in southeastern Indigenous cultures, identity, clan, and tribal belonging are passed matrilineally. Furthermore, land, property, and agriculture are under the purview of women. Non-Native men across the colonial record regularly sought to marry powerful Native women because they understood that with these women came access to tribal lands that were under their traditional control.[54] Thus, the tract that the French king "grants" the count was likely already his Choctaw wife's land. This piece of land becomes the future site of the titular Belles Demoiselles Plantation, which Cable brings us forward to in the early 1800s. Cable gives no indication that he was aware of the matrilineal association of Choctaw land. The plantation's eventual name comes from the seven daughters of the count's eventual descendant with his French wife, a son known in the story as "the Colonel." In a fittingly ironic twist, Cable creates two family trees where gendered associations with the land tangle and exchange places: one that begins with a Choctaw wife and an "orphan" Choctaw family "on the edge of the settlement," and another that begins with the count's French wife and their son and ends with their eventual descendants, the seven titular Belles Demoiselles.

Cable describes this land grant in the time of the original French count as "a long Pointe, round which the Mississippi used to whirl, and seethe, and foam, that it was horrid to behold."[55] He continues with imagery similar to Audubon's description of the shifting Mississippi River: "Every few minutes the loamy bank would tip down a great load of earth upon its besieger, and fall back a foot,—sometimes a yard,—and the writhing river would press after, until at last the Pointe was quite swallowed up, and the great river glided by in a majestic curve, and asked no more; the bank stood fast, the 'caving' became a forgotten misfortune, and the diminished grant was a long, sweeping, willowy bend, rustling with miles of sugar-cane."[56] Notably, like almost all one-crop plantation agriculture, the flat furrows of sugarcane fields can cause erosion of alluvial soils, particularly if the postharvest residue is not left to help mulch the land. According to soil scientists Ted Kornecki and James Fouss, "Historically, sugarcane residue has been burned following harvest, thus eliminating all benefits of the residue, such as organic carbon (oc) buildup and reduction of runoff and soil erosion."[57] Therefore, in addition to the natural erosive processes of the Mississippi that affected the Pointe in the past, the land of Belles Demoiselles in Cable's narration is probably eroding due to plantation agriculture.

Cable only alludes to the details of the plantation economy at work in the creation of the house. He writes, "Coming up the Mississippi in the sailing craft of those early days, about the time one first could descry the white spires of the old St. Louis Cathedral, you would be pretty sure to spy, just over to your right under the levee, Belles Demoiselles Mansion, with its broad veranda and red painted cypress roof, peering over the embankment, like a bird in the nest, half hid by the avenue of willows which one of the departed De Charleus,—he that married a Marot,—had planted on the levee's crown."[58] The cypress roof, indeed, must have come from the surrounding swamp forests, and tellingly, Cable attempts to create the impression that the plantation house is in the foreground of the visual frame, "like a bird in the nest," evoking an Audubonesque image. Moreover, *demoiselle*, in addition to meaning "young lady," is a type of European crane. However, unlike Audubon's indigenous egrets, these demoiselles are an invasive species.

With the house in the foreground, Cable leaves the details of plantation enslavement and agriculture in the background—present but distant: "The house stood unusually near the river, facing eastward, and standing foursquare, with an immense veranda about its sides, and a flight of steps in front spreading broadly downward, as we open arms to a child. From the veranda nine miles of river were seen; and in their compass, near at hand, the shady garden full of rare and beautiful flowers; farther away broad fields of cane and rice, and the distant quarters of the slaves, and on the horizon everywhere a dark belt of cypress forest."[59] This is the only mention of enslavement or the actual mechanics of the plantation that appears in the story, and the backgrounding of this labor is part of Cable's technique across his works. As Ieva Padgett notes, "Cable's gardens participate in this obliteration of other histories, especially because they are tended by ghost laborers otherwise omitted from the stories. A phenomenon observed in 'Belles Demoiselles Plantation'—where an impressive garden in the foreground outshines and obscures 'the distant quarters of the slaves' . . . who are presumably in charge of maintaining the garden—characterizes much of Cable's Creole writings."[60] Furthermore, in these details Cable makes the reader aware of how the plantation stands as distinct from the cypress forests in the background, which most likely signals that the land was once covered in dense trees like its surroundings. However, this story of land degradation never quite comes to the foreground until it creates a loss for the white protagonist, Colonel De Charleu. The numerous losses for everyone and everything else—the enslaved, the land, the Indigenous people—remain where audiences may or may not see them, present but out of focus along the margins of the narrative frame.

This difference between the background of Indigenous land and the pathos of white loss in the foreground comes into its most stark relief as the story progresses and the Colonel comes to interact with his distant relative, the Choctaw De Carlos, whom Cable troublingly refers to as "Injin Charlie."[61] The Colonel and Charlie have the original French count as their common ancestor, and their family trees diverge between the count's Choctaw first wife and his French second wife, both of whom, Cable notes, died young. The story sets up competing ideas of inheritance via French patrilineality and Choctaw matrilineality for claims to Louisiana land. Whereas the Colonel has held on to the "original" French land grant and turned it into the plantation named after his seven daughters, Charlie has maintained legal claim to a set of connected buildings in the city of New Orleans. The Colonel begins to covet how Charlie has managed his financial affairs. Charlie's old buildings are located in a part of the city that is starting to increase in property value, making Charlie relatively well off financially. To compound this resentment, over the years the Colonel's daughters begin to express their desire to have a place in the city. This becomes the impetus of the story's main plot, which is the Colonel's attempt to at first buy outright and later trade the plantation Belles Demoiselles for Charlie's property in New Orleans.

A lot of the short story and almost all of the resulting criticism focus on the language of blood affiliation and land negotiations between the Creole Colonel and the Choctaw Charlie.[62] While this line of inquiry is certainly important, it allows the larger concern of land degradation to float along in the background, as if Charlie and the Colonel are coming to the negotiation table with the same land attachment and stakes. Padgett approaches this understanding in her work on Cable's description and use of gardens, but she does not fully address the Indigenous claim to homelands that troubles Cable's logic. As she writes, "Cable's insistence on gardening as the proper way to engage with the national soil introduces a certain humility into one's relationship with the land.... Of crucial importance here is the requirement of attachment to the land *where one lives* as opposed to the land to which one feels entitled through the myth of filiation. Even such a position is in danger of leading to the all-too-familiar violence over 'grasping' the land (hence the risk). Thus, a genuine connection with the land is possible only through a careful and continual engagement with it."[63] I don't disagree that Cable is working to attach French Creole identity to land via his repeated use of garden imagery and metaphor, but this equation is incomplete without thinking through the specific Indigenous contours to the story. The fact is that Charlie's attachment to his homelands may be very different in spirit and materiality from that of his French Creole counterpart. His reluctance to part with his so-called property has a longer genealogy than Cable allows, one that calls up Robert Nichols's argument about how "in this (colonial) context, theft is the mechanism and means by which property is generated."[64] The point is that, for Charlie, this exchange isn't about property at all. The Colonel is trying to render the homelands Charlie has maintained into mere real estate. While Cable allows the Colonel the space and consciousness to read his identity through the land, he seems to struggle to give Charlie that same depth of character, despite the fact that it's certain Cable would realize that Native people, until very, very recently, had had legal and material claims to much of the land of southern Louisiana. So while I support critics who read the story in terms of what it says about race relations alongside questions of kinship and land affiliation in southern Louisiana, I find these inquiries lacking if they don't also consider the material condition of the land that precipitates Charlie and the Colonel's battle. This difference in perspective—Choctaw versus French Creole—affects how each character comes to the plot with a different understanding of temporality and scale for how the land shifts, holds on, erodes, and deposits.

Just as the Colonel is hopeful that he has convinced Charlie that a trade of the plantation for the spot in the city might be a good deal, he notices that the levee in front of his plantation property has cracked. A background splashing sound cues him to investigate: "He plunged down the levee and bounded through the low weeds to the edge of the bank. It was sheer, and the water about four feet below. He did not stand quite on the edge, but fell upon his knees a couple of yards away, wringing his hands, moaning and weeping, and staring through his watery eyes at a fine, long crevice just discernible under the matted grass, and curving outward on either hand toward the river."[65] As the Colonel witnesses this, he calls to God, and Cable tells us that "the tough Bermuda grass stretched and snapped, the crevice slowly became a gape, and softly, gradually, with no sound but the closing of the water at last, a ton or more of earth settled into the boiling eddy and disappeared."[66] Despite being one of the most common choices for American football field turf, Bermuda grass is an invasive species in the Americas. Colonial records indicate that it was used widely by 1807, and some scientists trace its introduction to the Americas to seeds that lodged themselves in hay bales on board ships that transported kidnapped and enslaved Africans.[67] Once the grass is established, its dense rhizomatic structure can quickly choke out other native plants, which may be more adept at existing long term in the brackish wetlands of southern Louisiana and thus better at preventing sustained erosion. Ironically, some agronomists today see Bermuda grass as one of the best hopes of preventing future levee failure, but again, as an already invasive species, its ubiquity and promise come at the expense of indigenous species of vegetation that may work better over time. In Cable's reference to the grass's futility in enforcing the levee in front of the plantation, we see both the colonial background and the foreshadowing of a plot of loss.

The Colonel knows this erosive development holds no good future for his estate, creating urgency to trade for Charlie's secure spot in the city. Somewhat oddly, the Colonel makes no effort to relocate his daughters out of the house that is sitting precariously close to an ever-failing riverbank before he goes again into the city to convince Charlie to accept his trade of city land for a plantation he knows is about to fall into the Mississippi. As he finally convinces Charlie that the trade might benefit both of them (an idea that Charlie repeatedly resists), the Colonel experiences a pang of guilt. He decides he wants to ride with Charlie out to the plantation so that Charlie can see it for himself, thinking, "If he chose to overlook the 'caving bank,' it would be his own fault;—a trade's a trade."[68] On the way to the plantation, they pass "hedges of Cherokee rose," another invasive plant species I discuss more in the following chapter, and the Colonel begins to have a crisis of conscience over trading the doomed plantation for Charlie's city home: "If he held to it, the caving of the bank, at its present fearful speed, would let the house into the river within three months; but were it not better to lose it so, than sell his birthright?"[69] In addition to his Job-esque logic, he worries about the ethics of betraying his distant kinsman. Importantly, however, the reality of the situation that perhaps even Cable fails to recognize is that this land isn't the Colonel's birthright beyond the colonial fiction of deeds and grants from far-off kings, a recursive property formation that negates what Indigenous homeland attachment might mean beyond capital. This homeland is Charlie's birthright. At best, Charlie's willingness to trade one "property" for the other is nothing but substituting one theft for a different one.

As the two men approach Belles Demoiselles, the Colonel repents, urging Charlie not to make the trade, which angers the normally patient Charlie, who has now entertained a host of negotiations and ride-alongs with his Creole kinsman. However, before the Colonel can explain, the story's grand climax comes into relief: "Both looked quickly toward the house! The Colonel tossed his hands wildly in the air, rushed forward a step or two, and giving one fearful scream of agony and fright, fell forward on his face in the path. Old Charlie stood transfixed with horror. Belles Demoiselles, the realm of maiden beauty, the home of merriment, the house of dancing, all in the tremor and glow of pleasure, suddenly sunk, with one short, wild wail of terror—sunk, sunk, down, down, down, into the merciless, unfathomable flood of the Mississippi."[70] Cable creates a prescient echo to Solnit and Snedeker's "unfathomable" claim about the historical depth of settler colonial Louisiana. Here the linchpin of the fathomable depends on a fantasy of Mark Rifkin's settler time, where the supposed unfathomable depth of history on these lands can only be (paradoxically) understood by those Indigenous peoples and communities who *can* fathom it.[71]

In this tragedy of the eroding Mississippi bank, the two old men become closer, as Charlie takes the now-homeless Colonel back to New Orleans and cares for him in his senile despondency. The tragedy Cable centers in his erosion story is the loss of the plantation and the white family. It seems clear that audiences are supposed to mourn for the Colonel's seven daughters and lost home. And in the foreground of the plot, Cable has Charlie mirror this mourning as he cares for the delirious Colonel in his New Orleans home. However, Cable weaves an odd detail into the background. A vine begins to creep into the Colonel's bedroom through the window as he convalesces. As Cable writes, "By the window came in a sweet-scented evergreen vine, transplanted from the caving bank of Belles Demoiselles. It caught the rays of sunset in its flowery net and let then softly in upon the sick man's bed."[72] Given Cable's description, the most likely identification of the vine is Carolina jessamine, an evergreen vine native to the US Southeast. Its close relative, swamp jessamine, would almost seem even more likely except that it produces odorless flowers. Both species, however, are known to grow in wetlands and work as effective ground cover to prevent erosion along steep banks.[73]

While Charlie nurses the Colonel, the vine continues to grow in the background, covering the window. The Colonel comes in and out of consciousness, worried about his debt to Charlie and shifting his concern: Did he and Charlie finalize their trade or not? Charlie adjusts his answer each day according to what he perceives as the sick Colonel's deathbed wish. In the middle of this, Cable again adds a small passing detail about the vine as Charlie works to draw the Colonel to consciousness: "Charlie wanted to see the vine recognized. He stepped backward to the window with a broad smile, shook the foliage, nodded and looked smart."[74] Cable does not spell out why the recognition of this transplanted vine is so important to Charlie. The fact that Charlie works to make the Colonel recognize this plant—this native species, previously growing along the caving bank with the invasive Bermuda grass—suggests that the vine holds a lesson if not even a comfort. It works similarly to the vines that Monique Allewaert discusses from William Bartram, whereby "the tendril that pulls a person into its spiraling motions joins the human will to that of plants, producing knowledge that changes human actions." This joining works to make connections, not unlike LeAnne Howe's theory of Indigenous tribalography that connects past, present, and future milieus. In a similar fashion Allewaert theorizes, "The grasping vines' desire for conjunction and collectivity with other tropical forces depends on the stretching outward of parts. The gracefully collectivizing movements result in huge strength."[75] Despite the Colonel's attempts to swindle him, Charlie doesn't give up hope that his kinsman the Colonel can pull a lesson from the background into the foreground. Like Allewaert's vine that "joins what it encloses in its delicate hold," this vine will hold the earth together; this is the life that the invasive grass disrupts as settlers try to control the winding life of the Mississippi River.[76] Notably, Allewaert points out that this ever-growing connectivity also poses a hazard for Bartram, who fears the "combinatory power."[77] In Cable's tale it seems the Colonel also struggles to make sense of this power of vining.

Ultimately, Cable seemingly sets up the tragedy of the story in the Colonel's death from the grief of losing his plantation and family. In the story's closing, even Charlie sheds tears for his relative. Perhaps Charlie does mourn his kinsman, but Charlie's losses are vaster. While a white nineteenth-century audience might see the loss of the plantation as tragic, reading the background of the story should allow readers today to ask if that is indeed the case. What if Charlie's tears signal the lessons unlearned? What if for his character the loss of the plantation is only tragic for the bad earth stewardship that it reveals in settler agriculture? He attempts to make the Colonel see something else on his deathbed: an indigenous plant that would have preserved the homeland that was lost. He tries to force the Colonel's eyes to focus on the background as he shakes the plant at him. As a Choctaw man, Charlie seemingly also understands the tragedy and folly of settlers who are increasingly trying to control the life of that river, building levees to protect sugar plantations and in turn creating even more erosion. So when Cable narrates that "two big tears rolled down" Charlie's face, I argue it is fair to ask who and what these tears are for.[78] The story repeatedly indicates that Charlie is a compassionate human, so perhaps one tear is indeed for his French Creole kinsman, but perhaps one of these tears is for himself, his people, his homelands—for all the loss that the white settler refuses to recognize.[79]

Cable's clear foregrounding of the pathos of white loss has been the standard of erosion discourse in the southern Mississippi River region since at least Audubon's time and remains so today. One must read between the lines to see the larger losses for Indigenous peoples in the works of both authors, and even present-day ecocritical pleas for the Louisiana coast tend to erase or glide over the specific Indigenous contours of loss in the region. The desire to stop the losses of plantation and farmland to the shifting Mississippi that both Audubon and Cable describe is the very thing that precipitates the "solution" that will spell devastation for southern Louisiana wetlands. Following the 1927 Mississippi flood, the US Army Corps of Engineers instituted a massive levee project that effectively forced the river to stay on course (by which, of course, they meant the settler-ordained course), to prevent tragedies such as the ones Audubon records and Cable dramatizes. Given that many Indigenous people in southern Louisiana had been pushed into the wetlands farthest south, which depended on the shifting deposits of the Mississippi for their continuation, these levee plans constitute a literal theft of homelands—not just a legal, paper title theft of property—from Native communities. The white settlers' experiences of naturally occurring erosion along the river produce the federal projects that offset that erosion to Indigenous communities farther south. Despite this offsetting, accounts of Louisiana land loss rarely engage in a sustained way with Indigenous perspectives on the situation, foregrounding instead the losses sustained by Cajun and Creole communities. Occasionally, there will be a passing mention of the Isle de Jean Charles Band of Biloxi-Chitimacha-Choctaw Indians' resettlement project as a "federally funded, first-of-its-kind effort that will offer resettlement options to current and former residents of Isle de Jean Charles in a safer and more sustainable community."[80] This often-quoted, but rarely engaged, and not entirely accurate narrative exemplifies how the Indigenous stakes of erosion in the region more often than not serve as background window dressing for a concern that seems largely still dominated by a foregrounding of white pathos. And notably, a lack of understanding from federal government officials about the divergent needs of different tribal nations and communities in the region has made these resettlement plans more difficult to execute than government planners considered.[81] This lack of consideration of tribally specific needs and plans represents both a cause and an effect of settler colonialism in southern Louisiana.

Regardless of this lack of awareness on behalf of the larger US public and authorities, Indigenous leaders and citizens from across southern Louisiana are working constantly in the realms of policy, activism, and art to account for their experiences. One prominent figure, active in virtually all realms, is the aforementioned Monique Verdin. Initially a photographer, later a filmmaker, and now an author and former tribal council member for the United Houma Nation, Verdin has emerged in the past decade as one of the key voices bringing awareness to the specific challenges facing Indigenous communities in southern Louisiana. In my previous work, I have discussed her film with Sharon Linezo Hong, *My Louisiana Love* (2012), which catalogs her family's and community's struggles from Hurricane Katrina in 2005 to the BP oil spill in 2010.[82] Weaving together issues from erosion to continued battles with pollution from the oil and gas industry, Verdin takes viewers into the intimate space of her own family with her aging, but ferocious, grandmother Matine; her charmingly witty father, who dies from cancer during the film; and her partner, Mark Krasnoff, who takes his own life toward the end of the documentary as a result of his depression from so many previous losses. The film chronicles loss without giving up, and it foregrounds survival without manufacturing a happy ending.[83] It forces the audience to consider the causes and costs of erosion and its contributing factors for Indigenous people at the end of the world.

In the remainder of this chapter, I turn to Verdin's book, *Return to Yakni Chitto*, to discuss her photography and writing and the way they put pressure on questions of endings and beginnings, backgrounds and foregrounds. In some ways, one might think of Verdin's work as bringing Indigenous communities to the foreground in the story of Louisiana erosion. I argue, however, that especially in her photography, Verdin does not simply invert a logic of the foreground. Rather, she constructs backgrounds that demand that careful viewers readjust their focus. While she consistently places her Indigenous family and community in the foreground of the shot, commanding a viewer's attention to their vital, expressive presence, she also creates backgrounds that adjust and confound perspective. The fact that she accompanies these photographs and photo collages with her own words and the words of her fellow Houma citizens and allies adds another layer of background to her use of the visual. Her book, then, follows a rich tradition in Indigenous women's writing, one that uses multivocal multimedia storytelling to construct what Choctaw writer and scholar LeAnne Howe calls *tribalography*. Howe's concept of tribalography encompasses and accounts for how Indigenous stories work in processes of relation, adding and multiplying connections between and among what might appear to outside audiences as diverse or divergent beings and perspectives.[84] In this way, Verdin's work joins other texts, such as Leslie Marmon Silko's *Storyteller* and Deborah Miranda's *Bad Indians*, that challenge Euro-American distinctions between "modern" literary and traditional forms of storytelling, between visual images and written narrative, and between history and memoir.[85] I argue that Verdin's ability to dwell in this liminality both mirrors the existence of Indigenous community at the border between land and water and models a deeper historical process for Indigenous people of the region who manage to live *with* the fluidity of the river over time, rather than the settlers who push *against* its shifting path. In other words, Verdin's use of background and foreground answers Nixon's call for new narratives that account for the past while pushing toward a future.

One of the key historical differences among Audubon's and Cable's and Verdin's texts—aside from the standpoints of Indigenous and gender identity, which cannot be overstated—is that Verdin's work comes after the oil and gas industry invades Louisiana in the 1940s. In addition to changing the economy of the region, this development fundamentally alters the landscape via the fragmentation of the wetlands to lay pipeline and via the pumping of oil and gas deposits out from under that land, exacerbating erosion.[86] Whereas Audubon and Cable are only tangentially involved or complicit in an emergent petroculture, Verdin's experiences are largely defined by it. There is certainly ambivalence around the emergence of the oil and gas industry along the Gulf Coast, and many people will outline how the industry brought a particular kind of economic uplift to a rural area where many people depended on subsistence agriculture. Even those who highlight the larger environmental costs of this economy also recognize their own or their family's immediate reliance on the jobs and products of the industry.[87] However, as with the plantation economy before it, an industry's ability to entangle entire communities does not render one's resistance to or critique of it simplistically futile or hypocritical. Verdin's work highlights this reality as she composes foregrounds and backgrounds to challenge her audiences' vision for the future of the Gulf Coast.

Return to Yakni Chitto offers multiple possible framing philosophies, but two in particular speak to the visual lesson of Verdin's compositions. The first is from the author of part 1, "Global Climate Change—a Houma Perspective," T. Mayheart Dardar, a United Houma Nation citizen and scholar. He writes, "In painting, the liminal zone is called *sfumato*. Leonardo da Vinci explored it in the hazy, ill-defined borders around his most famous portrait, the *Mona Lisa*. As much a philosophical principle as an artistic technique, it gives us a soft transition between colors and objects, as well as illustrating the frontier between wet and dry, stable and unstable."[88] He continues by tying this concept to the matter at hand: "Houma culture and lifeways were forged in this fluid domain."[89] This fluidity found in the borders of the image and in the focus describes many of Verdin's photographs across the book, as viewers must pause to take in the relationship between her ostensibly foregrounded subjects and the background she establishes behind them. This aesthetic and philosophical choice might be best understood in the second lesson that organizes the book, which comes from Verdin's father, Herbert Verdin. She recounts, "My father loved to remind me: 'People look, but they don't see. They hear, but they don't listen.'"[90] *Return to Yakni Chitto* is Verdin's attempt to draw the viewer in with the promise of *looking* at a twenty-first-century Indigenous community and instead making them *see* the realities facing life in the age of the Anthropocene.

Working in black-and-white photography, Verdin pulls from multiple genealogies in addition to the tradition of tribalography I outline above. Not only does Verdin quote from George Washington Cable's description of the "trembling prairie" of Terrebonne Parish, which appeared to glimmer and undulate as if simultaneously liquid and solid ground, but she also includes an archival photograph from Dorothea Lange's images of sugarcane farming from the 1930s Farm Security Administration. As I outline in the previous chapter,

Lange and Paul Taylor's evocation of erosion as metaphor and material reality for *An American Exodus* (1939) is continually bound up with legacies of settler colonialism. Verdin's own reuse of Lange's record of the plantation economy in Louisiana makes this connection all the more explicit. It allows Verdin to recontextualize the realities of Indigenous claims to homelands that always subtend considerations of plantation agriculture or environmental degradation even when those realities are not recognized as such.

Verdin frames the book in such a way as to center Indigenous womanhood as a vital perspective, undoing the gaze of Audubon, Cable, and many men who "looked" at this Louisiana space before her. The acknowledgments begin with the Mississippi River Delta. Verdin writes, "Without the life force of the Mississippi River delta, her sacred waters and unpredictable natural intelligence, these words and images, experiences and stories, would not exist. May we find new, and return to old, ways that honor and regenerate her dynamic system to heal past damages and to restore sustainable relationships for generations to come."[91] In other words, this is not Mark Twain's, William Faulkner's, or Oscar Hammerstein II's "Old Man River." And while some Western ecocritics might balk at gendering parts of the natural world as feminine, such critique fails to account for the myriad ways that Indigenous knowledges recognize power rather than subordination in such designations.[92] Verdin's acknowledgment alongside her opening thanks to "all the women" in her life re-centers an Indigenous reclamation and way of knowing the land, one that reorients land stewardship toward women, as found in traditional ecological knowledges and practices in southeastern Indigenous communities.

Verdin continues this structure by centering women in the text and in individual images. At the close of part II—the structural center of the text— Verdin offers two photographs of her grandmother Matine taken in 2000 at Pointe-aux-Chenes. The verso image includes Matine and her best friend, Jeanne, as they survey what is left of the lands where they grew up. Centered behind them is a live oak tree, also known as a *skeleton tree*, whose roots have been burned by the encroaching salt water of the Gulf and is now dead. Verdin explains of the two women, "They were shocked at the changes to the landscape."[93] The images, shot with a 35 mm Nikon FE2 with a wide-angle lens, sharply foreground the elderly women while the background of the bayou stretches behind, slightly out of focus. In the recto image, titled *Matine's Map*, Matine points to a live oak under which her mother, Celestine, buried her father after he died from yellow fever (figure 3.2). Matine's gesture and gaze direct the viewer back to the verso page, where a living live oak stands in the right background. This directed movement from image to image and from page to

3.2   Monique Verdin, *Matine's Map* | *Bayou Pointe-aux-Chenes* | *2000*. Photograph courtesy of the photographer and the Neighborhood Story Project.

page, left to right, and right to left to right again, forces the viewer to refocus on the spatial background and foreground as well as the narrative background of history and memory. Verdin's caption of the 2000 image explains that this live oak, where her great-grandfather is buried, is "one of the only surviving live oaks down on the bayou. The others have died from salt water intrusion," adding another level of narrative.[94] A present-day viewer must ponder if this last living live oak is alive today—after Hurricane Katrina, after the BP oil spill, and after even more acres of land have eroded away.[95] *Matine's Map*, then, offers an awareness of shifting time and space. In these centering images of the book, Verdin creates multiple layers of background and foreground, merging spatial and temporal considerations of how the materiality of the actual *ground* affects an engagement of Indigenous pasts and futures for the region. Rather than paper maps that fail to keep up with the land loss or digital maps that fetishize ever-shifting conceptions of modernity, *Matine's Map* offers a visual rendering of continuity and change drawn by Indigenous memory.[96]

While these images of Matine and Jeanne offer what might be thought of by some as traditional agrarian backgrounds—open bucolic land and nature in soft focus—Verdin also plays with the composition of the background in images of one of the youngest women in her work. Across *Return to Yakni Chitto*, Verdin features images of her cousin Allison, who readers can watch age (and sometimes move back in time) throughout the book. The photographs of Allison also act as a foil to the childhood photos that Verdin includes of her grandmother Matine taken in the late 1910s. Together, the photographs of the two young women separated by time create a narrative of continuity and change among the Houma communities along the bayou, illustrating Kyle Powys Whyte's articulation of "kinship time" that replaces the anxiety-laden doomsday linearity of the countdown clock with a concept of duration that foregrounds relationships and future possibilities.⁹⁷

In at least two images of Allison, one from 2000 and another from 2003, she stands in front of a large diesel truck. In both, one taken when she was three and another when she was six, the truck visually overwhelms the girl. Notably, the photographs are not arranged in chronological order. In the one taken in 2000, titled *Lil Black Chicken*, Allison is a fraction of the foreground, her head seemingly the size of the truck's headlight, and the grille of the machine, on which the viewer can barely read "International," almost seems to be in sharper focus than the girl herself. Although there are out-of-focus trees in the far background, the truck dominates the frame as foreground and background. Like the petroculture that has come to control the economy and existential threats of southern Louisiana, it is all-encompassing. The visually ironic reminder that this dependence on oil is "International" creates a visual tension where the audience must look at the young local girl and see the enormous global forces she must face down for her future.

The 2003 image of Allison, which serves as the cover image for part II, the titular "Return to Yakni Chitto," places her even more centrally in the world of petroculture (figure 3.3). She stands in the open door of the truck, which is clearly out of commission and serves as an example of a type of oil-industry detritus. In this photograph Verdin makes no obvious distinction of foreground and background; everything and nothing seems to be in special focus. The girl is simply among this reality, with her athletic shorts and shoes, offering a similarly defiant and proud expression to the one she exhibited three years prior. She exists as the person who must make a way in a fundamentally changed and ever-eroding world. As Dardar explains in the previous section:

> For over a century, unchecked economic development has ravaged our homeland. Its most visible result is coastal erosion brought about by the channeling of freshwater sources to serve the needs of commerce....

3.3 Monique Verdin, *Allison Rodriguez | Bayou Pointe-aux-Chenes | 2003*. Photograph courtesy of the photographer and the Neighborhood Story Project.

This same unchecked economic development has a global impact beyond our shores, significantly contributing to the causes of global climate change. The irony for us is that the fossil fuels harvested at great cost to our local environment continue to affect us directly as they are burned around the world, raising global temperatures and returning to us in the form of catastrophic weather patterns and rising seas.[98]

Verdin photographs Allison in the midst of this push and pull, with the diesel truck as both a product and a process of this industry where Indigenous children are caught in the double bind of local and global pasts and futures. The truck's side mirror, which seemingly reflects just to the right of the photographer—and by extension, the viewer—works as an evocative gesture to ask the audience about their own complicity in the scene before them.

Similarly to the questions posed by the framing of the images of Allison, Verdin also plays on the tension between background and foreground, and between looking and seeing, as she photographs her cousin Brent's high school graduation in the spring of 2005. She includes these photographs in part III, "Crossing Time," which is largely dominated by portraits of family and community members. Composed in verso and recto, like the photographs of Matine and Jeanne, two images of Brent work together to move the viewer's gaze and imagination between past and future. In the verso image, Brent stands on the trunk of an out-of-commission 1970s car (figure 3.4). A school bus that also appears to no longer operate stretches behind him. The depth of the image is difficult to determine, as Brent is almost entirely of the landscape around him, with everything (aside from the line of trees in the far background) of similar focus. Verdin offers this background information in the caption: "A few months before Hurricanes Katrina and Rita, I had the opportunity to photograph Brent Verdin's graduation. Even after decades of integration, high school graduation and college enrollment rates for Houma students remain below average. Twenty-two percent of Houma adults have not graduated high school."[99] Brent's figure among the broken-down school bus and car calls doubly on a community unduly burdened by petroculture economies and a lack of educational opportunities. His position, standing on top of the car and clad in his graduation robes, figures him rising above these conditions.

On the recto, Verdin offers a classic portrait of Brent, where he stands in sharp focus against a background of farm- and woodland. This portrait, titled *What the Future Holds*, works in conjunction with the verso image. The caption, however, tempers unmitigated enthusiasm, as Verdin narrates, "Shortly after graduating, Brent decided to look for other opportunities outside of

3.4 Monique Verdin, *Graduation | Grand Bois | 2005*. Photograph courtesy of the photographer and the Neighborhood Story Project.

Grand Bois. Instead of following in his father's and uncles' footsteps to work as a welder in oil fields and sugar refineries, he sought out another path and moved to Baton Rouge before eventually making his way to Texas."[100] This passage evokes an uneasy concern, as it can leave an audience wondering if young Houma citizens have to leave their homelands to find a way out of the conscripted economies of petroculture and plantation agriculture. Importantly, Verdin does not moralize or offer answers for this dilemma. Rather, she makes the audience see the stakes facing her community, and with her contributors, she ties those stakes to every human on the globe. Or as Dardar offers, "As with all Indigenous Peoples, our existence and identity is tied to the lands and waters that have given birth to us. As the avarice of capitalism continued to devour our world, we wonder with our brothers and sisters around the globe, what will be left to pass on to our descendants?"[101] Together, Brent, Allison, and Verdin herself are these descendants recorded and recording their existence, leaving and returning to Yakni Chitto.

As I note in the introduction, *to erode* in its earliest etymology had a meaning "to eat," and in this plantation-turned-petroculture capitalism, the devouring consumption continually pulls humans into cannibalizing the earth. Verdin makes her own entanglement in this dynamic plain. She notes how after an exhibit of her work in 2017 at the Historic New Orleans Collection, an older woman told her how the family who endowed the collection gained their wealth as owners of the F. B. Williams Cypress Company. As Verdin explains, "In the early 20th century, the company dredged canals through thousands of acres of swampland to log cypress trees until the wetlands were almost completely deforested. The company then turned their investments towards the oil and gas industry. While trying not to look like I just got hit over the head by a 2×4, I managed to say, 'Oh, really!?'"[102] The poetic flair she offers—being hit over the head with a lumber product as she must engage the legacy of clear-cutting trees from her peoples' lands—makes clear the entanglements that turn over between histories and products, producers and consumers. She continues to explain, "'Oh really, not surprising!' would have been a better response. Most arts and culture in south Louisiana are brought to you by name-your-oil-and-or-gas company," and she follows with a list of examples.[103] The erosion of the Gulf Coast has a deep historical background, and it's not hard to imagine that the family who eventually endowed the space where Verdin exhibits her work is the same cypress-logging family Audubon describes in "The Squatters of the Mississippi." In this erosion epoch, it can be hard to parse circles of life from vicious cycles.

This difficulty is all the more true for Indigenous communities who don't fit non-Natives' preconceived ideas of Native people or who exist in places already considered "sacrificial zones" by the larger US popular imagination. While in 2017 many Americans were captivated by the protests against the Dakota Access Pipeline (DAPL), Indigenous activists and allies in southern Louisiana were trying to raise opposition to the Bayou Bridge Pipeline, which is the other end of the Dakota Access. This action did not receive the same media attention. Perhaps this is because for many non-Native people in the United States, Indigenous peoples of Louisiana don't fit their idea of who a Native person is or what they look like or how they sound. If only one Indigenous movement is going to get widespread attention, most Americans—even "well-meaning" ones—have been trained by Hollywood to gravitate to the images of the Great Plains, with horses, headdresses, and the vast open lands of their favorite Hollywood western. It matches the background they already "know." Perhaps Native people speaking French and living by the rhythms of the water are too anomalous, too complex for the larger non-Native imagination. It's a background too out of focus, too blurry for them to make out. Or perhaps it's easier for people to *look* at Louisiana—Mardi Gras, Cajun food, shiny plastic beads—than to *see* how the state is disappearing under their eyes. While Verdin notes in an interview with Kirstin Squint that the NoDAPL protests did bring increased activism to southern Louisiana for action against the Bayou Bridge project, the movement never quite gained the same cultural traction as its Dakota counterpart.[104] Even ecocritics who ponder the relationship between energy resource extraction and the Gulf South seemingly prefer to cite Standing Rock rather than the Bayou Bridge action, or they discuss the BP *Deepwater Horizon* spill separately from their discussion of Indigenous community.[105] The sfumato that Dardar describes and Verdin illustrates perhaps makes it difficult for audiences to hold all of these things in view as they consider the land before them.

Ultimately, erosion in southern Louisiana occurs in the background of everything. Ironically, its hyperquotidian ubiquity makes it hard to see. The foreground catches the eye like Audubon's birds and Cable's plantations. The loss of these foregrounded objects evokes a pathos of loss often attached to whiteness in multiple manifestations, whether in the conservative, plantation-worshipping Confederate memory or in the progressive, eco-conscious birder. And yet the material ground of this foreground and its attendant background is *always Indigenous ground*. This material disappearance of homelands is perhaps the final theft of earth staged by colonialism, and it doesn't happen by way of a sudden event, proclaimed policy, or legal document. Erosion occurs through insidious, constant inaction right under a viewer's eyes. And speaking of occurring under a viewer's eyes, let's return to the background of this very chapter. You have read just over fourteen thousand words, and in that time a football field's worth of Yakni Chitto has eroded away.

4

Gullies and Removals of the Plantation South

IN 1940 the *Chicago Daily News* ran a short piece titled "Soil and Sanctuary" that outlined the serious problems facing the country's topsoil. The author uses the soil crisis in the United States to caution against criticizing Europe for being at war, noting that while the Europeans may be destroying some aspects of their societies, they were at least practicing good stewardship of their land even amid the killing. Near the middle of the article, the author also gestures toward the height of American popular culture at the time: "A nation is composed of two elements—folk and soil. Occasionally we have an admission of that in our country, but we go on wrecking our lands. In 'Gone with the Wind,' Scarlett's father reminds her that the land is the root of all, and the picture closes with a pretty 'back to the land' climax—but the facts are that more than 100,000,000 acres of that very soil have long since gone with the wind, sacrificed to one-crop soil mining under the lash of usury.'"[1] Even though Margaret Mitchell's 1936 novel and the later 1939 film and the crisis of the Dust Bowl during the 1930s were clearly contemporaneous, virtually no critics have considered what *Gone with the Wind* (GWTW) has to say about erosion. The most obvious evidence for considering Mitchell's opus within the frame of 1930s soil science is that the very title *Gone with the Wind* is an erosion metaphor—a fact not lost on some of the book's and film's initial audience, as demonstrated by this piece in the *Daily News*. Mitchell wrote and revised the novel during the height of the Dust Bowl, when accounts of surface erosion precipitated by windstorms in the southern Great Plains saturated news coverage across the entire nation.[2] Moreover, Franklin D. Roosevelt signed the Soil Conservation Act on April 27, 1935, and the amended version, the Soil Conservation and Domestic Allotment Act, in February 1936, just a few months shy of GWTW's June publication.[3] Thus, it's hardly a stretch to say that Mitchell herself would have been keenly aware of her title's resonance for her 1930s audience.

Taken directly from Ernest Dowson's 1897 poem "Non Sum Qualis Eram Bonae Sub Regno Cynarae" (I am not as I was under the reign of the good Cynara), the title emerged as Mitchell negotiated details of the novel with her editor, Harold Latham.[4] Dowson's third stanza reads:

> I have forgot much, Cynara! gone with the wind
> Flung roses, roses, riotously with the throng,
> Dancing, to put thy pale lost lilies out of mind;
> But I was desolate and sick of an old passion,
> Yea, all the time, because the dance was long:
> I have been faithful to thee, Cynara! in my fashion.

The images here call up the Confederate Lost Cause pathos of Mitchell's novel that emerges from the flowery youthfulness of Scarlett's escapades on the eve of the Civil War. The lines ending in *wind* and *mind* are the only instance of slant rhyme across the poem's four stanzas. If read for perfect rhyme, *wind* (the movement of air) must become *wind* (the action one does to tighten the spring of a timepiece), or *mind* (linked to the organ lodged in one's skull) must be pronounced something like *mend* (to repair that which is broken). All of these possibilities are intriguing. Together they gesture toward a slippage in the logic of Mitchell's title. Most provocatively, to render the noun *wind* into the verb *wind*, one must pause to think about the temporality that Mitchell's novel calls up. It is a novel ostensibly about the "tornado" of the Civil War and

Reconstruction, but as many other critics have noted, it is so fully embedded in its 1930s era of composition that it is impossible to understand it without considering the Great Depression landscape.[5] *Gone with the Wind* attempts to wind the clock on Georgia, to make the temporality of the novel's era of publication match up with the historical moment it represents, to set the clock to work—to keep time—again. In this way, Mitchell's work serves as a core example of the struggle of orientation in what Mark Rifkin terms *settler time*, whereby non-Native stories attempt to cast settler colonial temporalities as the single frame by which the world operates.[6] So when I say that Mitchell's novel seeks to wind a clock in order to keep time, I mean both a process of naturalizing a settler plantation landscape in the Southeast *and* a possessive act—to keep time for white supremacy. As Nicholas Mirzoeff articulates, working from Katherine McKittrick's concept of "plantation futures," "From the plantation, there was a compound temporal projection from past conquest into present and future domination."[7] *Gone with the Wind* attempts just this type of temporal projection, trying to wind a plantation past into a 1930s white settler fantasy that does the work of settler memory. Yet as a close reading of the novel's geological registers reveals, the temporality of Mitchell's Georgia landscape could not be restored to her pre–Civil War fantasy as easily as winding a clock, or contained as a possession. It was indeed gone with the wind, or rather, in the case of plantation Georgia, the water.[8]

Medora Perkerson of the *Atlanta Journal Sunday Magazine* asked Mitchell in a 1936 radio interview, "The title of your book, *Gone With the Wind* means that the ante-bellum civilization was swept away by the tornado of war, doesn't it?" To which Mitchell responded, "Yes, Medora, that is the meaning of the title, naturally I would be glad if people thought that the book did tell the story of the whole South. But that isn't the kind of book I tried to write. It is a book about Georgia and Georgia people,—especially North Georgia people."[9] Despite the interviewer's assumptions, Mitchell doesn't spend any time on the mechanics of the *specific* wind destruction of the title—tornado or dust storm—so it seems possible to say only that the title evokes wind damage. Almost certainly the single largest damaging wind event of Mitchell's life, the one that dominated popular consciousness in the 1930s, was the Dust Bowl. What Mitchell does focus on in her answer, however, is how the damage of these forces impacted Georgia specifically. In other words, *Gone with the Wind* is a tale of a highly particular place, not all of the South. The focus of Mitchell's novel is so particular, I argue, that we can read it for what it says about the literal particles of the soil as they eroded away under the plantation economy and its aftermath.

Given my focus on erosion, then, one might wonder about my choice to discuss *Gone with the Wind* at all. It may seem more logical to turn to a text such as Erskine Caldwell's *Tobacco Road* (1932), which depicts white sharecroppers amid hopeless land degradation and their anxieties that losing any more proverbial ground will render them "below" their Black southern counterparts. However, I argue that there is something useful in looking to the pervasive soil science of Mitchell's novel and engaging with the way the novel's seemingly lowbrow popularity has forwarded a romanticized agrarian South without any ecocritical scrutiny. Rather than offer a warning about environmental destruction, as does *Tobacco Road* or even Caldwell's *God's Little Acre* (1933), GWTW attempts to make soil exhaustion romantic—as it likewise and troublingly does with racism and rape. This romance *is* the danger of Mitchell's novel, for as long as these messages float on the surface and are free from scrutiny, the fantasies of white lives on red earth overdetermine the history and popular understandings of the US South.

Much of the plantation South's erosion occurred due to water runoff from farms, creating massive gullies across the landscape, a phenomenon that esteemed soil scientist and "father of soil conservation" Hugh Hammond Bennett explored in detail and divided into three types: gully erosion, rill erosion, and sheet erosion.[10] A specific examination of this water erosion in what we currently call Georgia reveals how the temporality of the landscape in Mitchell's novel mirrors the physical effects wrought on the geological space of the state by plantation agriculture. The perfect example of this is Providence Canyon, a massive thousand-acre site just outside of Columbus, Georgia, near the Alabama border (figure 4.1). Named for a nearby Methodist church, which was established in 1832–33, Providence Canyon is colloquially referred to as the Little Grand Canyon. Its name is misleading, however, as Providence Canyon is technically a gully. At its deepest point, the gully is over 150 feet deep, and it reveals approximately thirty million years of the earth's history in its strata. The canyon's strata showcase red iron-laden clay, alongside deposits of white clay known as kaolin, down to the bottom, where there runs a small creek. Given this history, the canyon has long served as geoscientists' premiere example of the dangers in unchecked gully erosion.[11]

The very temporal scale of Providence Canyon's creation boggles the mind and sets clock arms spinning. In just under fifty years, thirty million years of the earth's geological record were revealed as agricultural runoff in the area quickly eroded the topsoil and sediment in Georgia's Coastal Plain geological province. Taking Providence Canyon as one of its central texts alongside *Gone with the Wind*, this chapter examines how narratives of settler colonial plantation legacies often result in Anthropocene scars that later become reimagined as "natural" spaces, just as Mitchell's novel attempts to naturalize a white Confederate Lost Cause pathos to the region's history. I begin with an overview of Providence Canyon, followed by a close reading of the corresponding soil science concerns and the work of settler memory embedded in *Gone with the Wind*. I close the chapter with a return to a reading of Providence Canyon via artist Elizabeth Webb's 2019 installation *For the Mud Holds What History Refuses (Providence in Four Parts)*, which draws from African American poet and scholar Thomas Jefferson Flanagan's 1940 work *The Canyons at Providence (The Lay of the Clay Minstrel)*. These texts, individually and together, engage questions of temporal strata that create meaning for how settler colonialism, the plantation economy, and ongoing white supremacy manifest in discourses of erosion. While some might imagine that these three areas of inquiry—invading settler colonialism, followed by a resulting plantation economy and then continued in ongoing white supremacy—are layers of historical processes built atop one another, this chapter engages in conversations begun by scholars such as McKittrick, Mirzoeff, and Robert Nichols about how all of these processes exist in continual recursivity and can be read neither wholly diachronically nor synchronically.

If one is thinking with a linear timeline, it's significant that Providence Canyon emerged as a gully resulting from the destructive agricultural practices between the years of Muscogee Creek Removal in the 1830s and the US Civil War. Whereas historians such as Paul Sutter have examined the nineteenth-century creation and 1930s "preservation" of the canyon, this chapter considers the site within its literary landscape. Similarly to how Mitchell attempts to retell the nineteenth century in her 1936 novel, local residents attempted to retell Providence Canyon—and by extension the nineteenth-century agricultural practices that created it—in their tourist boosterism efforts of the 1930s. In a way that evokes Providence Canyon's creation following Muscogee Creek Removal, Mitchell frequently references the historical context of Removal that precipitates the creation of the famed O'Hara family plantation, Tara, whose name, as other critics have pointed out, is obviously reminiscent of *terra*.[12] Together, GWTW and Providence Canyon demonstrate how the geological turn illuminates moments when historical and agricultural contingencies quickly calcify into naturalized narratives of place. These erosion narratives become "sense of place" formulations within regional studies that produce troubling ongoing affinities with settler colonial practices, including theft of Indigenous homelands and Black enslavement. They work in the way that Mirzoeff describes how "white supremacy has a natural history. It is the history of how

white dominance was made natural and in turn claimed nature as part of its domain."[13] While Flanagan's poetry approaches the conversation from a different perspective, it nonetheless grapples with romanticized ideas of Indigenous loss alongside the legacy of the (in)visibility of Black enslavement as it relates to the erosion that creates Providence Canyon. It works akin to what McKittrick explains as how "through the violence of slavery, then, ... the plantation produces black rootedness in place precisely because the land becomes the key provision through which black peoples could both survive and be forced to fuel the plantation machine."[14] When combined in analysis, these texts illustrate what K. Wayne Yang argues as how "the logic of the plantation, when spelled out, illuminates the present of settler colonialism, but also the underground of Black survivance."[15] In the case of my argument, this underground is the underearth exposed in the massive eroded gullies of the plantation South, which illuminate the landed violence of settler colonialism and, as Flanagan's poetry shows, also serve as physical reminders of Black survivance on the land despite violence and enslavement.

Somewhat coincidentally or significantly (depending on how one thinks of these things), the concept of erosion within a geological context shares part of its history with Providence Canyon. As I note in the introduction, Oliver Goldsmith first uses the word *erosion* in his *History of the Earth and Animated Nature* (1774) to describe earth events.[16] Following this, the first instance of the word *erode* and an extended consideration of the phenomenon as having to do with geological features appears in Charles Lyell's three-volume work *Principles of Geology*, which began its publication in 1830, the same year that Andrew Jackson signed the Indian Removal Act. Notable for its forwarding of uniformitarianism (the theory of change as slow, gradual, and constant rather than catastrophic) in earth sciences, *Principles* outlines erosion as an example of the power of the sustained effect on the earth's features from the subtle and slow action of external forces such as wind and water. Thus, the very phenomenon that Lyell lays out in *Principles* is illustrated almost immediately in the formation of the Providence gully. However, rather than demonstrating Lyell's theory of gradual change, the formation of Providence Canyon moves at a lightning pace for geology, simultaneously and paradoxically proving and disproving Lyell's ideas about the pace of changes in the earth's physical appearance. In other words, the gully both made visible the

4.1   (*overleaf*) Providence Canyon in Stewart County, Georgia, 2017. Photograph by the author.

layers of slow, gradual change, supporting Lyell's hypothesis, and was itself created in rapid fashion. In this way, Providence Canyon and Mitchell's novel share a kind of temporal register where their meaning lies somewhere between the time of their creation and the projection of narrative onto their layered histories of southeastern Native Removal and plantation agriculture based on Black enslavement.

Lyell notes several gullies of a similar nature across Georgia during his travels in North America, which he records in the two-volume *A Second Visit to the United States of North America* (1849). Lyell's texts prove a compelling complement to GWTW, as he describes much of the landscape that Mitchell later tries to capture in romantic terms in her novel. As Sutter explains, Lyell recorded "substantial gullying" in locations ranging from the then capital of Georgia, Milledgeville (where the earth formation came to be known as the Lyell Gully), to just north of Providence Canyon in Columbus and on to Tuscaloosa, Alabama. In the second volume, Lyell writes of his future namesake gully:

> Twenty years ago it had no existence; but when the trees of the forest were cut down, cracks three feet deep were caused by the sun's heat in the clay; and, during the rains, a sudden rush of water through these cracks, caused them to deepen at their lower extremities, from whence the excavating power worked backward, till, in the course of twenty years, a chasm, measuring no less than 55 feet in depth, 300 yards in length, and varying in width from 20 to 180 feet was the result. The high road has been several times turned to avoid this cavity, the enlargement of which is still proceeding, and the old line of road may be seen to have held its course directly over what is now the widest part of the ravine.... In another place I saw a bridge thrown over a recently formed gulley, and here, as in Alabama, the new system of valleys and of drainage, attendant on the clearing away of the woods, is a source of serious inconvenience and loss.[17]

While as Sutter points out, later geologists would indicate that the Milledgeville gully was not quite as big as Lyell had suggested, it was still one among many pieces of evidence that settler agriculture was contributing to rapid gully erosion.[18] Lyell does not indicate exactly what has been lost in the massive gullying he observes, but it is important to consider that this awareness of vanishing land for the plantation economy—this sense of loss—seems to exist prior to what will become familiar to the region as the Confederate Lost Cause.

Furthermore, Lyell makes a direct connection between Removal and the eroded landscape, writing first that "the last detachment of Indians, a party of no less than 500 quitted Columbus only a week ago for Arkansas, a memorable event in the history of the settlement of this region and part of an extensive and systematic scheme steadily pursued by the Government, of transferring the aborigines from the eastern states to the far west."[19] He continues, "Here, as at Milledgeville, the clearing away of the woods, where these Creek Indians once pursued their game, has caused the soil, previously level and unbroken, to be cut into by torrents, so that deep gulleys may every where be seen; and I am assured that a large proportion of the fish, formerly so abundant in the Chatahoochie, have been stifled by the mud."[20] It seems possible that Lyell slightly exaggerates his temporal proximity to the "last detachment" from Columbus, but, nonetheless, he clearly finds that Removal and gullying are connected. Despite his awareness of the land destruction from the expansion of plantation agriculture, Lyell resists critiquing the plantation system based on enslavement as it connects to the loss of land. He continually waxes offensively poetic about how "contented" enslaved people seem to be, and any reader looking for abolitionist sentiment in his works will be sorely disappointed. In other words, his work is simultaneously a subtle condemnation of land damage by the plantation and an apologia for slavery.[21] His refusal to make the connection between the abuses of land and human labor demonstrates the mechanisms by which this violence later comes to be disassociated, as Tiffany King and Mark Rifkin discuss in their critical work on Black and Indigenous peoples' intersections of experience.[22] When read together, Lyell, Mitchell, and Providence Canyon writ over time offer a constellation of how the process of erosion via the settler colonial plantation moves from a question of soil science to a vehicle for the metaphors of the white southern pathos of the Confederate Lost Cause that renders Black enslavement and theft of Indigenous homelands unseen and often disconnected despite their direct relationship.

Through its quick uncovering of thirty million years of the geological record, Providence Canyon asks us as earth readers to hold roughly a hundred years in the foreground: from 1830 to 1930, or from southeastern Native Removal to the Great Depression. The creation of Providence Canyon—as a geological feature and later as a Georgia state park—occurs almost directly as a result of these two sociopolitical events. The Indian Removal Act made possible the large-scale forced removal of Muscogee Creek people from the area in southwestern Georgia that is now Stewart County. Following this, the land was opened for large-scale plantation agriculture dependent on enslaved labor. As David Montgomery argues in *Dirt*, the southern plantation economy exhausted soils and slowly moved westward across the American continent. The movement followed the opening of Indigenous territories for settler farmers. Even white citizens in Alabama noted that the movement of the plantation model of agriculture (high-yield, single-crop) produced rapid soil exhaustion, creating in turn the need for more land. In this way, erosion both precipitated the desire for Removal in the opening of more available lands and then exhausted those lands as a result of the removal of Indigenous peoples and, by extension, their knowledge of land management in the region. Within this larger setting, the gully that becomes Providence Canyon appears on the landscape. By the 1930s, when soil exhaustion became a feature of Works Progress Administration and New Deal responses to the Great Depression, Providence Canyon was serving as the poster child for land abuse, even appearing as the very first photograph illustrating extreme erosion in Bennett's 1939 landmark textbook *Soil Conservation*.[23]

Despite the canyon's frequent appearance in tracts discussing the United States' "southern problem" of rural poverty and soil degradation, Stewart County residents touted it as a possible economic incentive to gain tourist money in the emergent automobile traveling culture. In this move, they transitioned the gully from one scale of human geological effect in the nineteenth-century plantation economy to another in the rise of the individualized fossil fuel–based travel economy. Interestingly, some residents attempted to claim that the canyon was "natural" rather than the effect of human activity over the past hundred years. And not surprisingly, they relied on the canyon's stunning appearance to do so.[24] As Dudley Wilmeth described in a piece of clear boosterism in the *Atlanta Journal* in 1939, "Whew! It's a sheer hundred feet from where you stand to the tops of the tall trees growing in the bottom, straight as marble shafts, for no wind blows there to bend them. In contrast to the soft green of the trees a white-capped peak of limestone juts up from the canyon's floor, a long, slim finger pointing to heaven as its maker. On the wall of the canyon across from you the striated blending of white, yellow, orange, red, and purple makes a canvas of chronomatic splendor. No sun-set ever had more colors."[25] Admittedly, the canyon *is* aesthetically appealing. However, even in this romantic construction, there is a slight nod toward the problem of soil erosion. As Wilmeth closes his piece, "Soil erosion, a plague to most states, has created a thing of beauty from the innards of Georgia. It's an ill rain that brings nobody good."[26] Sutter discusses the canyon's co-present messages as "ironic," but as a text, I would argue that it more closely approaches sublimity.[27] It's the horror and awe of thirty million years of the earth turned inside out within a century due to terrible human actions. In a way, it is both epic and epoch, and the boosterism that sought to market it as a location worth seeing is very much of the agrarian nostalgic and romantic land attachment trend of 1930s literature emerging from the US South, including Mitchell's novel.

It is little surprise, then, that Providence Canyon becomes a potential site of local pride in the 1930s—the heyday of what Michael Kreyling calls the "invention of southern literature."[28] The decade witnessed a proliferation of attempts by white southern writers to claim their own providence from the past century, exemplified most explicitly in the Southern Agrarian manifesto *I'll Take My Stand* (1930).[29] Similarly, the revisionist logic of Mitchell's novel centralized white obsessions with southern soil and attempted to read the previous hundred years into a providential future where a white southern hierarchy survives. Certainly Mitchell is not alone. Though more dubious of nostalgia projects such as these, the Agrarians' favorite, William Faulkner, repeatedly turns toward narratives of land abuse during the second half of the decade in Sutpen's wrenching of his own plantation from the swampy soils of the Mississippi Delta in *Absalom! Absalom!* (1936) and his retelling of the 1927 flood in the "Old Man" section of *The Wild Palms* (1939). As a geotext within this literary landscape, Providence Canyon acts as a similar site of nostalgia, depending on the audience.

All of these texts by white southerners foreground loss as beauty, and they all participate to different degrees in what Charles Reagan Wilson has termed the "civil religion of the Lost Cause" and Robert Jackson calls "the southern disaster complex."[30] For Wilson, the southern civil religion of the Lost Cause looked toward a history of righteous defeat that "offered confused and suffering Southerners a sense of meaning, an identity in a precarious but distinct culture."[31] As Jackson explains in a slightly different way, the US South and its authors have long engaged in disaster narratives where "the full sense of tragedy that so many Southerners have perceived in their culture seems a particularly apt response to living in an environment where disasters of whatever origins occur with alarming frequency."[32] Jackson's articulation of the southern disaster complex describes the aesthetic and psychological uses of tragedy in examining the Confederate Lost Cause. Additionally, Jay Watson argues that like southern studies, environmental studies needs to interrogate not simply the language of loss or lost objects but the ontological and epistemological frameworks embedded in different conceptions of lostness.[33] This is all the more true if we are to think ourselves out of the continued destruction wrought by Anthropocene projects, many of which are themselves extensions of settler colonialism in the Americas. For as Jackson explains, "This native Southern discourse has succeeded in imagining a history that becomes not just a resource for environmental protection, a political tool, a portentous spiritual and aesthetic wellspring, but a restorative environment in itself. Welcome to Dixie."[34] As the tone of his closing sentence suggests, such projects should be read skeptically for the old white southern ideologies they uphold through embedded narratives of so-called traditionalism. Put simply, I want to think critically about the *conserve* in southern *conservatism and conservation* when engaging the uneven terrain of Anthropocene discourse that scholars such as Kathryn Yusoff and K. Wayne Yang outline. When understood through the logic of settler plantation systems, these critical approaches reveal an uneasy tension between white pathos and what we might think of as the "liberal" politics of earth preservation. Indigenous peoples—from around the globe, both at home and in diaspora—have experienced far different effects of, and entanglements with, the Anthropocene as they have borne the brunt of various iterations of the global imperial project and its attendant losses. In the US South this local imperial project largely took the shape of the settler plantation—often discussed as the *plantationocene*—that enslaved Black people and stole Indigenous homelands while naturalizing the resulting damaged landscape as providential.[35]

For its part, Mitchell's novel has been linked to ideas of plantation modernity and concepts of the plantationocene, particularly in the work of Amy Clukey.[36] I would like to extend these conversations to examine GWTW's engagement of Indigenous Removal in Georgia and its frequent exposition on the nature of material southern dirt. Patricia Yeager ever so briefly touches on the iconic scene of Scarlett holding a handful of dirt in David Selznick's film adaptation in her germinal study *Dirt and Desire*. She compares Selznick's melodramatic shot of Vivien Leigh, as Scarlett, clutching the fistful of earth with the rather banal scene in the novel. Yeager writes, "Scarlett's fantasy points *Dirt and Desire* in two directions. First, it evokes the tragic history of dirt in the South, the way that dirt as property, as money-making machine, is mingled not only with desire but with blood. Second, these bodies point to the genesis of the southern grotesque within a particular locality—the back-breaking labor needed to establish the white southern home."[37] However, this is the last mention of the novel, or the film in Yeager's study. I take this as a departure point for my reading of the novel, given that, one, the "home" in question here is stolen Indigenous homelands, and, two, an omnipresent anxiety of Mitchell's story is that the particles of dirt that make up those stolen homelands are quickly eroding away.

I also take as a cue Yeager's argument that Mitchell's novel does not lend itself to the act of deconstructive unearthing. Rather, the whole meaning is right on the surface, evinced by its very title. As Yeager asserts, "*Gone with the Wind* explores—without quite intending to—the cult of fetishized, never-seen surfaces, what is hiding in plain sight, the preoccupation not with 'under' or 'beneath' or 'depth,' but with the cult of 'besides,' of what is proximate, next-to, and therefore invisible."[38] In other words, at its most formal level, GWTW is a novel of the topsoil. It opens with gestures to both Removal and erosion in its first pages. These are the things right on the surface of the entire story, and this is the anxiety of the white South: in addition to stolen labor, the entire system of the plantation is built on stolen homelands, and that land is disappearing.

In the opening paragraphs of the novel, Mitchell explains, "Life in the north Georgia county of Clayton was still new."[39] Of course, life in the space currently known as North Georgia was decidedly *not new*. We know that Mitchell knows this, because she goes on to refer to Removal multiple times throughout the novel. Beyond this, when Mitchell first describes the setting of Tara and the surrounding area, she refers to both evidence of harmful land degradation (seemingly unwittingly) and efforts to prevent erosion (directly):

> Spring had come early that year, with warm quick rains and sudden frothing of pink peach blossoms and dogwood dappling with white stars the dark river swamp and far-off hills. Already the plowing was nearly finished, and the bloody glory of the sunset colored the fresh-cut furrows of red Georgia clay to even redder hues. The moist hungry earth, waiting upturned for the cotton seeds, showed pinkish on the sandy tops of furrows, vermilion and scarlet and maroon where shadows lay along the sides of the trenches. The whitewashed brick plantation house seemed an island set in a wild red sea, a sea of spiraling, curving, crescent billows petrified suddenly at the moment when the pink-tipped waves were breaking into surf. For here were no long, straight furrows, such as could be seen in the yellow clay fields of the flat middle Georgia country or in the lush black earth of the coastal plantations. The rolling foothill country of north Georgia was plowed in a million curves to keep the rich earth from washing down into the river bottoms.
>
> It was a savagely red land, blood-colored after rains, brick dust in droughts, the best cotton land in the world.[40]

While Mitchell may have imagined the predominance of red imagery would nicely highlight her protagonist Scarlett's connection to the reddish earth, what she depicts is the clay-heavy, iron-rich ultisols of the Georgia Piedmont that had been so exhausted by cotton plantation monoculture that their normal red hue had intensified into crimson.[41] As Sven Beckert explains, "Ever newer cotton frontiers replaced one another, motivated by the unrelenting search for land and labor, as well as soils that had yet to escape the ecological exhaustion that so often came with cotton growing."[42] This exhaustion resulting in southern clay was described by Bennett in the 1930s as he developed his theories of sheet erosion:

> We noticed two pieces of land, side by side but sharply different in their soil quality. The slope of both areas was the same. The underlying rock was the same. There was indisputable evidence that the two pieces had been identical in soil makeup. But the soil of one piece was mellow, loamy, and moist enough even in dry weather to dig into with our bare hands. We noticed this area was wooded, well covered with forest litter, and had never been cultivated. The other area, right beside it, was clay, hard and almost like rock in dry weather. It had been cropped a long time. We figured both areas had been the same originally and that the clay of the cultivated area could have reached the surface only through the process of rainwash—that is, the gradual removal, with every heavy rain, of a thin sheet of topsoil. It was just so much muddy water running off the land after rains.[43]

Here Bennett almost perfectly describes the exhausted red clay soils of Mitchell's romantic setting as resulting from the exact farming that she waxes poetic over through Scarlett's eyes. Bennett's word choice of *removal*, while likely more coincidence than political acknowledgment, is also telling. Compounding this depiction of soil exhaustion, Mitchell directly describes that the plantation owners must practice contour plowing, creating "a million curves" in an effort to prevent the little topsoil they had left from eroding into river bottoms—in effect what had caused massive gullies such as Providence Canyon and gained Lyell's attention during his surveys of the 1840s. Right away, then, readers are keyed to the fact that the novel understands the erosion of the landscape it seeks to establish. To return to Yeager, Mitchell's meaning lay right on the surface.

In addition to the explicit description of soil degradation and erosion prevention, Mitchell describes an attending anxiety of the novel: "The plantation clearings and miles of cotton fields smiled up to a warm sun, placid, complacent. At their edges rose the virgin forests, dark and cool even in the hottest noons, mysterious, a little sinister, the soughing pines seeming to wait with an age-old patience, to threaten with soft sighs: 'Be careful! Be careful! We had you once. We can take you back again.'"[44] For one, the description of the forests as "virgin" relies on a settler colonial fantasy of untouched land, but even more tellingly, the passage signals the nervous precarity of the plantation against anthropomorphized forests that look to reclaim that which has been stolen and abused in the form of the earth. Mitchell's use of the common American gothic trope of a threatening haunted forest is certainly not new. In authors since at least Nathaniel Hawthorne, this device has signaled a settler anxiety over Indigenous homeland claims. These opening passages tie together the two central components of my argument in their clear awareness of the dangers of erosion and their direct appeal to the anxiety of white—cotton, people, supremacy, American—disappearance.

The connections to an eroding landscape and Indigenous Removal continue throughout the beginning of the novel. Just a few pages later, Mitchell describes the setting as Scarlett travels to the Wilkeses' Twelve Oaks, where she hopes the object of her affection, Ashley Wilkes, will propose to her rather than, as rumored, his cousin Melanie: "Along the roadside the blackberry brambles were concealing with softest green the savage red gulches cut by the winter's rains, and the bare granite boulders pushing up through the red earth were being draped with sprangles of Cherokee roses and compassed about by wild violets of palest purple hue."[45] Again, the novel describes the red earth as "savage," and in this case, it attaches the adjective to "gulches" (another word for gullies), signaling the serious erosion problems surrounding Tara. Furthermore, the addition of the "Cherokee rose" alludes to the apocryphal tale that the flower bloomed along the Cherokee Trail of Tears, with each white petal representing a tear that a Cherokee person cried and the yellow center of the flower representing the gold that was stolen from the tribe near present-day Dahlonega.[46] The flower, an invasive species in the US Southeast, was introduced by settlers in the late eighteenth century.[47] As readers, we know that Scarlett's own path will be paved with tears, as virtually any reader can guess that Ashley is not going to choose her over Melanie. Furthermore, thanks to historical hindsight, readers will certainly anticipate that loss looms for the white Confederate characters in a novel set on the eve of the Civil War. More important than this, however, is the fact that Mitchell uses a specific symbol of romanticized Indigenous loss to forward her own heroine's impending fall. She merges a white Confederate pathos of loss with an Indigenous one. Draped underneath is the evidence of an ever-eroding landscape to which virtually no one seems to pay any heed. Likewise, there can be little doubt that GWTW depicts enormous human abuse in its portrayal of enslaved and convict labor. The novel is also a veritable catalog of land abuse. This needs to be made explicit not because one supersedes the other but because the two are irrevocably linked: there is no land abuse that does not also abuse humans, and many human rights violations also violate the earth.

While numerous important scholarly works outline the myriad ways that Mitchell's text abuses its Black characters, there also exists so much direct textual evidence of land abuse in Mitchell's novel that it would take almost this entire chapter to record every instance that appears in the text, suggesting the ways the two abuses are imbricated. I offer a sampling of passages that I won't consider in detail but that nonetheless record the soil degradation of the novel: the Tarleton twins travel across red furrows in the land in the opening pages; Scarlett stares down a "blood-red" and "gashed" road waiting for her father after a rainstorm; she looks out her window at "freshly turned red earth" before leaving for Twelve Oaks; she chastises her father for kicking up "a heap of dust" on her on the road; she notes the "muddy-red roads" at Atlanta's emergent Five Points intersection; she watches red dust over the city from her window; she observes a courier with news of Tara kicking up more red dust in the street; she has "red dust thick upon her ankles" as she realizes Atlanta is under Union siege; she notes the "deep gullies" along the road as she escapes to Tara out of Atlanta; and Mitchell describes even more "red gullies" along the road near the Flint River once Scarlett reaches Tara. To continue, there are "weary red miles" that isolate Tara following the war, and despite all this prior evidence of soil exhaustion, the "spring plowing" commences in full force the next season. Then later there are the even redder, muddier streets that greet her on her return to Atlanta; Scarlett also expresses her desire to plant "miles and miles of cotton" at Tara even though the entire first half of the novel depicts land that is already exhausted.[48] Shortly after this, Scarlett observes that the earth is desperately trying to go back to seedling pine and brambles, which I'll add is not shocking, given the years of land abuse the reader has already witnessed.[49] Finally, there's Scarlett's own Reconstruction-era lumber business, which is most certainly clear-cutting even more trees from the land, leading to what will certainly be even more erosion across the landscape. And this is just a selection of the scenes I am *not* considering in detail in my argument.

Indeed, Scarlett does have anxiety about the land's productive efforts and her attachment to it. As Richard Gray narrates in his foreword to *South to a*

New Place, Mitchell recounts this as the "southern curse": "The curse I refer to is loving land enough to give everything you've got to get it. Never would I own a foot of it, city or country land. If I had spare money it would stay in the bank or the stock market but never in red clay. Then about two years ago when I set out to write the great American novel I was confronted by the fact that whether I liked it or not, it was a story of the land and a woman who was determined not to part with it."⁵⁰ However, despite this supposed love for the land, Scarlett does not notice that *she* is the one destroying the land. The land is disappearing because of her actions and those of her father. This diagnosis is not simply upstreaming later recognitions of soil science back to the temporal setting of Mitchell's novel. As Bennett explains in his 1939 textbook, "In 1832, the farmers of Georgia and South Carolina were advised that only manure could save the sterile field from 'our wretched system of agriculture,'" and he quotes from an 1860 pamphlet from N. T. Sorsby that calls the region's agriculture "the murderous system."⁵¹ Thus, what we see in *GWTW* is not the love of the land itself but the love for control over and ownership of that land. When Scarlett panics about her loss of this land, it isn't land loss that gives her trouble. It's the loss of white supremacy.

In addition to the sheer number of passages that I could select to support my reading of *GWTW* within a consideration of erosion, it also bears mentioning that many of the scenes that have become iconic in the larger public imagination surrounding the novel involve descriptions of exhausted soils. In these scenes audiences see the tension between Scarlett's relationship to the earth and the novel's pathos of white loss. As Yeager rightly points out, Selznick's iconic scene of Scarlett raising her fist in the air and proclaiming "I'll never go hungry again" is less triumphant in print, as in the novel Scarlett lies in the dirt and vomits a turnip.⁵² Following this scene, Mitchell again returns to her favorite earth trope: "Her love for this land with its softly rolling hills of bright-red soil, this beautiful red earth that was blood colored, garnet, brick dust, vermilion, which so miraculously grew green bushes starred with white puffs, was one part of Scarlett which did not change when all else was changing. Nowhere else in the world was there land like this."⁵³ The red clay imagery again signals the exhausted soil that pervades the entire novel. Scarlett imagines her love of this landscape as "unchanging." Meanwhile, the land itself had been drastically changed in the past thirty years, as illustrated by Mitchell's own descriptions. However, Scarlett envisions the land attachment going on ad infinitum: "These were the only things worth fighting for, the red earth which was theirs and would be their sons', the red earth which would bear cotton for their sons' and their sons' sons."⁵⁴ However, this land only belongs to her family under the fiction of land title, and their supposed love for it hasn't stretched back for generations, as readers know that Scarlett's father was among the first generation of settlers to clear and plant these acres with cotton. She is fighting for a white plantation futurity that *she* has rendered onto the land. This is a process akin to the way theft generates property, as described by Robert Nichols, inverting the terms by which we understand dispossession and its legacy for Indigenous landed sovereignty.[55] And this plantation future Scarlett desires is the thing that will cause this land to disappear. The reader sees her caught in a paradox of her own destruction. Mitchell describes her as "a woman to whom nothing was left from the wreckage except the indestructible red earth on which she stood," thinking, "'We'll plant more cotton, lots more.'"[56] This attitude is indicative of what Bennett diagnoses in his textbook when he notes that the US South before the Civil War was "still enslaved by a one-crop system," drawing an association between enslaved labor and soil destruction. Even as Scarlett worries about what will happen to the land, she's planning to do the very thing—double down on a massive one-crop system maintained by her abuse and enslavement of Black humans—that will continue to ruin it via soil exhaustion and eventually sheet erosion and gullying.

This paradox is completely bound up with the loss of white supremacy, as Mitchell describes how Scarlett's plan for continued land abuse fundamentally depends on enslaved Black labor:

> "In another year, there'll be little pines all over these fields," she thought and looking toward the encircling forest she shuddered. "Without the d[——], it will be all we can do to keep body and soul together. Nobody can run a big plantation without the d[——], and lots of the fields won't be cultivated at all and the woods will take over the fields again. Nobody can plant much cotton, and what will we do then? What'll become of country folks? Town folks can manage somehow. They've always managed. But we country folks will go back a hundred years like the pioneers who had little cabins and just scratched a few acres—and barely existed.
>
> "No—" she thought grimly, "Tara isn't going to be like that. Not even if I have to plow myself. This whole section, this whole state can go back to woods if it wants to, but I won't let Tara go."[57]

While Scarlett vows to plow Tara herself, it seems unlikely that anything about the system of land use she desires can be achieved without enslaved labor. She works against a loss of land to the forest, but in her plan to plant even more cotton, she will cause even more destruction, more sheet erosion, more gullying, and less and less hope for human survival on the land. And this is the point: Scarlett's anxiety isn't about land; it's about whiteness and white supremacy.

With all of this talk of planting more and more and exploiting the land for every nutrient the plantation can leach out of it, one might think that Scarlett, via Mitchell, just simply wasn't aware of basic farming techniques and soil science. Yet the novel makes it plain that Scarlett *does* understand the basics of soil exhaustion and regeneration. After she returns to Tara for her father's funeral, she wakes up to gaze across the land as she had done in the opening pages of the novel. Through Scarlett's eyes, Mitchell describes Tara as a place that is healing precisely because it has ceased to be a massive plantation operation:

> Scarlett's heart swelled with affection and gratitude to Will who had done all of this. Even her loyalty to Ashley could not make her believe he had been responsible for much of this well-being, for Tara's bloom was not the work of a planter-aristocrat, but of the plodding, tireless "small farmer" who loved his land. It was a "two-horse" farm, not the lordly plantation of other days with pastures full of mules and fine horses and cotton and corn stretching as far as eye could see. But what there was of it was good and the acres that were lying fallow could be reclaimed when times grew better, and they would be the more fertile for their rest.[58]

Almost miraculously, Scarlett, who has spent the previous two-thirds of the novel only desiring more planted land despite the obvious signs of soil damage around her, suddenly recognizes the need for crop rotation and small-scale land management. Even though the novel and its characters acknowledge crop rotation and land management, the novel does not demonstrate an awareness that the pervasive red gullies across the landscape indicate all bad things to come for land use—even without a civil war. This unawareness appears despite the fact that soil scientists even in the nineteenth century were addressing these very problems.

However, I argue something else is going on here in this tension around soil exhaustion. It's the place where the novel and the novelist know what is to come. In the soil-to-soil comparison of the mid-nineteenth century with the 1930s, we see the temporal registers of the novel come into locked relief, the clock winding to pull them into a recurrent cohesion of settler time where the concept of land as property continually enacts manifold thefts against Black and Indigenous people in the region. The land problems of plantation Georgia are the land problems of the Great Depression. Scarlett is the farmer that Bennett worries about. Indeed, Bennett is concerned for the land and its demise, but there's also a concern here for settler-owned land—for what losses white people sustain when they have abused the land into no longer sustaining them. Bennett makes this plain. In the 1930s he specifies exactly which Americans are to blame: "The white inhabitants of this country, in their 'conquest of the wilderness' and their 'subjugation of the West,' piled up a record of heedless destruction that nearly staggers the imagination."[59] As a novel very much of its time, GWTW isolates the moment of anxiety when white southerners have lived and burned through the surface topsoil and are left alone with the hard red clay of their actions. Rather than change their behavior, these white southerners make the evidence of their destruction—hard red clay—a central figure of their identity via the pathos and co-optation of loss, forming a key symbol in their white agrarian fantasies of the region.[60] In this process white southerners attempt to erase the histories of theft of Indigenous homelands and of Black enslaved labor from the landscape, as Scarlett does in her vision of Tara becoming a "two-horse farm."

Scarlett's retreat into a myth of agrarianism is in line with the novel's larger attachment of an anxiety over waning white supremacy to a loss of the soil. Abby Goode argues that these fantasies of agrarianism work as "agrotopias," which are "characterized as much by racial purity and reproductive stability as they are by small farming and independent labor."[61] She continues, "American literary fantasies of agrarian perfection did not come out of nowhere; they emerged in reaction to imagined, racialized threats to the nation's demographic and environmental stability."[62] Mitchell's novel weaves together the threads of settler memory in its forwarding of the fantasy of an emergent agrarian resurgence. Scarlett's sudden desire to prevent erosion, then, has less to do with saving the soil than with saving whiteness.

Toward the end of the novel, the main characters become increasingly aware of the tension between what happens on the surface and what anchors into the earth. As Scarlett reflects on the increasing numbers of new people flooding into Atlanta, she worries in explicitly earthly terms: "They didn't care, these people from God-knows-where who seemed to live always on the surface of things, who had no common memories of war and hunger and fighting, who had no common roots going down into the same red earth."[63] Deep roots are the very thing that often prevents soil erosion, and the irony here is rich, given that Scarlett has made her money in the Reconstruction South via clear-cutting forests for her lumber mill, only to now be worried about "roots." She isn't alone in this concern. Following the death of their daughter, his rape of Scarlett, and the accident where he causes her to fall down the stairs and miscarry, Rhett, too, becomes concerned about roots. He has left Scarlett and Atlanta only to return with a resolve that he must once and for all move on from their relationship. Shortly before uttering his most famous line, Rhett explains to Scarlett, "I'm forty-five—the age when a man begins to value some of the things he's thrown away so lightly in youth, the clannishness of families, honor and security, roots that go deep—."[64] This white nativist desire for roots is the very thing that leads him away from Scarlett and Atlanta, yet the question that every character seems to float past with ease is, How is it possible for them to have deep roots after—at best—only a few generations in a place?

Scarlett and Rhett yearn for an impossible thing, as there can be no deep roots on stolen land. They cannot prevent the erosion of the earth because their very presence—with its absence of literal and figurative roots—is the very cause. The novel at least tacitly seems to recognize this, as much earlier Mitchell writes, "She could not desert Tara; she belonged to the red acres far more than they could ever belong to her. Her roots went deep into the blood-colored soil and sucked up life, as did the cotton."[65] This description, where Scarlett sucks life out of the soil like the exhaustive soil-degrading crop, represents the crux of her character development. Her roots, however, are only as deep as her father's generation, which benefited directly from the theft of Muscogee Creek homelands. Her and Rhett's sudden desire for life beyond the surface comes only after they have realized that the surface has been washed away.

Importantly, Indigenous Removal precipitates the entire plot. In the very beginning of the novel, readers learn how Gerald O'Hara came to establish Tara: "Then the hand of Fate and a hand of poker combined to give him the plantation which he afterwards called Tara, and at the same time moved him out of the Coast into the upland country of north Georgia."[66] This "Fate" was, in truth, likely the fraudulent Treaty of Indian Springs, which resulted in the forced Removal of the Muscogee Creek Nation from the state and a lottery to divide the remaining lands among Georgia settlers.[67] Mitchell narrates Gerald's experience as follows:

> It was in a saloon in Savannah, on a hot night in spring, when the chance conversation of a stranger sitting near by made Gerald prick

up his ears. The stranger, a native of Savannah, had just returned after twelve years in the inland country. He had been one of the winners in the land lottery conducted by the State to divide up the vast area in middle Georgia, ceded by the Indians the year before Gerald came to America. He had gone up there and established a plantation; but, now the house had burned down, he was tired of the "accursed place" and would be most happy to get it off his hands.[68]

Gerald eventually manages to win the land lot from the man in a drunken game of poker. It seems that Mitchell attempts to absolve the O'Haras from being *direct* beneficiaries of Georgia's genocidal actions by noting that the Creeks had "ceded" the land the year before Gerald arrived in the United States. However, in the same scene, readers see Gerald directly figure the scope and pace of Removal when considering his own strategy for acquiring land: "From the stranger's description, his plantation was more than two hundred and fifty miles inland from Savannah to the north and west, and not many miles south of the Chattahoochee River. Gerald knew that northward beyond that stream the land was still held by the Cherokees, so it was with amazement that he heard the stranger jeer at suggestions of trouble with the Indians and narrate how thriving towns were growing up and plantations prospering in the new country."[69] From this passage we know that Gerald (and Mitchell) is thinking very specifically about the effort to steal Indigenous homelands in the region and its effect on his own self-interest.

This awareness follows the plot along, as Mitchell also attaches Atlanta's founding to Removal. She writes, "When Gerald first moved to north Georgia, there had been no Atlanta at all, not even the semblance of a village, and wilderness rolled over the site. But the next year, in 1836, the State had authorized the building of a railroad northwestward through the territory which the Cherokees had recently ceded."[70] What Mitchell marks as a cession is in fact the fraudulent Treaty of New Echota. Furthermore, in both the Cherokee and the Muscogee Creek Nations, there were already multiple plantations dotting the landscape that were operated by Indigenous people. The wilderness was not wilderness at all but rather a place being transitioned into a destructive plantation economy. Despite these actions by some Native families, however, the rampant soil degradation notably does not appear until the forced mass Removal of Native people from the region. This isn't to argue that an Indigenous-owned plantation that enslaved people was somehow better in its treatment of the earth or of humans. However, the scale of erosion due to the disregard for the land via rapid clear-cutting and monoculture likely could not have happened without forced Removal across the state, as the sheer amount of monocrop agriculture that emerged in the aftermath accelerated soil exhaustion to a near limit point.

Repeatedly across the novel, Mitchell perhaps unwittingly offers up reminders of the previous land stewardship of Native people. In fact, we learn that the eponymous twelve oaks of the Twelve Oaks plantation aren't some feat of the Wilkes family but rather a feature seemingly purposefully planted by Muscogee people. When Scarlett approaches the plantation following her escape from the siege of Atlanta, she notes, "There towered the twelve oaks, as they had stood since Indian days, but with their leaves brown from fire and the branches burned and scorched. Within their circle lay the ruins of John Wilkes' house, the charred remains of that once stately home which had crowned the hill in white-columned dignity."[71] There are old, deep roots here in those oak trees, but they aren't the roots of Scarlett and her people. Such a fact, however, does not preclude her or any other white character from mapping their own sense of loss onto the Indigenous landscape that surrounds them.

The relationship of theft of Indigenous homelands, Black enslavement, and white supremacist futurity via the plantation becomes explicit in one particularly telling scene. After Scarlett and Melanie's return to Tara from war-torn Atlanta, Mitchell describes how Melanie's baby nurses at the breast of the enslaved Dilcey. Tellingly, it is in this moment that Mitchell returns to the plot point of Dilcey as Native—her Black identity not negating this fact, as one can be both Black and Native in both traditional and contemporary understandings of kinship and belonging. The reliance on enslaved Black labor that sought to codify distinctions around Black and Indigenous identities in the region further served to buttress white supremacy. Likewise, the language of Black Native people being "part Native" or reduced to the language of "blood" serves only to divide Black and Indigenous communities in white supremacy's service. This understanding likely would have been far beyond Mitchell's understanding or recognition. She writes of Dilcey in an offensive way that relies on phenotype:

> The door opened softly and Dilcey entered, Melanie's baby held to her breast, the gourd of whisky in her hand. In the smoky, uncertain light, she seemed thinner than when Scarlett last saw her and the Indian blood was more evident in her face. The high cheek bones were more prominent, the hawk-bridged nose was sharper and her copper skin gleamed with a brighter hue. Her faded calico dress was open to the waist and her large bronze breast exposed. Held close against her,

> Melanie's baby pressed his pale rosebud mouth greedily to the dark nipple, sucking, gripping tiny fists against the soft flesh like a kitten in the warm fur of its mother's belly.[72]

Scarlett remarks on her surprise and gratitude that Dilcey elected to stay at Tara despite Emancipation. Dilcey says her decision was because of how well she was treated by Gerald O'Hara. In Mitchell's ignorance, this scene brushes over a key fact that exists outside the narrative: if Dilcey is Native, then these are very likely her homelands, which would go a much longer way toward explaining her desire to stay than some account of "good" treatment under Gerald's enslavement. This unnamed detail would almost certainly *not* have been Mitchell's intention. Rather, she imbibed and disseminated a romantic fantasy that enforced the filiation of white supremacy and Black enslavement. The larger context of this scene that Mitchell does not understand, however, is one of Black and Indigenous kinships and desires for homelands. In this moment with Dilcey, Mitchell unwittingly creates and records what she cannot or will not see: a Black Indigenous woman's resistance to being removed from her own homelands despite the harshest of circumstances. This scene demonstrates the means by which the milieus of settler colonialism and enslavement are joined and made invisible by narratives of white loss attempting to define the region on its own terms. Moreover, just as Scarlett sucked the life out of the soil like cotton, in this scene the imagery is the same: the tiny white son of Twelve Oaks is sucking life out of the enslaved Black Indigenous woman's body.

One of the single most significant factors in the emergent global plantation economy, and in the US Southeast in particular, was the relentlessly hungry theft of Indigenous homelands and Indigenous life.[73] However, many popular accounts of Removal in Georgia often focus on the Dahlonega gold rush, even though this much later event does not begin to compare to the money generated by cotton plantations.[74] The plantation system's drive for more and more land produced more and more negative effects on that land. As Providence Canyon historian Paul Sutter explains, "The environmental destructiveness of southern agriculture was the product of settler society pushing relentlessly westward and clearing land."[75] While he is careful not to romanticize all Indigenous practices as 100 percent perfect environmental stewardship, his conclusions are direct: "Settler agriculture, as the cutting edge of an emerging global capitalist economy, was driven by a land hunger and market orientation that [Native] agriculture never expressed. . . . [T]obacco and, later, short-staple cotton . . . were both commercial crops feeding transatlantic markets, and they tore through fresh land with a pace and magnitude that [N]ative agriculture never matched." And he ultimately concludes, "At every step, though to varying degrees, accelerated soil erosion corresponded with—and was the product of—expanding settler agriculture."[76] His conclusions match nearly all available historical evidence, even down to the material from Lyell, solidifying the connection between plantation gullying and Removal. Thus, as we have seen in each of the preceding chapters, settler colonialism not only stole Indigenous homelands through the rendering of property relations and land titles but also stole the physical earth via its destructive capitalist practices.

Sutter specifically goes on to link the erosion of Stewart County and Providence Canyon directly to Muscogee Creek Removal, writing, "While Creek agriculture was sophisticated, dynamic, and syncretic, and while it reshaped portions of Stewart County's landscape, the massive erosion that soon scarred Georgia's western Fall Line Hills was a unique legacy of settler agriculture."[77] He outlines the dates and details of Providence Canyon as thoroughly as the archive allows, using records from the nearby Providence Methodist Church: "Congregants began clearing the land of its forest cover and farming and grazing the territory surrounding the church in the early 1830s. It was not long after that the land began slipping away from them. There is little solid evidence about exactly when or how this happened, but we do know that by 1859 gully erosion had become so bad in the vicinity of the church that the congregation had to move to a new house of worship on the north side of the road to avoid sliding into the expanding chasm."[78] The dates here are significant, for they almost perfectly mirror Gerald O'Hara's own movement into Creek territory and the eve of the Civil War, when the novel begins. Simply put, both Tara and Providence Canyon were born from Removal.

Like Mitchell, who attempted to repackage this history in the 1930s, Stewart County residents attempted to repackage the history of Providence Canyon in the same decade. In this campaign the massive gully was often rendered as "natural" to the landscape and linked to Manifest Destiny, as Sutter describes in detail via his extensive archival work on local newsletters and public addresses. In many cases, boosters leaned heavily on the "Providence" name to suggest that God himself had opened up this wonder. Additionally, at least one local forwarded the incredibly odd theory that the gully was a result of trails left by Native people—as if they had worn down through thirty million years of the geological record by walking. Other local historians later repeated this story. They occasionally amended these stories to say that Creek people had walked away only *some* of the land and that it left slight ravines, where the water runoff from settler agriculture quickly channeled.[79] This theory is, obviously, preposterous. It signals a settler desire for absolution for the destruction wrought by the plantation economy—a clear instance of the work of settler memory. These mythical origin stories where Indigenous people somehow serve as the foundation of the region's whiteness permeate much of southeastern and US "history." In the case of Providence Canyon, these stories combine with the geological landscape to produce an uncanny monument to the settler plantation economy. While other monuments in the region attempted to naturalize white supremacy, efforts to make Providence Canyon a national park attempted to naturalize settler agriculture (and by extension white supremacy) into the very earth for tourism profit.[80]

While I appreciate Sutter's historical work on the canyon, I am also troubled by his own lapses into myths of southern exceptionalism to imagine this confluence of factors. He writes in his introduction:

> When I read about the poor farming practices that had allegedly created the canyon, I found it hard to get past the irony of the place: the seeming incongruity and even the dark humor in a park that preserves the scenic results of an environmental disaster. As a northerner new to the Deep South, I also found myself thinking, "how southern," by which I meant how exemplary of the South's renowned boosterism to create a park from such an ignominious example of the region's land use history. It seemed to me a particularly southern gesture to turn a scar into a point of pride.[81]

Here Sutter clearly likens the canyon to an earthen monument to the Confederacy, enslavement, and the southern civil religion of the Lost Cause. Moreover, as a "northerner," he maintains a regional exceptionalism that seemingly lets the rest of the nation off the hook for their investment in these same ideological roots. He goes on to modify his initial assessment but only somewhat, noting, "As for its southernness, I sensed that that too was up for debate. I remain convinced that there is much that is distinctly southern about Providence Canyon and that one can learn a lot about the soils of the South and the region's larger environmental history by staring into its multihued abyss," even though he also states that "Providence Canyon should not . . . be seen as only a southern site, for it can teach us both smaller and larger lessons about conservation and environmental history."[82] As the previous chapters of this book have illustrated, erosion—and its attendant white settler pathos—is not exceptional. It affects every corner of the continent. Perhaps Sutter sees something particular in the confluence of the plantation economy and romantic narrative recovery in the canyon, but these anxieties about land loss are only that. They are not an exception. Just as white supremacy and pathos surrounding a perceived loss of white identity pervade nearly every corner of the nation, so too do the concerns over ecological damage. Regardless of region, the anxiety of white disappearance is repeatedly mapped onto the loss of the soil across time.

Despite its peak popularity in the 1930s as a site of both nostalgia and destructive agriculture, Providence Canyon continues to have resonance with artists today. In 2019 Elizabeth Webb staged the exhibit *For the Mud Holds What History Refuses (Providence in Four Parts)*. The multimedia exhibit explores a number of ways that Providence Canyon was represented over the twentieth century, in materials ranging from Robert Flaherty's documentary film *The Land* (1942) and Works Progress Administration photographs of Stewart County to an oral history interview with poet and scholar Thomas Jefferson Flanagan and Webb's own film footage. Webb layers these objects across the four parts of the exhibit, and together they mirror the palimpsest nature of the canyon itself. The centerpiece of the exhibit is a roughly four-minute video that layers archival footage and audio with Webb's own contemporary visual work. She frames the entire exhibit with a quote from the same southern sociologist and critic of sharecropping that Dorothea Lange and Paul Taylor quote in *An American Exodus*, Arthur Raper: "Gullies are the region's receipts for the 'bargains' the system got out of virgin soil, slavery, and farm tenancy combined."[83] Significantly, here the myth and rhetoric of "virgin land" creeps in when one attempts to make sense of the southeastern landscape and the horror wrought by the plantation economy. Nonetheless, Webb's artistic work is compelling for it attempts to imagine the twenty-first-century legacy of this particular erosion disaster.

The layering of sound and image asks the viewer to imagine their own place in the strata of the landscape in question. In an audio loop of an archival oral history recording, Flanagan narrates his own experience of racism as a Black child when he and his white neighbor friend would have to "divide" at the canyon in Stewart County to go to their separate schools, placing the canyon as directly symbolic of another social history in addition to an ecological one. The sound of Flanagan's voice with the crackling audio of the oral history interview recorded in 1973 and the intentional "projector" clicking sound of the 16 mm film pulls together several eras into one auditory space. When Webb later narrates her own twenty-first-century experience of the canyon, noting trees falling over the edge after rainstorms and the need for the State

Park Service to move fences farther back from the ever-shifting edge, she continues this layering of meaning. She personifies the canyon as having an ever-gaping mouth from which its own "voice" rises. The aural composition of these documentary moments induces the listener to consider what the canyon itself might be saying, and I am drawn to the possibility that such sound composition might effect. However, there remains something troublingly romantic about imagining the canyon's voice. As an observer of the canyon, I am unsure if it speaks for itself or if audiences and interpreters such as Sutter merely hear the echo of their own presumptions about the region when they listen at the edge of the abyss.

Webb also layers film from Flaherty's *The Land* with her own footage. One can see multiple film stocks running over one another, so both are visible at the same time rather than as a consecutive montage. Just as the multiple temporal registers of the sound work to give the viewer a sense of the same earth layers present in the canyon itself, the visual elements also mimic layers of meaning wrought from the landscape. While Flaherty's film has sound of its own, Webb chooses to use only his moving images. Such a choice is interesting, as the writer of *The Land*'s catalog description on the Internet Archive notes, "The only pity is that Flaherty, at heart a silent film director, wouldn't let his subjects speak."[84] While Flaherty attempts to speak for the earth and his subjects, Webb in turn silences his narrative voice in favor of his moving images. Webb also zeroes in on the distinct color contrast of the canyon's walls as Flanagan narrates how the difference in skin color between him and his childhood friend created their eventual estrangement. The layering of the present-day color images with the black-and-white images from Flaherty works to suggest that the world has moved beyond "black and white" logic but still bears the vivid color scars of a previous era.

Despite this careful artistic rendering of Providence Canyon's symbolic meaning, Webb's work never mentions two key things: erosion or Indigenous homelands. In some ways, it might be clear from the images of water running through that canyon that erosion is certainly the overarching issue that does not need to be named. She makes erosion plain in virtually every other element of the film. The absence of a named Indigenous history, however, is more concerning. Just as the framing quote about "virgin land" from Raper suggests, Webb sets up a Providence Canyon that floats free from the very specific act of Muscogee Creek Removal. Her omission of this illustrates in many ways what Tiffany King argues in *The Black Shoals*. King offers that Black and Native studies (and history) have largely been thought of as disconnected, and she imagines the shoal as a geological formation that can tie together

Black diasporic metaphors of the water (e.g., *The Black Atlantic*) with Native metaphors of the land as the "fulcrum" of the field. The shoal, as both land and water in dynamic negotiation, brings these two fields together.[85] In the case of Providence Canyon, I would argue that the process of gully erosion, in which water runoff cuts through the earth, works in a similar vein. One cannot thoroughly consider the contours of the canyon without understanding the exploitation of both Black and Indigenous people in the plantation South at the same time.

I do not want to fault Webb in particular for this oversight. Rather, I want to posit this oversight as a moment when, as I have argued in previous chapters, erosion merges with erasure, even if unintentionally. As King writes, "U.S. racial discourse tends to be organized by a White-Black paradigmatic frame that often erases Indigenous peoples. When U.S. Black studies has engaged Indigenous thought and politics, the field has been less likely to articulate Black-Indigenous relations through a discourse of settler colonial relations until recent, twenty-first-century scholarship."[86] Webb's work, in its provocative layering of Black and white relationships via the intersection of the eroding canyon, manages to miss the crux of how the theft of Black life and labor worked in conjunction with the theft of Indigenous homelands as each was converted to units of property for white supremacy. In other words, even smart, beautiful attempts to make meaning of erosion in the Southeast today still struggle to imagine how the eroding landscape remains bound up with unspoken anxieties over the disappearance of stolen Indigenous homelands. Even when layering multiple histories, settler colonial narratives based on racial reconciliation continue to promote Indigenous erasure.

In one notable detail, however, Webb makes a subtle gesture. Indeed, it's so subtle that it is difficult to tell if it is a gesture at all. She mentions at the end of the short film that the plumleaf azalea grows in the bottom of the canyon, narrating that "a rare flowering azalea known as the plum-leaf thrives. Their roots hold these disparate layers together, blooming away from the canyon walls, outward and upward toward the sun."[87] Unlike Mitchell's Cherokee rose, the plumleaf azalea is native to the southeastern United States. Perhaps Webb imagines in her evocation of its roots holding the layers together that as an indigenous plant it works to hold the earth in place. This is an evocative image. Along with the sepia-toned film stock—which the exhibition notes is "pigmented by red clay soil [Webb] harvested from her family's former plantation land in Eastern Alabama"—layered over the black-and-white archival footage and photographs, it enables a viewer to stretch their sight and imagine a gesture toward Indigenous history.[88] However, this gesture is ultimately unsatisfactory, for an indigenous flower is not an Indigenous person. Its roots are not control over homelands, and its presence does not heal the scars torn across the earth by Removal. Indeed, it seems Webb is striving to layer meaning from her own experience with the history in question, but unfortunately the exhibit never quite reaches the level of acknowledging the theft of Indigenous homelands. This does not entirely undermine her other attempts to make sense of the racial injustice and conflict around Providence Canyon, however.

In one provocative piece of the exhibit, she includes a case with pages of Flanagan's *The Canyons at Providence (The Lay of the Clay Minstrel)* (1940) spread out under glass. Underneath and around the pages, she has added red soil from her family's land, which the exhibition notes is her attempt to "[implicate] her own family's land practices."[89] This coupling of Flanagan's poetry with the physical land, in the same case, asks the viewer to process how different people might regard the scenes of earth degradation via soil exhaustion—themes that Flanagan explores in his poetry. Flanagan grew up in Stewart County, Georgia, near the canyon and later attended Atlanta University in the city. He worked as a mail clerk and for the railroad while writing poems for publication. Later in his life, he was the minister at the Antioch African Methodist Episcopal Church in Decatur, Georgia, and he contributed regularly to the *Atlanta Daily World*. His poetry often focuses on questions of place, and his work on the canyons adds a complicated piece to the story of reclaiming the canyon beyond its ecological destruction. Webb's use of the manuscript images of these poems brings Flanagan's understudied work into the contemporary conversation about land use, the US South, and memory. Canyon historian Sutter also attempts a reading of Flanagan's collection. Although I agree with some of his conclusions about the poems' ambivalence, Sutter's attempt at literary analysis is perhaps limited by his training as an environmental historian, as he somewhat simplistically attributes the poem's meaning to Flanagan's "intentions" and conflates the speaker of a poem with its author, as if they were always coterminous. His interpretation also remains at the level of content, neglecting to consider how—especially for poetry—form can offer just as much, if not more, insight into what a poem offers. And like Webb, who neglects Indigenous history in her artistic consideration of the canyon, Sutter seemingly misses the gestures toward questions about Indigenous homelands in Flanagan's work.[90]

For his own part, Flanagan opens the short collection *The Canyons* with a foreword that wonders why Providence Canyon has not been lionized previously in literature. The opening poem, the title poem of the collection, begins with the following lines:

> LOW in the volumes of the caverns I come,
> Beating tom-toms on the red sand drum.[91]

These first two lines allude to a style of drum stereotypically associated with Indigenous roots, and Flanagan continues with these vague romantic references to a possibly Native speaker:

> Oh, I love the canyons, for as deep as their goal
> Sinks in the soil are the pits of my soul,
> Eroded by the tears I have wept for my clan,
> From the world's foundation—the forgotten man.[92]

While *clan* can signal simply family, the reputation of Muscogee clans, the reference to "tears" amid the ubiquity of the Trail of Tears, and the midcentury pathos around supposedly disappeared and forgotten Native people support a reading where Flanagan gestures to the connection between the canyon and Indigenous peoples. In this version of the canyon's genesis, Flanagan attributes its beginning to the Native speaker's tears, making the Trail of Tears an agent of physical erosion on the landscape. Through this opening Flanagan links in verse what other artists of the period seem reluctant to accept: Removal precipitated the large-scale land damage of the Southeast.

In the opening section to the next poem, titled "The Erosion March," Flanagan works with irregular meter and falling identical rhyme to mirror the landmark he describes. His highly formally complex poem demonstrates the extent to which mere questions of content are unsatisfactory for addressing the architecture of systemic white supremacy that has, in this case, materially embedded itself in the geological structure of the earth. He composes four-line sets that mostly end with identical rhymes of *them* and occasionally with *you*. Each line ending with *them* or *you* works as a falling rhyme, where the last syllable is unaccented. Instead, the penultimate words before *them* and *you* all follow a pattern of perfect rhyme. In this way, Flanagan motions the reader down into the canyon, falling farther and farther with each line. One set of lines breaks from this pattern, ending with the unusually accented *roomless* and *groomless*, where the closing syllable *-less* receives the emphasis.[93] In this way, Flanagan paints a portrait of the canyon with an ever-descending meter punctuated by a single moment, representing in verse the few notable standing pillars of limestone that rise up from the canyon floor. Notably, the emphasis on *-less* accents the very negative-space quality of the eroded site, where the profound presence of less and less *is* the defining feature of the canyon. Thus the opening section of the "March" does not simply fall spatially into the earth but also works to drop the reader back in time, moving backward through the time of Christ into the time of Jupiter. As Flanagan moves down into the geological layers of the canyon, he references the layers of earth's human history. His poem works as a marker of historical content and also at the level of material form. This meaning at the level of form in Flanagan's work demonstrates Yang's point that "the plantation is both architecture and the story we tell about it—place and plot."[94] In Flanagan's specific case, however, it is both place and *poetics*, a merging of a rhythm of Black life and survivance with the effects wrought on the earth by the legacy of the plantation. Unlike the boosters who attempted to turn Providence Canyon into a "naturalized" space, Flanagan attempts to dwell in the ambivalence the site calls up.

At the level of content, Flanagan makes four distinct narrative moves in the content of the "The Erosion March": one, he begins by referencing an old Indigenous presence with the canyon; two, he mourns the enslaved Black laborers who might have worked the land over the years; three, he notes that the ever-increasing canyon flattens the distinctions between rich and poor, enslaved and enslaver; and, four, he contrasts the canyon with other "wonders" of the world, noting that while they were built with enslaved labor, Providence Canyon was built by God and nature. His first point is more romance than reality, as the canyon does not date back to the period of Muscogee homelands in the region. His second point is far more grounded in fact, as there were undoubtedly "mute, inglorious Miltons" who never found the full promise of their potential under enslavement or the sharecropping system. His third point is likewise true. The avalanches of soil from along the canyon walls certainly do not care whose land they take under. However, Flanagan's fourth point remains difficult to reconcile. It is known that Providence Canyon almost certainly exists because of the land abuses and human abuses of enslavement. Indeed, enslaved people did not build Providence Canyon in the same manner as enslaved people built the pyramids. However, the canyon does exist because of Black enslavement and destructive plantation agriculture.

Given Flanagan's championing of the dignity of his enslaved ancestors, a reader might expect him to compose a poem more critical of the plantation enslavers' abuse of the land, which ultimately formed the canyon. As a contemporary reader, I find it difficult not to desire a reading of Flanagan that lays the destruction of erosion at the feet of the plantation economy. Sutter similarly admits to wanting to find a kind of "irony" in Flanagan's work.[95] Toward the end of "The Erosion March," Flanagan writes:

> Unmarred by biting chronicles of broken bone and sweat
>   We inherit a coliseum no mortal hand has set.
> No blistered ant-swarm beneath unrequited toil
> Raised these stalwart pyramids and fashioned from the soil
> These nameless busts that peer on time out of crumbling loam
> That tints their base from carpet spread to swinging chapel dome.[96]

One cannot fault Flanagan's speaker for seeing a kind of majesty in the canyon itself. The lines suggest that the plantation system, with its own "chronicles of broken bone and sweat," did not contribute to the earthen scar that is the canyon. Indeed, enslaved people were not used to "build" the canyon, as they were for the pyramids, but the canyon is certainly still a product of the plantation based on enslavement and monocrop agriculture. However, Flanagan's work certainly does not wallow in a pathos of loss about the canyon that waxes anxious about the loss of whiteness or white supremacy. Rather, he attempts to recover the canyon as a reminder of a great equalizer of erosion among humans, taking a more circuitous path of critique that suggests God made the canyon to deliver equalizing justice. In this way, he makes the gaping chasm—itself a kind of absence—into a profound site of visibility for the legacies of Black enslavement.

In a similar way, he writes earlier in the "March" of this leveling evoked by the canyon:

> The road was the dividing line betwixt the black and white
> Placed in last encampment ere this leveling blight
> Contending that all dust is dust and there is neither rank
> Nor creed beneath the challenge of the master of the flank:
> Hamp Mayo died a chattel but fond freedom knows his ghost
> And sawing on his fiddle, he surveys the haunting host
> Cap'n McGinty, whose estate was all the lower ranch
> In disgust claps his palsied hands at each avalanche.[97]

Flanagan works in verse to break racial hierarchy, reminding the reader that at the end of the day, erosion—like death—does not care who it takes under. There is power in this assertion even as Flanagan's poem perhaps lets the plantation enslavement system off the hook for creating the "leveling blight" of the canyon itself. Near the beginning of the poem, the speaker owns his bias toward the canyon as an aesthetic object that blends "beauty and tragedy." Flanagan writes:

> But a strangeness of beauty and tragedy blends them
> And though here's the boneyard of the farmer and his farm,
> The art-thirsty eye, admiringly defends them
> For their romantic glory and rerety of charm.[98]

Perhaps in this silver-lining reading from the defending "art-thirsty eye" there lies a reclaiming of more than the canyon itself. Flanagan's poem offers something redeemable that has emerged from something destructive, a futurity in the region not based on white supremacy.

Flanagan doesn't ignore the tragedy in the landmark, but he also does not seem to be composing this praise in a verbally ironic fashion. He does, however, complicate what appears to be his sincere appreciation for the aesthetics of the canyon. It seems likely that he would have been aware of the contemporaneous analysis of Providence Canyon as a direct result of Indigenous Removal followed by the destructive farming practices based in the plantation and later sharecropping economies. However, the poem upholds the canyon as a net positive, because it renders permanently visible that labor on the land—notably, the same labor that Lyell's accounts of plantations and Mitchell's agrarian dream of returning to yeoman farming attempt to erase. One cannot simply ignore Providence Canyon, nor can one forget the people whose abuse and dispossession made this landmark possible. It is as if Flanagan posits that the land itself remembers. The Confederate Lost Cause South might be able to manipulate erosion narratives and histories into its own versions of plantation futures, but it cannot erase the eroded canyon from the landscape. The "less" and "less" it created will remain accented every time one gazes at the canyon. It cannot make this violence against the land and people unseen in the work of settler memory. Flanagan's poetry reclaims Providence not for the abusers but for the survivors of the South's plantation erosion legacies.

Ultimately, what does a geotext such as Providence Canyon do for our thinking about erosion within a southern literary landscape largely overdetermined by popular images taken from novels such as *Gone with the Wind*? Providence Canyon itself seemingly layers Removal, plantation destruction, and the attempted recovery of this narrative as naturalized nostalgic space. While Indigenous peoples in the Americas constructed elaborate earthworks that scholars have discussed as literary texts, or as earth-writing, Providence Canyon itself appears as an unwitting earthwork, written on the landscape by the destructive human agency of the settler colonial plantation.[99] As Flanagan suggests in his writing, this eroded space cannot be erased. It disrupts the "compound temporal projection" that the nostalgic and systemic racism of plantation futures depends on. It is the "leveling blight" of erosion where contemporary audiences must choose what narrative of southern space they seek to remember for the future.

5

Littoral Cells and Literal Sells of the Atlantic

THE SUMMER I TURNED SIXTEEN, I woke each morning to haul PVC pipe, several five-gallon buckets, a metal rod, and a sledgehammer to the northernmost end of Shell Island in the town of Wrightsville Beach, in present-day North Carolina. My friend and I would then take core samples of the shoreline along the edge of Mason Inlet in the immediate shadow of the Shell Island Resort, a 169-unit complex where individuals own condominiums that are rented out most of the year as lodging accommodations through a central rental program. I sometimes look back and wonder what people must have thought: two sixteen-year-old girls hammering PVC pipe several feet into the sand and then extracting, cataloging, and bagging the resulting soil samples. It was clear to nearly everyone that Mason Inlet was moving quickly toward the resort, eroding the northern end of the island (figure 5.1), but people don't always enjoy being reminded of disappearance during a day out at the beach. Our research that summer was part of a program called North Carolina Summer Ventures in Science and Mathematics, which selected high school students from around the state to spend the summer at various college campuses, conducting independent research projects. I had applied because I thought I wanted to work in estuary ecology, but after an introductory course on coastal geology, I was hooked. We learned that the Outer Banks and other barrier islands are all moving south through their own natural processes. Essentially the current moves sand from the northern end of the islands to the southern end, causing inlets to shift south. Unlike many other instances of erosion one might imagine, this erosive cycle of the barrier islands is for all intents and purposes what one might think of as "good." This process of sand erosion and deposit is entirely natural, as part of the littoral cell system of the Atlantic barrier islands I had grown up on and loved.

The problem, however, was that real estate developers had long been allowed to build far too close to the northern ends of the islands despite the moving inlets. North Carolina has strict laws prohibiting the building of hardscape, including seawalls, to slow or prevent erosion, as it has been regularly proven these structures cause far more harm than good.[1] However, as the Shell Island Resort found itself closer and closer to the inlet, the pressure was on the state legislature and leaders to protect the building at all costs. Meanwhile, the southern end of Wrightsville Beach was growing as the sand continued its own process of redistribution. Perhaps in many ways, Shell Island is where this book began—long before I would ever imagine my life as a literature professor, of all things. I was struck by the questions of a deep time that I could *see* in the layers of the soil and of the politics of class that made 169 individuals' real estate investments a priority over laws meant to protect public beaches. This, combined with the knowledge that the earth worked in ways so incredibly unconcerned with my own teenage life left an indelible mark on me that summer. And the story I've learned since then about the Atlantic Seaboard has only grown richer and more complicated. This closing chapter is largely about the dynamism of place as a geological phenomenon and as a site where humans struggle to cope with a changing earth that is often not of their own making.

These triangulated concerns of a natural geological process, climate change, and the loss of human-built structures inform much recent representation of the southern Atlantic shoreline. Because this chapter deals in the narratives and geological conditions of an ecosystem determined by shifting sandbars and dynamic islands, it likewise moves and shifts across places

5.1 Shell Island Resort behind increasingly eroded dunes caused by the approaching Mason Inlet, 1996. Photograph by the author.

and texts that may initially seem disconnected from one another by water. However, these places are in actuality *connected* by that water and the littoral cell system that controls the way sand erodes, flows, deposits, and sinks. The chapter, then, moves from the southern tip of North Carolina north to the Chesapeake Bay, zooming in on individual islands all affected by the conflict between shifting sands and capitalist investment. The earliest histories of English colonial invasion subtend all of these spaces. In many ways the Eastern Seaboard works within an Algonquian linguistic continuum that long predates US fictional regions of "North" and "South" and endless debates on where in Virginia or Maryland the South begins or ends.[2] Here the waters connect. Although these aqueous connections are often overlooked in this space, they are no less important than ideas of land to understanding the stakes of erosion. In my formal approach to this chapter's organization and analysis of texts, I attempt to show how one might perceive the discrete parts of the littoral cell system as fragmented but how often-unperceived currents connect island to island, and histories to presents.

I begin in the town of Wrightsville Beach, which serves as the main beach area for Wilmington, North Carolina, site of the 1898 racial terror plot that effectively broke the emergent political and professional power of Black communities in the city and ushered in decades of Jim Crow policies. Wrightsville Beach appears as one setting in Charles Chesnutt's novel about the events of 1898, *The Marrow of Tradition*. Notably, as I discuss later in this chapter, the same individual white supremacists who stoked the animosity toward Black people and participated in the violence themselves were also the key landowners and early developers of Wrightsville Beach. Nearly one hundred years later, Ellyn Bache published her short story "Shell Island," in which questions of family, white masculinity, and futurity brush up against early signs of climate change, demonstrating how this one location has long served as a kind of magnet for white anxieties of disappearance. Around the same time as the publication of Bache's story, North Carolina was in the middle of a rare bipartisan campaign to save the state's iconic Cape Hatteras Lighthouse from shoreline erosion. The eventual relocation of the entire lighthouse, dubbed "The Move of the Millennium," demonstrates just how far state and federal entities *can go* in saving something they deem important—something that is, perhaps not incidentally, a literal beacon of protection for those who would come ashore to settle in the early United States.

Recently, farther north up the seaboard, the fate of Tangier Island received special attention as journalist Earl Swift covered the isolated Virginia Chesapeake community in *Chesapeake Requiem*. As Swift notes, even though the island is on the front lines of climate change, as it loses more land and suitable crabbing grounds every year, the residents are almost uniformly skeptical of climate science and enthusiastically supportive of Donald Trump. The islanders do not talk in terms of a warming planet or sea-level rise, preferring instead to refer consistently to their central problem as "erosion." The elision of scale from global to local informs the way I frame this chapter's argument about particular regions' depictions as foregone spaces of loss. These representations follow well-worn narratives of regional exceptionalism, where some spaces serve as sites of psychological quarantine where problems can be set aside from larger national patterns of politics and crisis. Moreover, the figure of the island amplifies this phenomenon. As Andrew Lipman argues, "By favoring ground over water, [historians such as Frederick Jackson Turner] actively created the myth of American exceptionalism."[3] Through this creation of a grounded American exceptionalism, the island as a figurative location becomes a site of exception to American exceptionalism. While Swift represents the people of Tangier Island with nuance, the larger media depiction of Tangier frames it within long-held discourses of rural, southern spaces as "out-of-time," backward areas that cannot fathom their own existential threats. Notably, famed English colonial invader and Disneyfied legend John Smith recorded exploring the waters around the island in his early writings on the region. He refers to the place that would become Tangier and the surrounding isles as "Limbo." This descriptor offers an eerily prescient diagnosis of the in-between condition of anxiety that has come to define life under apocalyptic climate change. Of course, there might not be a direct line of thought about whiteness and anxiety between Smith and the present-day residents of Tangier, but so often historical coincidence reveals meanings that are difficult to ignore. The narratives of southern exceptionalism and American landed exceptionalism crash together in this space, demonstrating the perpetual tensions between dynamism and stasis in a United States invested in settler real estate.

At stake here is the question of shifting, dynamic land and who gets to claim it for what ends. Pricey real estate ventures, infrastructure, and investment all contribute to a capitalist understanding of place that needs the land to be static rather than dynamic. It is, after all, easier to sell land that doesn't move about of its own will. As coastal geologists Wallace Kaufman and Orrin Pilkey explain, "No erosion problem exists until people lay out property lines and build."[4] Within this consideration there is a powerful correlation to the questions of race and Indigenous identity along the Eastern Seaboard. As other scholars have demonstrated, American capitalism based on enslavement and land theft needed racial codification in order to maximize its profits. However, because of the nature of colonialism on the Eastern Seaboard, with its attendant issues of epidemics and enslavement alongside shatter zones and subsequent recoalescence of Indigenous societies, the nature of Indigenous and racial identity along the Atlantic coast has been difficult to recognize and codify for outsiders interested in pinning down what it is. Just as capitalist settler colonialism needs the land along the Eastern Seaboard to stay in place, it also needed and needs Indigenous identity to be fixed and recognizable in supposedly stable categories, which often fall along the lines of imposed Black-white binaries in the United States. The dynamic littoral cells of the Atlantic—where erosion is a process embedded in the survival of the entire ecosystem—have long been a problem for settler colonialism, just as have the dynamic Indigenous communities who have continually changed, adapted, and sometimes even moved in this region in order to survive.[5]

Monacan poet Karenne Wood captures these questions of recognition and dynamism in her work. Her collection *Markings on Earth* (2001) traces the shifting soils and phenomenologies of Indigenous belonging from Jamestown to her own present in Virginia, reminding the reader that the perception of appropriate land use and the perception of Indigenous presence are deeply intertwined. The collection opens with a poem of the four directions and then proceeds in the first section to offer poems of human-animal-plant kinship prior to the appearance of English colonists, including John Smith. Notably, Wood resists the romantic backward glance, instead focusing on moments when Siouan and eastern Algonquian people found themselves working, reluctantly or necessarily, in tandem with the colonial forces that were quickly reshaping their material world.

Much of this material reshaping exists in relationship to the soil. For example, in the poem "Orinoco," named for the variety of tobacco that dominated the early colonial plantation economy, Wood writes:

> Virginia's alluvial bottomland greened—we cleared it all, even hillsides, not for food crops but tobacco. Dark gold it was, and we gave our land to it.[6]

Thus, she outlines the collective Indigenous speakers' own participation in the clear-cutting and upturning of a once rich alluvial soil for a sacred plant rendered a colonial cash crop. The poem then seems to almost speed through time to explain:

> A gift to us, it seemed, this money and land enough for all, until the earth itself failed us, its richness spent, and the topsoil drifted away. We saw the children hungry in a drought of our making, the plant a new form of destruction.[7]

Here Wood directly references the damage wrought on the land by the monocrop tobacco agriculture, and she alludes to the rise of cancer (perhaps a nod to her own battle with the disease that would ultimately claim her life far too soon, in 2019, at the age of fifty-nine). Significantly, this destruction is linked directly to the erosion of the topsoil through the farming of tobacco. However, the tragedy is more than the imposition of colonial plantations taking and destroying the land; it signals the entanglement of Indigenous peoples bound up with the exploitative farming of their own sacred medicine turned cancerous danger. There is a change here, and this change links the earth and people in ways that trouble easy distinctions about land use under colonialism. This is not a poem of foregone, romantic Native people unencumbered by the pressures to participate in colonial exploitation. Yet Wood explains that some people tried to hang on: "We went by moonlight to the fields, hacked stalks / down, held ceremonies, prayed, but greed spread like blight, and others / took the money, then the land."[8] No process of change is uniform and totalizing, nor is any past free from those who one wishes might have made other, better decisions about what compromises to make with increasingly land-greedy settlers. Wood reminds the reader that one must hold all of this in view as they consider the legacy of agriculture and Indigenous complicity versus resistance along the Eastern Seaboard.

With the homelands under others' control, her collection continues to recount changes in people and soil emerging from the founding of Jamestown. In "Jamestown Revisited," Wood juxtaposes the dynamics of Indigenous peoples having been invited to a gathering at the site of the 1607 settlement, where descendants of Jamestown hoped to apologize for all the wrongs that proceeded from the founding of the colony. Wood takes an appropriately exasperated tone:

> Here you come again,
> asking. Do you see
> we have nothing
> to give, we have given
> like the ground, our
> mountains rubbed bare
> by hybrid black poisons
> concocted from tobacco.[9]

Again, she connects the people to the earth itself, which has been exhausted by tobacco farming. The enjambment following "nothing" is amplified by the white space on the page to the right, leaving the reader to linger on just how much has been taken. The caesura between "ground" and "our" emphasizes the rupture that has emerged from this soil and human exhaustion. Moreover, as with the sped-up temporality of "Oronoco," Wood pushes the settler audience to see the (dis)connect between their actions and apologies:

> You could repent to us,
> weep into your robes an
> emotional, talk-show-like
> moment to absolve almost

>   four hundred years, then
>   go home to mow your lawn.¹⁰

The juxtaposition of historic tobacco-based land exploitation with present-day lawns forces the reader to see that the soil (and human) abuse is not merely a past wrong that can be forgiven. Instead, it is an ongoing nonrecognition of how the literal management of the soil is connected to continuing frameworks of even well-meaning settler colonial "apologies." The performative "talk-show-like" nature of this apology calls up images of middle-class viewers consuming *Oprah* or *Dr. Phil* while they experience a fleeting catharsis for colonial wrongs. The closing return to one's "lawn" shows the reader that the earth destruction that fueled tobacco speculation buttresses the casual land and water abuse of sprawling suburban development. It signals how apologies and recognitions will ring hollow as long as Indigenous homelands remain in the possession of settlers and their mortgage companies.

Wood connects this casual nonseeing to how settlers fail to see the Indigenous people before them *as* Indigenous. Alluding to Virginia and the East Coast settler legacy of refusing Indigeneity to mixedblood people, she states:

> We are
> nameless, named by others:
> *mulattos* and *mongrel*
> *Virginians*.¹¹

In reversing this gaze of the settler impulse to prescribe Indigenous identity through what *they* can see, Wood closes the poem with a call to a dynamic Indigeneity that exists within itself and is made up through multiple genealogies of relations that reveal any so-called phenotypic markers of Indigeneity as fallacies of understanding. She writes:

> Only our eyes look around.
> Earth-toned eyes, forest
> eyes, thunderhead eyes,
> eyes flecked with gold, eyes
> like obsidian, eyes that are
> seeing right past you.¹²

The lines that connect eyes with earth tones, the forest, a thunderhead, gold flecks, and obsidian tie the gaze of Indigenous peoples back to the questions of land use. These connections also represent a range of phenotypic markers in how any given Indigenous person's eyes might appear. Regardless of what an Indigenous person looks like to an outsider, an Indigenous person's eyes can see past the lies of colonialism, reversing the gaze of Indigenous people as objects observed to subjects who do the observing. In this way, "Jamestown Revisited" echoes and complicates the closing of the earlier poem "Oronoco," where Wood remarks that her people now

> stood in the orchards at breaktime, smoked our ready-rolls, coughing a little—we scuffed the ground with our feet when we spoke and did not see each other's eyes.[13]

Through this repetition of eyes in the two poems, Wood shows how the question of seeing Indigenous identity remains tied to the land uses and abuses across the centuries. Through these poems, she asks at least two key questions: Who sees the Indigenous dimensions of the Eastern Seaboard? And because of these centuries of change and adaptation to survive, do Indigenous people of this place see each other across the histories of settler colonialism, white supremacy, and anti-Blackness that obscure their relationships to one another and to the land? To return to Herbert Verdin's quoted pronouncement in chapter 3 of this book, "People look, but they don't see."[14]

This question of phenotype, recognition, and land connection carries through to the poem "Colors," where Wood outlines the legacy of colorism surrounding eastern Indigenous nations. She begins by recalling her mother finding her

> in the
> basement, seven years old with new
> watercolors, painting my arms
> Indian red because I couldn't
> wait for summer.[15]

This longing for a change in skin color from the summer's rays carries through as she recounts scenes of discrimination against darker-skinned children left behind by the school bus or refused from the Cub Scouts. Her invocation of "watercolors" brings together the connectivity and fluidity of the water and the connectivity and fluidity of her Indigenous community across phenotypic traits. She then reflects on her own daughters' paler skin and curly hair. Indeed, these questions of phenotype are uncomfortable ones, and on the surface it might be difficult to see how this relates to the land abuse that Wood calls up earlier in the collection. However, the poem's close leaves little doubt about the relation between the range of physical appearance for Indigenous peoples and their connections to this specific physical place on the earth. She writes of her children and others:

> They haven't painted
> themselves yet, but in Aberdeen,
> Seattle, Syracuse, or Richmond, a seven-
> year-old with black hair or blonde will be
> scrubbing or painting her hands or just
> waiting for summer.
>   And I say
> let her be a moonflower
> unfolding to the night. Let him be
> Carolina's blue-shadowed black waters.
> Let their children be
> every shade of the soil that feeds us.[16]

This closing invocation of Carolina's blue-shadowed black waters and of shades of soil calls together the littoral cells of dark swamp soils and pale oceanfront sands as all pieces of the larger puzzle of Indigenous homelands *and* homewaters that span the continent. It is this relationship of soil to people and people to recognition that ties an Indigenous presence to the Eastern Seaboard even when some outsiders may think it is absent or too obscure to see. Like the dynamic shifting soils of barrier islands, Indigenous people of the Eastern Seaboard have also had to survive through their ability to shift and move. It is only when a hardscape is built, when a line is drawn in that proverbial sand, that these connections become harder to see, and the phenomena of movement—of needed erosions and deposits—start to seem like ruptures rather than necessary continuities. Karenne Wood's poetry poses a challenge to the settler mentality that needs things to stay fixed in place, not to adapt. Colonialism, built on extraction, and settler colonialism, built on both extraction and the illusory permanence of identities, each desire stasis as a means to reap maximum profit. Lands must be fixed for property "development" and cash-crop agriculture; human identities must stay static to determine who is legally or socially obligated to perform the labor that makes these extractive dreams possible.[17] However, the lands along the East Coast have other, very old geological plans. The Indigenous peoples who know these places as their homelands and homewaters have worked with these dynamic factors over millennia. Erosion was not necessarily a crisis because it was understood in relation to other deposits. As Helen Rountree explains, "The people native to the Carolina Sounds region inhabited a natural world of interlaced land and waters, much of the land being wetlands that did not seem like land at all. In this world they lived by hunting, fishing, foraging for plants, and farming, all of which required them to know exactly what they were doing in order to survive. Over many millennia they had learned to cope expertly. But any Europeans who became stranded there were in danger of starving, unless they could tap the locals' knowledge."[18] Comparatively, as Lipman offers, "Americans have a long national tradition of imagining the oceans around us as a protective moat and assuming our past is always grounded ... fetish[izing] the moment colonial feet touched dirt."[19] Despite the English eventually "flood[ing] out" the eastern Algonquians (in Rountree's words), it seems that colonists never let go of that initial anxiety of the fluid, shifting landscape they stole. The next four hundred years of the Eastern Seaboard become a story of white settler anxiety about change and the erosion of white supremacy via capital investment (not soils) at the first site of English colonization.

For example, while at the turn of the twenty-first century Shell Island stood as a symbol of literal erosion anxiety, the history of the island reveals other anxieties regarding land use, real estate, and questions of racial belonging. Before 1965 Shell Island was, as its name implies, a separate island north of the island (and town) Wrightsville Beach, divided from it by Moore's Inlet. Wrightsville Beach had been incorporated in 1899 and developed by Hugh MacRae of Wilmington. MacRae is perhaps best known for his role as one of the Secret Nine, who invented and disseminated the false instigating rumors about imminent violence by Black people that ultimately resulted in the Wilmington massacre of 1898.[20] He was also likely responsible for material acts of violence against Black citizens during the massacre, and he benefited directly from the burning of Black Wilmington, as he was also in the business of real estate development. One of his main real estate projects was the development of Wrightsville Beach as a vacation and leisure location exclusively for white people. In the early 1920s Shell Island, to the immediate north, became an exclusive resort for Black beachgoers from all over the eastern United States.[21]

Notably, Shell Island was developed by white business associates C. B. Parmele, Thomas Wright, and Robert H. Northrop. Parmele was a son of another member of the Secret Nine, Edgar Parmele, who had been installed as police chief of Wilmington following the massacre and successful white supremacist coup of the local government. Parmele was the president of the Wrightsville

Beach, North Carolina Home Realty Company, which was enmeshed with the finances of MacRae, who owned the title to Wrightsville Beach itself.[22] These white developers wanted to capitalize on a wealthy Black community that desired access to coastal leisure. The fact that the uninhabited Shell Island already served as a bathing site for Black beachgoers going back many years offered a convenient reason to focus on the island for this business venture. The initial plans called for fourteen streets and 270 individual lots alongside plans for a pavilion and restaurant as well as other infrastructure, including a streetcar spur from the Tide Water Power Company (of which MacRae was president). The grand opening of Shell Island as a resort community on May 30, 1924, was according to all known accounts a success, with over a thousand attendees.[23]

Despite what should have been a successful real estate venture in the development of Shell Island as a Black resort, the invisible hand of the market has always seemingly had a visible skin color. Historian Andrew Kahrl notes, "The presence of black professionals enjoying their own leisure and entertainment at Shell Island sent whites in Wrightsville Beach into a panic."[24] The familiar currency of rumor once again took hold, just as MacRae had orchestrated in 1898. However, this time, rumor seemed to be working against MacRae and his associates' business interests. Ahead of the grand opening in 1923, the *Wilmington News* assured readers that "racial lines are still observed" and that "the waters of the Atlantic Ocean still beat against [Shell Island's] borders on all side[s], divorcing it completely, entirely, and eternally from those fictitious drawbacks which the fertile imagery of Dame Rumor has woven."[25] As Kahrl notes, white residents who had been sold an oceanfront real estate dream on Wrightsville Beach by MacRae "feared that [Shell Island's] opening would wash all that they had worked to build, and the image they hoped to project, out to sea."[26] This anxiety of Black proximity for white people on Wrightsville Beach seems all the more absurd and acute given that, in addition to being separated from Shell Island by Moore's Inlet, the north end of Wrightsville Beach, which was closest to Shell Island, wasn't even developed at this time. Land *and* water separated these anxious white residents from their Black counterparts. These anxieties of the erosion of the trappings of exclusive white wealth—of being "washed away," in Kahrl's words—threatened by the proximity of Black wealth were ultimately not assuaged by assurances that Shell Island was quite literally *separate* from the island of Wrightsville Beach. Rather, it was the *equal* access to leisure that appeared to threaten a supposed erosion of whiteness.

After the first summer, several prominent Black businessmen formed the Shell Island Beach Resort Development Company to work with the Home Realty Company on the development of the island. However, local historian Marc Farinella concluded from his research that "the developers were not making money as they had hoped, and there was probably no chance of them making money as long as there was a high-profile African American resort on the island."[27] This lack of profit likely promoted the sale of the island to Charles W. Bannerman—an employee of Northrop's Insurance Agency—for fifty dollars in late 1924. Oddly, though, Bannerman sold the island back to Parmele and associates just nine months later for a hundred dollars—a 100 percent increase in price—even though no records indicate that Bannerman had sold any additional lots or made any additional capital improvements to the island during his ownership of it.

In the summer of 1926 a fire that has been described by different sources as "mysterious" and/or "beginning in a dining room" ultimately leveled Shell Island.[28] Details on the fire are scant, aside from the fact that it occurred in June and that the results were so devastating as to not warrant investors making any tangible effort to rebuild the resort. As Farinella has stated, "By the time the fire occurred, no one seemed to have any interest in continuing to develop a Black resort on Shell Island. The fire seems to have been very convenient for the developers."[29] Notably, when I was researching the history of Wrightsville Beach and Shell Island in the archive, I found a July 12, 1926, letter sent to Nelson MacRae, care of Hugh MacRae and Company, indicating that on June 18, a development deal was closed with Sanford and Brooks for the development of 206 lots on the north end of Wrightsville Beach, for which MacRae and Company would enjoy a 29 percent return on investment.[30]

I hesitate to propose conspiracies, but I want to make the timeline clear: Shell Island burned to the ground at the *exact same time* MacRae was working to extend real estate development for white families to the north end of Wrightsville Beach. Moreover, this all occurred under the contextual umbrella of existing white residents of Wrightsville Beach expressing their anxiety about proximity to an exclusive Black resort *and* MacRae's associates, all of whom worked in real estate and insurance, not seeing a sustained return on their investments due to the racist climate of the coastal South. And lest anyone forget, twenty-five years earlier, MacRae himself had been involved in rumor-instigated arson attacks on Black citizens in nearby Wilmington.

The fact that Shell Island was to become a site of erosion anxieties at the other end of the twentieth century might again stand as more coincidence than correspondence. However, when one looks at the concrete history of place, it's difficult not to see how for the littoral cells of the Eastern Seaboard, there have long been anxieties born of the literal sales of real estate development. As Wood's poetry demonstrates at the opening of this chapter, questions of development illuminate how land, race, and belonging have been entangled on these islands since the beginning of settler invasion. The eventual reliance on enslaved labor that sought to codify distinctions around Black and Indigenous identities in the region would further serve to buttress white supremacy. In 1965 the Army Corps of Engineers filled in Moore's Inlet, which was already beginning to fill in from natural erosion and deposits, and joined Shell Island and Wrightsville Beach into one continuous landmass. Up until that point Shell Island had remained largely undeveloped since the fires of the 1920s. After the constructed merger of the islands, real estate development continued north onto the formerly separate Shell Island; however, this development was largely contiguous with the established whiteness of Wrightsville Beach. The early twentieth-century history of Shell Island eventually faded from most of the public consciousness, but the island itself remained a site of erosion and anxiety.

In late November 1985 (the same year that the town of Wrightsville Beach officially annexed Shell Island), writer Ellyn Bache clipped an article from the *Wilmington Morning Star*: "Hot Weather Has Turned Fishing Season Upside Down."[31] The short piece outlined how the waters off Wrightsville and other North Carolina beaches had stayed from the summer, topping sixty-nine degrees Fahrenheit in some places. These conditions were damaging the fishing prospects, as the larger migrating fish were not arriving while the water remained ten to fifteen degrees hotter than normal. In the mid-1980s it was rarer to see occurrences such as these attributed to climate change or global warming. The article never mentions anything related to larger concerns but instead stays at the scene of the local anxiety about the fish who would not appear as long as the water remained so warm.[32]

Bache, a moderately known fiction writer, went on to compose a short story drawn from the seed of inspiration embedded in this short news article. The resulting piece, "Shell Island," combines the unsettlingly warm winter with a larger backdrop of anxieties about family, masculinity, and race. The short story depicts a white father and son: Eban, a recently divorced professor who has moved from western Maryland to the area for a new job, and his son, Alex, a woefully indifferent and seemingly depressed preteen. Rather than live in Wilmington, Eban opts to rent a house on the Shell Island end of

Wrightsville Beach. Of the unseasonably warm 1985 weather, Alex laments to his father, "It's not normal."[33] His father reminds him, "We have no way of knowing what's normal. We've never been here before in fall."[34] And the narrator tells us that Eban "rather liked the idea of a climate without autumn."[35] Even though Eban tells his son there is no evidence of what is normal, the narrator tells us that Eban is seeing the same news coverage that inspired Bache to write the story, suggesting that Eban is professing a certain denial of the local knowledge that directly indicates that the temperatures are far from usual for the fall. This slow creep of rising temperatures and the cascading effects it brings continue throughout the short story. There is a close call with a late-season hurricane, and the sticky, hot air causes Eban distress even while he tries to imagine that he enjoys the new "normal" he is making with his son. Notably, however, across the entire story Eban seems awash in anxieties about his failed marriage, his own aging, and his inability to connect with his son in any meaningful way. The rising temperatures serve as a slow background behind Eban and his own anxious white masculinity after the collapse of his nuclear family.

Even while the story does not seem to engage directly with the racial history of Shell Island, it positions the larger context of race, anxiety, and disappearance at nearly every turn. The narrator tells us of Eban and the island: "The scenery struck him as exotic, and eased his pain."[36] More than by this scenic exoticism, however, Eban is drawn by the island's disconnect from the rest of the world. The real estate agent tells him as they look at rental properties early in the story, "Of course out here you'll have all the little problems of living on a barrier island."[37] But she assures him, "Don't worry, these places have been here for a while. They've witnessed a few storms, they don't blow down that easily." The narrator then tells us Eban "was beginning to like the idea of living on a *barrier island*."[38] Later, when he contends with the drawbridge that crosses the intercoastal waterway separating the island from the rest of North Carolina, we are reminded that "he rather liked the bridge blocking passage so effectively—making the island a barrier from the mainland as well as the sea: impermeable: safe."[39] This barrier quality puts Eban in a place where he can feel isolated from the world of the mid-1980s. This bridge to Wrightsville Beach had a long history of serving as a patrolled space for Black people. As Kahrl explains, Wrightsville Beach had a "pass" system for Black people attempting to come onto the island: the bridge keeper would call the white person whom the Black person was attempting to visit, to confirm the visitor had business on the island.[40] Bache's story gestures toward this history, even while it doesn't name it so directly.

In one moment, however, readers learn that Eban has opted to send Alex to a private school (likely a segregation academy) rather than a desegregated public school that had a "bussing order for racial balance."[41] In one of their infrequent phone conversations, his ex-wife challenges him on this decision about their son's schooling: "I never thought you had it in you to be a bigot," and chastises: "Don't tell me it never entered your mind."[42] The narrator tells us that the issue of segregation had not occurred to Eban, but clearly the accusation of racial bias against a desegregated school strikes a nerve with Eban's own white masculinity even if it had not been at the forefront of his conscious mind. He retorts, "Well, I never thought you had it in you to be an adulteress."[43] His reaction at having his liberal whiteness questioned forces him into a stance of barbed and wounded masculinity, linking questions of race to questions of gender, all under the guise of one white couple's acrimonious divorce. The fact that his wife has left him for a less refined, seemingly more masculine loading-dock manager on the family's apple farm haunts Eban throughout the story. When he escapes with his wounded masculinity to an island that had been reclaimed for whiteness and now serves as his own impenetrable barrier of safety, readers are led to connect his anxieties over his own white masculinity to the larger anxieties of disappearance and "abnormal weather."

Following an eighty-three-degree Thanksgiving day, Eban agrees to let Alex return to Maryland for Christmas. After taking his son to the airport, he notices that the first frost may have occurred, rendering the island something like a barren landscape. He sits alone in his house, feeling "empty" yet "almost relieved." His colleague Dennie, a widow with whom he has developed a kind of aromantic relationship, invites him for dinner so that he will not "brood." His melancholia at sending his son back to Maryland permeates his mood. The narrator tells us that after dinner Dennie surprises him when she looks at him in what seems to be a "come-hither way," yet before they shift their friendship to a more sexual arena, they stand and examine her shell collection: "Among the cowries was a long strand of flattish beige disks, almost like segments of a sand-colored lei."[44] He identifies them as whelk eggs, which Dennie confirms. She tells him how they wash up onto the beach in rows and rows during the summer and shows him: "With a long fingernail she slit open one of the compartments and emptied it onto the coffee table. Hundreds of tiny, whitish whelk shells fell out, perfectly conical, perfectly detailed miniatures even to the spiky nobs near their tops."[45] Dennie remarks, "You can imagine how many eggs there must be altogether. . . . And each one has—what? Maybe a hundred of these little compartments?" On examining the mass of tiny white shells that partially replenish Shell Island each summer,

Eban "touched his index finger to his tongue and then to the table, so that several of the little whelk shells adhered to it. He raised them to his eyes. They were small and white, each one capable of a new life, and they struck him as distinctly hopeful."[46] There is an ambivalence in this supposedly hopeful ending. The thing that brings Eban comfort deep within his white liberal professorial soul is the sight of thousands of white whelk babies. He sees within this spread of tiny white shells a symbol of his ideas of perfection, capability, and replenishment. When this is combined with his noted tacit support of maintaining segregation, his desire for barriers, and his own sense of beleaguerment, a pattern emerges that shows Eban's anxieties to be all tied to his own white masculine futurity.

Indeed, the whelk shells are not the same type of whiteness as Eban's constructed white masculinity. These are shells from a species that is supposed to live in and contribute to this space. Moreover, whelks are hermaphrodites that reproduce on their own. However, in this moment Eban maps his constructed human white masculinity onto this other animal, projecting hope onto a whiteness wholly unlike his own. Even though Bache never notes an actual scene of erosion on Shell Island, her entire character study presents a protagonist deeply anxious about the erosion of white manhood as granting him the privileges and ease to which he feels entitled. The hope he feels in the closing lines results from a confirmation that the white sands of Shell Island will be replenished naturally. Rather than ending with an erosion anxiety, the story gives the audience the inverse: one man's confirmation of continuing whiteness. However, in addition to the rising temperatures, the other constant background of the story is the "supply trucks hauling building materials to the northern tip of the island, where next summer's condominiums were progressing at a rapid pace."[47] These condominiums are indeed none other than what would become Shell Island Resort, itself a sign of foolhardy capital investment in spite of a warming, rising tide, built on the northern end of an island destined by geology to disappear.

During the 1980s, when Bache was writing about points farther south along the coast of present-day North Carolina, another immediate coastal erosion saga gripped most of the state. The famed Cape Hatteras Lighthouse (the tallest brick lighthouse in the country), located on the easternmost point of the Outer Banks and itself a pervasive symbol of the state, was imperiled by coastal erosion. As Stephen Fletcher writes, "Public awareness of the lighthouse's imminent danger surfaced in an Associated Press Reports article published by *The Charlotte Observer* on October 14, 1980. The report noted that local residents feared the lighthouse 'might not make it through the winter' because tides were 'gnawing away at its foundation' and its steward, the National Park Service, did not have a plan to save it."[48] The superintendent of Hatteras National Seashore, William Harris, estimated that less than ninety feet of sand remained between the structure and the ocean. Describing the erosion as "very acute," he also noted to the AP that moving the lighthouse inland would be the cheapest option—an option that was precluded by the fact that "we have no money."[49] The news of the lighthouse's peril and the options for its survival return to the logic of the *Los Angeles Times*' "The Ocean Game," which I describe at the beginning of chapter 1: hardscapes versus beach nourishment versus managed retreat.[50] The debates over which of these strategies would be the best to save Cape Hatteras Lighthouse continued across the next two decades and saw the confluence and unlikely partnership of several influential North Carolinians: the famous actor Andy Griffith, the staunch political rivals Democratic governor Jim Hunt and Republican senator Jesse Helms, and the famed North Carolina photographer and activist Hugh Morton, who also happened to be Hugh MacRae's grandson. Together, these men would work to save the lighthouse from an ever-encroaching tide, but Morton in particular remained opposed to a plan for its managed retreat that would ultimately be its savior.

Like every other place in the Americas, Hatteras Island has a long history for Indigenous peoples. For many years, many have thought that Hatteras Island was the location where refugees from the failed English "Lost Colony" fled to coalesce with their ally Manteo's community of Croatoan (as the island was known among sixteenth-century English settlers). Over the years, Indigenous towns were recorded at locations such as present-day Buxton and Frisco.[51] In the early 1700s, the Native Hatteras community sided with the English during the Tuscarora War, and in 1756 Hatteras leader Thomas Elks complained to the colonial governor about English encroachments on Indigenous homelands. Even though he had gained a land grant, it seems that two years later he retreated to Mattamuskeet, on the mainland. However, up until at least 1788, when the Elks sisters (presumably relatives of Thomas Elks) sold a two-hundred-acre tract, Indigenous people owned part of the Hatteras cape. Even though the historical record of Indigenous political decisions, allegiances, and life on the island remains scant, there is no doubt that Hatteras represented a stronghold of Indigenous homelands on the Atlantic coast during early English colonization. It's entirely likely it even represents one of the first places that Indigenous peoples of North America offered assistance and refuge to the struggling English.[52]

The first lighthouse at Cape Hatteras was authorized by Congress and secretary of the Treasury Alexander Hamilton in 1794—only thirty-five years after the Elks sisters sold their land—to help mitigate the dangers of Diamond Shoals, an area known popularly today as the Graveyard of the Atlantic. Geologically speaking, these shoals occur because of the confluence of the nearby Gulf Stream and the Labrador Current. This yields numerous shifting sandbars that have proved perilous to approaching deep-bottomed ships since the sixteenth century. Although some apocryphal accounts say that Hamilton himself was endangered by the shoals near Cape Hatteras and thus was motivated to appropriate funds for construction, it's more likely that building lighthouses was viewed as a good investment by the early US leaders, as trade was needed to bolster the early nation's economy. In fact, the ninth law passed by the US Congress was the Lighthouse Act, which also became the country's first public works program, taking control over local lighthouse projects away from the states and giving it to the federal government.[53] In many ways, then, lighthouses are powerful symbols of settler colonialism. They were designed and built to make approach safer for deep-bottomed European ships trafficking in settlers and enslaved peoples. So while, on the one hand, lighthouses are symbols of public safety and the public good, one must also pause to ask for whose protection and profit these structures were conceived. In addition to serving as a sort of surveilling observation tower, lighthouses are symbolic of a particular historical moment facing a particular geological "problem" of settler colonialism—how to make the eighteenth-century growth of the settler nation safer and easier despite the Eastern Seaboard's geological resistance to the invaders. Perhaps this nearly unconscious symbolism of Cape Hatteras Lighthouse is what made saving it such an appealing bipartisan cause.

In August 1981 renowned North Carolina photographer Hugh Morton took a series of aerial photographs of Cape Hatteras, showing how dangerously close the lighthouse was to the pounding waves of the Atlantic. Morton had previously used his photography to spur action on issues, including the preservation of Grandfather Mountain in the western part of the state and the relocation of the USS *North Carolina* to the Cape Fear River waterfront in Wilmington.[54] Morton had photographed the lighthouse from the ground in the late 1950s, but his later aerial shots captured an existential threat that ground photography could never have demonstrated. In contrast to John James Audubon's portraits of birds, Morton's photography gives the viewer the bird's-eye view—a surveillance view of the object of surveillance. One color photograph

in particular from the early 1980s shows the lighthouse from above at a near forty-five-degree angle. The tide is seemingly going out, and a shallow pool of water remains dangerously close to a rock riprap wall built around the base of the structure (figure 5.2). Another aerial portrait of the lighthouse from the west depicts the water coming almost toward the viewer, suturing the audience's point of view to the endangered structure. It was this picture that Morton would later photograph Governor Jim Hunt holding at the kickoff of the "Save the Cape Hatteras Lighthouse" fundraising campaign.

Unlike the photography that I examine in previous chapters—Lewis Baltz's flat horizons, Dorothea Lange's portraits, Monique Verdin's combined landscape and human portraits, and Elizabeth Webb's archival palimpsests—Morton's attempts to look at the issue of erosion from above. The aerial perspective not only gives a kind of impersonal, detached, and scientific feel to the scene of erosion but also zooms out on the human conflicts that make dealing with issues such as erosion and climate change so fraught. In this bird's-eye or god's-eye view, all that is left is an image of imminent threat. It creates a sense of urgency as the aerial view offers the viewer a kind of objectivity. However, aerial photographs are not geological maps. Morton adjusts the angle, plays with perspective, and situates a frame to evoke this sense of both objective danger and clinical detachment. Perhaps this sensation is what allowed Morton to begin to assemble a group of unlikely allies to save Cape Hatteras Lighthouse. To say that the Democrat Hunt's vision for North Carolina was antithetical to that of the long-term Republican Senator Helms is an understatement. At the time, their battle for the US Senate in 1984 was one of the most contentious and expensive campaigns ever held in the country.[55] And yet Morton's visual depiction of the imperiled lighthouse brought these two foes together to fundraise and build a plan that would save it from coastal erosion. Tying the danger of erosion to an iconic symbol of the state—and of settler colonialism—temporarily closed the gap between the mainstream American left and right. It encouraged citizens to rise above their disagreements to confront a particular crisis.

The early campaign to save Cape Hatteras Lighthouse focused on a plan of beach nourishment via sand fences, sandbags, and an underwater structure composed of synthetic seaweed. This structure was intended to capture

5.2   Hugh Morton, *Cape Hatteras Lighthouse* (aerial view), early 1980s. Courtesy of Hugh Morton Photographs and Film, Wilson Special Collections Library, University of North Carolina at Chapel Hill.

more migrating sand deposits in order to build out the shore right in front of the structure. At first, this plan seemed to work. Morton took a number of new images to illustrate the success of this plan to extend the beach. He also took photographs of Hunt and Helms at the landmark. One image in particular depicts Hunt holding one end of a tape measure at the edge of the ocean while Helms holds the other end at the base of the lighthouse, showing the two "measuring" the success of their efforts. In addition to these photographs, Morton made a portrait of famed North Carolinian Andy Griffith holding two of Morton's aerial images of Cape Hatteras, showing the enlarged beach in front of the lighthouse. Even though structural plans like sand fences and beach nourishment can slow or even potentially reverse the effects of coastal erosion, these benefits, as "The Ocean Game" I examine in chapter 1 explains, have temporal limits. Littoral cells operate by long-standing physical geological principles. The sand is going to move. Moreover, as the effects of sea-level rise from unchecked climate change continue, the tide is going to continue to rise.

Thus, just as "The Ocean Game" attempts to teach audiences today, and as Superintendent Harris of the National Park Service noted in 1980, for human-made structures the solution to sea-level rise is managed retreat. The retreat of individual humans is of course a different enterprise than moving a two-hundred-foot-high brick lighthouse inland. The National Park Service, acting on the advice of the National Academy of Sciences, decided that the only way to save Cape Hatteras Lighthouse was managed retreat. Morton and the others adamantly opposed this plan. As Jack Hilliard explains, "Ironically, both sides wanted the same thing—to save the Lighthouse—but differed on how to achieve it. Morton's committee listed three reasons for not moving: (1) the committee was convinced the structure could be saved for a little over 2 million private dollars, versus 12 million in taxpayer dollars; (2) the movement could cause tremendous stress on the old brick structure, and (3) taking the lighthouse away from its historic seaside location would destroy its promotional and scenic value."[56] None of these are necessarily bad reasons to oppose moving the lighthouse, but at the end of the day, none of them was entirely realistic on an ever-warming planet or within a littoral cell with certain physical principles of erosion and deposits. As for the first reason, sea-level rise is going to cost money, and the two-million-dollar plan would have likely needed repeating ad infinitum, eventually costing as much as managed retreat. The second reason is also a valid concern, but if the lighthouse stayed in place, its demise was also nearly certain. The third reason is perhaps the most interesting as it is based more on the affect and aesthetics of a highly specific place than on money, physics, or geology. In this formulation, managed retreat from erosion becomes a symbolic battle rather than a material crisis.

This third factor is perhaps the most interesting piece of the so-called Battle of the Beacon. Cape Hatteras Lighthouse was built to make the coast more hospitable to settler colonialism. This isn't to argue that it should have been allowed to simply crumble into the ocean but rather to provoke an honest conversation about what the settler state views as worth saving from erosion. If leaders such as Hunt and Helms and their political inheritors were interested in protecting the nation from actual erosion, then we would see a real material investment in legislation to slow and reverse the effects of climate change. Instead, the settler state wants to protect the symbol of protection rather than invest in protection itself. In this we see that for many people, the dangers of erosion have always been symbolic rather than material or even existential. Even though the managed retreat of Cape Hatteras Lighthouse was eventually successfully completed in 1999, Morton never entirely made peace with the retreat. In 2003 he asserted that he still thought the move was a mistake: "There are three prominent United States coastal landmarks in a class by themselves that are known around the world: the Statue of Liberty, the Golden Gate Bridge, and the Cape Hatteras Lighthouse. . . . Now, our lighthouse is half a mile away [from the ocean], back in the bushes by a cell tower."[57] Although Morton's remark evokes the sense that the lighthouse is out of both harm's and historical value's way, the truth is that even this move will likely not last. The US Geological Survey Coastal Change Hazards Portal places this entire section of Hatteras Island within a "red zone" that in over seventy-eight years will still see an average shoreline loss of two meters a year.[58] Saving a symbol of protection from material erosion seemingly remains more about the preservation of the symbolism of the settler state than about caring for the physical land in all its dynamism.

As one of the first places that the English attempted to colonize in the Americas, the Atlantic Seaboard from the Chesapeake Bay to barrier islands such as Hatteras has long been bound up with narratives of Indigeneity, settler colonialism, vulnerability, and permanence. Many of these islands and sounds were mapped by John Smith, giving England a kind of first look at the landscape they would come to imagine as indelible to their own identity. Farther north up the coast among this land and waterscape was a place then called Russels Isles, a set of islands in the Chesapeake Bay that includes what is now known as Tangier Island. This cluster of small islands, which also may have included present-day Smith Island and Watts Island, is a claimed homeland of the Pocomoke peoples of present-day Virginia, who were known as the Wighcocomoco people in Smith's account. The name of the river Pocomoke has been translated by some to mean "broken (or pierced) ground," which may seem historically ironic or telling given that the land in question is quickly being broken by anthropogenic sea-level rise.[59]

Smith described the isles as "uninhabited." However, archaeological finds on the islands suggest a history of human occupation, and nearby oyster middens evince long-term fisheries management in the area.[60] On encountering the isles, Smith writes of "an extreame gust of wind, rayne, thunder, lightening [and] unmercifull raging of that Ocean-like water."[61] As he and his crew sail around the bay, thrown by the inclement weather, it's a bit difficult to chart exactly where he is when he recounts that their ship was nearly lost and for "two dayes we were inforced to inhabit these uninhabited Isles which for the extremitie of gusts, thunder, raine, stormes, and ill wether we called Limbo."[62] He then goes on to call this area of isles Limbo two more times, firmly suturing this region of the Chesapeake and its isles with the first circle of Dante's Hell.

Like the isles that Smith records, Limbo is inhabited.[63] While it has several iterations in Catholic theology as thinkers from Augustine to Thomas Aquinas attempted to prove its existence, Limbo is notably the place of the virtuous pagans, those who did not have the opportunity to know Christ due to the inconvenience of linear time.[64] Christ's "harrowing of hell," where he retrieves important Old Testament figures for transport to a post-salvation heaven, is a key event of Limbo only tacitly referred to in canonized books of the Bible but accounted for in the apocryphal Gospel of Nicodemus and expanded on in the Middle Ages.[65] Given Smith's clear account of the various Native peoples that he does encounter in this so-called uninhabited space, readers might pause on the parallel he creates between the Indigenous peoples of the Americas and the virtuous pagans. It is as if Smith either sees the place as a kind of Limbo—where the souls that inhabit it are indeed good but deprived of a so-called true Gospel—and he is the underworld explorer Dante sent to pity them or (perhaps more likely for Smith) imagines himself as a kind of Christlike figure sent to harrow.

If it is the second, Christlike parallel that Smith imagines calling on, then one must consider that *harrow* works on multiple levels as a verb that comes to us from Old and Middle English. In one iteration, it means "to harry, rob, spoil," as Christ does with hell in his descent to Limbo to gather the prefigures of his teachings in the New Testament.[66] In another definition, *harrow* means "to draw a harrow over; to break up, crush, or pulverize with a harrow."[67] In other words, like a now-obsolete meaning from Richard Stanyhurst's 1583

English translation of Virgil's *Aeneid*, *harrow* means "to plough." This meaning of *harrow* is consistent with its noun form, also derived from Middle English, which means "a heavy frame of timber (or iron) set with iron teeth or tines, which is dragged over ploughed land to break clods, pulverize and stir the soil, root up weeds, or cover in seed."[68] Thus, in this comparison of the Chesapeake islands to Limbo, Smith evokes coterminous ideas of salvation, robbery, spoiling, and that most dangerous activity for any soil's longevity of place: plowing.[69] Perhaps coincidentally, the relatively contemporaneous usage of *harrow* as "to plough" emerges from none other than Dante's guide through hell, Virgil, himself a resident of Limbo. In this way, it's not particularly surprising that there seems to be some correlation between Limbo as a harrowed space and the possible meaning of the word Pocomoke as "broken (or pierced) ground."

These islands are a space where fragmentation seems to sit front and center conceptually. Such a reading is consistent with philosophical readings of the space of Limbo as signifying a paradox of perennial crisis.[70] As a region, Limbo often figures as a literal border of hell itself, consistent with Smith and his crew's interpretation of the extreme weather they encountered in their voyage. As Kristof Vanhoutte outlines, Limbo has also long been associated with walls, doors, and gates—as the barrier between one state and another.[71] Limbo is where otherwise enlightened and virtuous souls such as Virgil recognize that they are doomed to a kind of permanent crisis of nonsalvation. They can see the lack of hope before them, understand it even, and yet they are powerless to change their condition. Or as Vanhoutte explains, according to Alexander of Hales, this is an "epistemic" condition where residents of Limbo were "well aware of their being separated and disconnected from God. Their state was thus one of anxiety and of sadness, but also hope of being 'one day' saved."[72] They would need a Christ figure to come save them. However, Christ himself has been and gone, and their condition remains the same.[73] Multiple fragmentations of land and spirit define Limbo. It is a place of perpetual anxiety.

However, despite this being "cut off from hope" and "liv[ing] on in desire," as the character Virgil describes in the *Divine Comedy*, Limbo is for Dante also a relatively nice place.[74] As Vanhoutte explains, "The luminous and fresh place reserved for the greatest minds of antiquity . . . does seem somewhat surprising—especially considering they are in Hell."[75] Their hell is, as mentioned before, epistemic. It is the knowledge of not knowing salvation that is the torture, despite the lack of overt punishment and the illuminated tranquil fields that make up the geography of Limbo. Vanhoutte asserts, "For

Dante... Limbo thus seem[s] to constitute the paradigm of the paradoxical place."[76] It is this paradox that animates contemporary life in the Chesapeake Bay islands initially described by Smith. As Vanhoutte explains, "Living in perennial crisis is analogous to living in Limbo; or to use Agamben's language of the signature, the concept of the idea of the perennial crisis is the secular signature of Limbo."[77] He goes on to consider Pope Benedict XVI's 2007 suspension of Limbo as a category as a possible recognition "that [Limbo] is not a category of the afterlife but the nature of our present life, today, down here on earth."[78] This shift in theology speaks to the present-day crisis of living on an ever-warming planet, facing an ever-rising sea, from the oceanfront towns of California to the "hot weather" that spurs Bache's "Shell Island."

Accordingly, the residents of present-day Tangier Island are locked in their own contemporary Limbo as they wait for help against a rising tide and a land broken by what they consistently refer to as "erosion." As Earl Swift explains, "The lower Chesapeake's relative sea-level rise—the one-two punch of water coming up and land going down—is among the highest on Earth, and of all the towns and cities situated on the estuary, none is as vulnerable, none as captive to the effects of climate change, as Tangier."[79] However, as he interviews resident after resident, he hears a similar refrain: "Sea-level rise, that might be occurring, but it's small scale next to the erosion," and "Tangier's demise is going to be erosion."[80] Swift reminds his readers, though, that "erosion and sea-level rise are no either-or proposition: They are inextricably linked, for the bay's erosive power grows as its level climbs. Accelerating erosion is a symptom of a global phenomenon, not a product of local winds alone."[81] Despite this truth of the daunting double whammy facing the Island, Swift quickly points out that the residents' distinction between sea-level rise and erosion is likely more "political... than biblical."[82] He quotes one resident, Carol Moore, early in the book as she explains, "If the government officials insist that it's sea-level rise, what can you do about sea-level rise?... Nothing. Not a thing. And if that's what they see this being, then they won't want to spend any money to try to stop it." Moore closes her remarks to Swift by noting, "If we don't get help, we're going to be history. The end."[83] This tension—between an apocalyptic event too big to fight, in the form of global climate change, and the local-based issue of "erosion" that requires only a savior or saving plan—constitutes the paradox that leaves Tangier Island in a state of limbo.

Just as the knowledge of nonsalvation creates the anxiety of the souls in biblical Limbo, Swift seeks answers in Tangier to the same sort of epistemic crisis of life on the edge of climate change, "joining its watermen on their boats, absorbing its odd and long-standing customs to discern what we'd lose with its demise, and plumbing its collective anxiety over what the future holds."[84] In his account of the island, Swift reveals the ways that a secularized limbo of the crisis moment manifests. Living under the crushing weight of an anxiety of disappearance leaves the residents of Tangier looking for a savior that might perform a contemporary harrowing and save them from the crisis. Paradoxically, they don't want to be transported to another place in an attempt at managed retreat. They want to be saved—perhaps understandably—*in place*. Rather than be swooped up from Limbo, they want Limbo itself saved from a sea hell-bent on taking them under.

In 1989 Tangier Island received assistance from Congress to build a 5,700-foot-long riprap wall that was designed to curb erosion along the island's western edge. To demonstrate their own investment in this project, residents of Tangier voted unanimously to increase their real estate taxes fivefold. As Swift explains, "The armoring stopped the erosion fast. Not an inch of ground has slipped into the bay there since. But it did not halt the Chesapeake, which took aim at other parts of the island with new ferocity."[85] This constant push and pull of fighting the Chesapeake has led many on the island to see permanent sea hardscape as the solution to saving the island, and yet, as with the case of erosion on the West Coast, seawalls, riprap, and jetties only deflect consequences elsewhere. As Orrin and Keith Pilkey explain, "Beach-destroying seawalls in the United States are under construction or repair in an attempt to postpone the date when the occupants must flee."[86] In fact, almost all coastal geologists agree that hardscape plans represent only deferrals rather than solutions: "Seawalls, bulkheads, groins, and house foundations reduce the flexibility of the system to respond to changes in the dynamic equilibrium," and "jetties ... have been responsible for the disappearance of whole towns."[87] Nonetheless, residents of Tangier Island began campaigning for a new jetty project that might slow the erosion elsewhere on the island.

Swift records a curious kind of psychological phenomenon that began to take hold among residents in the early 2000s about this proposed jetty: "The jetty's size and form shifted in the minds of many Tangiermen. No longer was it seen for what it was—a simple and rather limited ploy to stave off shoaling and protect the harbor from wind-driven waves. No, now it was conflated with other, grander schemes.... The jetty was reshaped in the public imagination into a seawall not unlike that guarding the airport—a seawall that promised the island's survival."[88] The desire for this barrier in the form of a seawall seems not unlike the concept of the barrier that Eban finds comforting in "Shell Island." Each represents a kind of isolationism in the face of crisis. For

Tangier residents, the seawall becomes a subject of prayers and vague hopes, an "amorphous product of wishful thinking, a mass confabulation of what has been considered, planned, hoped for, and promised for years."[89] The residents of Tangier face the ironically glacial pace of bureaucracy and funding, which might now be moving much slower than the actual glaciers that are melting and adding to the erosion of the island.

As might be expected, the imagined saving wall begins to connote other types of anxieties in the present political moment. Thus, Tangier Island finds itself battling erosion anxieties in multiple contexts. In the 2016 election, 221 Tangier Islander votes out of the available 253 were cast for Donald Trump—a candidate who is, *at best*, indifferent about the effects of climate change. Swift quotes one reverend from the island telling his congregation, " 'We are at an anxious time in our country politically' " and encouraging them not so subtly to vote for Trump.[90] Evoking anxiety, the reverend encourages people to vote out of their own existential dread. However, given the racist and xenophobic nature of Trump's campaign, it's unclear exactly what kind of anxiety—geological or ideological—the reverend hopes will spur his congregation toward the far right. Three days after election day, Swift uncharacteristically breaks his journalistic fourth wall and corrects a group of fishermen claiming that Trump will be the first Christian president in years. Of course, his argument about Barack Obama's and Hillary Clinton's Christianity fails to land with these islanders. They believe that Trump has been saved, spiritually, and that in turn he is going to save them, physically, by building a seawall.

During this postelection context, CNN arrives on Tangier Island to talk to the residents who are being taken under by sea-level rise and yet believe that a president who has called climate change a hoax will help them. When Jennifer Gray interviews the assembled residents, they insist that it is erosion—not sea-level rise—that plagues their continued existence. When asked what they might say to the new president, one resident proclaims, "Build us a wall!" The mayor follows up, "I love Trump," and continues, "as much as any family member I got," and finally concludes, "Donald Trump, if you see this, I mean, anything you can do—we'd welcome any help you can give us."[91] The mayor's appreciation of and appeal to the forty-fifth president results in a barrage of online vitriol and disbelief from many across the country. Understandably, the larger public finds this situation nearly impossible to understand and even harder to sympathize with. Stephen Colbert even makes a joke about Trump making the ocean pay for the seawall. The entire scenario seems nearly tailor-made for the proverbial indecision about whether to laugh or cry, as here exists an island that has voted almost unanimously against not just their best interests but their very material existence. And as Swift explains, the residents of Tangier are shocked that these outsiders are shocked by them.

The external pressure starts to reveal a remarkably old logic of colonial anxieties and so-called saviors. Within this mix, all of the public opinion commotion gains the attention of former vice president and climate activist Al Gore, who attempts to discuss the macroscale issue of sea-level rise and the microappearance issue of erosion with the mayor of Tangier Island in a town hall hosted by Anderson Cooper. Gore tries several strategies to pull the ends of these events together for the mayor and the larger CNN audience. However, the mayor digs in: "Our island is disappearing, but it's because of erosion, and not sea-level rise.... Unless we get a seawall, we will lose our island." When Gore again attempts to prompt him to see the two as connected, the mayor responds, "This erosion's been going on since Captain John Smith discovered the island and named it."[92] The callback to the colonial invasion of Indigenous homelands and the place that Smith himself called Limbo signals a kind of uncanny time loop. In some ways, the mayor is right. The littoral cells of the Eastern Seaboard have been shifting sands, creating and disappearing islands for millennia. It's only the very settler presence on the island that the mayor seeks to save. The erosion of the land is itself part of a dynamic system, but this dynamism challenges the fixed property that settler colonialism depends on. In the mayor's reference to John Smith, we see just how proximate the anxieties of erosion are to the anxieties of the white settler colonial state.

The anxiety of erosion is also an anxiety of national disappearance. In addition to the thick colonial residue coating their homes, residents of Tangier are living in the contemporary psychological space of Limbo as described by Vanhoutte. The tacit awareness of existing in a state fundamentally defined by the very absence of the hope of salvation causes people to seek a savior—even a highly improbable, if not impossible, one. Significantly, as Vanhoutte outlines via Wendy Brown in *Walled States, Waning Sovereignty*, this profound condition of anxiety is connected to the very image of the walled Limbo as akin to the walled nation. As Brown puts it, "Rather than resurgent expressions of nation-state sovereignty, the new walls are icons of its erosion."[93] Campaign slogans that advocate for walls speak to anxieties of the erosion of a static nation-state. In the case of Trump and the political right, this anxiety is about an erosion of the white nation-state. However, when those from the left bemoan the erosion of the country or of democracy at Trump's hands, as uncomfortable as it may be for them to admit, they are also appealing to an anxiety of disappearance of a nation-state founded on the theft of Indigenous homelands.

When we compound this Tangier anxiety with a tradition that emerges from John Smith in Limbo, we return to the questions of dynamism that animate Karenne Wood's poetry about the Eastern Seaboard. As an Indigenous woman, Wood does not write from the compounded anxiety of US disappearance, focusing instead on the ebbs, the flows, the washes, and the returns. This isn't simplified resilience; instead, it is an assertion of a kinship time created for survival. It's not akin at all to John Smith's Limbo. As Wood writes in the poem that closes *Markings on Earth*, "First Light":

> At this hour, who could discern where land ends,
> or water, where creek becomes bay, bay becomes
> river and stretches across to a blue verge
> of Maryland, all the way black now, invisible.[94]

The enjambment that follows both "becomes" and "verge" accentuates this continued act of becoming and fluidity over being stuck in place or time, walled in. Rather than offering a conventional view of the Chesapeake Bay that might render Indigenous people as background, Wood reverses this logic as possible Tangiermen appear in the distance:

> From the marina around the bend, two crabbers set out.
> Their diesel chugs reverberate as prows cut new waves.[95]

The diesel engine calls up a climate paradox of dependence on the things set to destroy these crabbers' very way of life. The end stops of the lines signal a cutting off of possibility or becoming or living in the dynamism that surrounds the bay and defines the littoral cells of the Eastern Seaboard.

This dynamism and a healthy respect for how it structures a sustainable life in this place carries throughout the poem. Wood continues the second half of "First Light" with a question and answer that evokes perpetual movement. She writes:

> What matters? At the end, we become what we have
> loved, each thing that transfixed us in the rapture
> of its moment, its grace of its own making, ours the same.
> We grow around the land as it grew around us, and
>
> dawn crosses over us, whether asleep in nests or
> berths or in the ground becoming life again. Here is

> the moment: here, among herons, ospreys, morning,
> river. I believe in *this* light: it is the light of the world.[96]

She ends the third stanza with an enjambment that begs for connection with the closing conjunction "and." She pushes the reader to continue on to see the connectivity and movement that propel the space forward past the space between the two stanzas—perhaps not unlike those blank white spaces marked as "unimportant" on the 1934 erosion map that spurred me to begin this project. The land exists in reciprocal growth and change with the people who are willing to live with its changing face—to see the lands *and* the waters, the lines *and* the spaces, all as *important* parts of the connected whole.

The rising light of dawn in this poem connects species to species and past to present to future. Wood answers in verse Rob Nixon's call to think prospectively in addition to retrospectively, and she illustrates a version of LeAnne Howe's relational tribalography that forwards Kyle Powys Whyte's concept of a kinship time of continuance and becoming over a settler time of linear anxieties of disappearance and finality.[97] This is not a darkened or dimmed Limbo that John Smith must traverse or harrow, occupied by those who cannot know salvation. It is Indigenous homelands and homewaters, allowed to be dynamic, peopled by those who exist in *this* world. Wood places all of her "us" in the rapture, defined only by what this collective group loved. If one loved the land, it transfixes them. If one loved only the things they built on that land, then they are likewise enrapt by that. Significantly, Wood does not punctuate this rapture with anxiety. Rather than "erosion," her poem focuses on "growth."

This insistence on not disappearing distinguishes the perspective of the Chesapeake Bay and Eastern Seaboard from one that desperately needs to hold the earth in place. Wood's poem argues that survival requires an awareness of flux. She pushes her reader out of John Smith's imposed Limbo and back into the world before them to look around and see a dawn rather than the end times.

Dawn has long oriented life along the Eastern Seaboard of the Americas. It is the place of the first light that has drawn and sustained human beings since time immemorial. I myself close this chapter writing from an early morning on Shell Island—a place my mother and I return to every Christmas holiday. We stay in the very resort that gave me so much consternation as a teenager, full with the knowledge that where we look out at the Atlantic should be smack-dab in the middle of a migrating inlet, now reshaped by a multimillion-dollar effort to give us humans more time on this piece of land.

One of my mother's dear friends refuses to stay at the Shell Island Resort, opting instead to stay a few blocks south at the Holiday Inn, which stands on the old Moore's Inlet—"the Holiday Inlet," my mother and I quip every year. "Six of one, half a dozen of the other," we agree in this age of late capitalism, where every choice is a compromise. I must confess, though: I do deeply and guiltily love this island; nowhere else makes me feel so poignantly anchored in my own insignificance. Or perhaps it's a pull toward what I know are my family's faraway, but no less real, roots across that ocean—my own desire to undo something that should never have been done.

However, this cathartic smallness and wistful longing is perhaps the very problem of this place. Having this feeling—this draw to our transient, fleeting existence—is not enough. Knowing the layered history—measured in the soil samples of my early amateur forays into erosion research and later confirmed by my learning about the history of settler colonialism and white supremacy that stole land and life from Indigenous peoples and made dreams of Black wealth and leisure in the US South largely untenable—is not enough. Wistfully imagining undoing my own implication in settler colonialism is but a cop-out if I am not committed to direct action. Writing this book is also not enough. I sit on a balcony of a resort that should not exist, that should have disappeared years ago, looking at an ocean and inlet that need to keep moving. To the south, the current is depositing sand, building land in a process that long predates and will likely outlast any of us. Should this erosion give me anxiety, or should it evoke another feeling? Is my desire for the earth a desire of care or a desire of my own settler existence in this space? To love, to survive our own existence on this earth, may have always been and will always be a crisis about letting go of the things that have made that very existence possible.

Conclusion: What We Talk About When We Talk About Erosion

AS I WAS REVISING THE MANUSCRIPT OF THIS BOOK, I happened across a curious mistake. At several different points across the text, my word processor's autocorrect had "corrected" words such as *ecocritic* and *ecocritical*, and even mistyped instances of *erosion*, changing them to the word *erotic*. Upon first noticing this, I chuckled to myself, imagining how confused a reader might be at the sudden foray into "erotic readings" and "erotic problems" when only a sentence before I had been railing about issues in soil science. However, my second thought was to pause on this AI-generated error masked as a solution to my supposed bad spelling and sloppy typing. (I am admittedly guilty on both counts, to be fair.) The words *erode* and *erosion* are not etymologically related to the Greek *eros*, after all. However, this was the first time it occurred to me that regardless of this fact, the word *eros* had been sitting right there in front of me this entire time, obviously present in the very title of this book: *Erosion*.

Despite the words being unrelated in a denotative etymological sense, I began to think about the concept of eros as a kind of all-consuming desire for something. As opposed to the aims of philia or agape, where love takes on a more selfless posture of care or mutual reinforcement, eros has long been associated with a kind of mania, a burning desire resulting from a wound. Eros often yields unfortunate if not disastrous outcomes for those affected. It both consumes and is all-consuming. In this way, it seems not so far off from the earliest sense of *erosion*, as a kind of cankerous eating away of the body's tissue. Eros is that which can destroy what it purports to love.

I am not the first to make this association. In 1971 Austrian avant-garde performance artist Valie Export staged a scene titled *Eros/ion* where she rolled naked in broken glass and transferred the evidence of the wounds onto a glass plate and then a paper screen, enacting the wounding of the body followed by the printed replication of the evidence of those wounds. While aside from the title Export seemingly does not make any explicit connection to the geological phenomenon of erosion in her work, she nonetheless forces her audience to think about desire and wounding as a kind of spectacle that creates a visual archive.[1] I don't want to imply here that when we talk about the earth, we are always talking about our own human bodies. Rather, my intention is to query the divergence between a care for the earth and a desire for the land that ultimately destroys that same land. This erotic, erosive desire is often mistaken for love even as it wounds. Export's work pushes the audience to see how this eros/ion is projected, duplicated, printed again and again—a replication of the wounded absence mistaken for presence. This process is akin to the way theft generates property as described by Robert Nichols, inverting the terms by which we understand dispossession and its legacy for Indigenous landed sovereignty.[2] Export makes visible the eros/ion that renders the body as property through its wounding, which forms a cartography of desire in its duplication and printing. Similarly, the wounds of settler agriculture on the land, as seen in spaces such as Providence Canyon, signal Indigenous dispossession and Black enslavement through their hypervisibility of erosive absence.

Understanding this eros/ion in its associatively capacious form offers tools for thinking about the settler desire for land—a land that settlers claim to love even as they destroy it with their very presence. This is the eros of John James Audubon toward birds; it is the "love" Scarlett feels for Tara; it is the desire for a walled-in existence on Tangier; it is perhaps even the feeling I have on Shell

Island. This is the eros that creates what Deborah Miranda describes when she asserts that "the loss of land is a kind of soul-wound."[3] Indigenous peoples are the wounded here, bodies tormented by the consuming desire of colonialism for more and more of their homelands. Kristina Lyons echoes this when she asks, "How do soils—what may or may not be conceived as an object called 'soil'—harbor the irreparable wounds and tracks of violence and germinations of transformative proposals and alternative dreams?"[4] This question returns us to Miranda's queries via Thomas King from chapter 1: "Isn't it time to pull off the blood-soaked bandages, look at the wound directly, let clean air and healing take hold?"[5] Miranda, by way of King, argues against the walling off of the wounding eros, asking instead for healing through other forms of love based on care. In this way, the healing comes in words reminiscent of Lynn Riggs and Ramon Naya's Vine Theater, which desired a community that would "pour into it a fertile and continuously growing life spirit."[6]

This is not to say that Indigenous peoples and other nonsettlers are immune to the problem of eros/ion. Riggs himself offers a cautionary tale of assimilation and allotment in Julie and Clabe's mutually held desire for one another and desire to be one another in a world upended by the imposition of settler gender norms. Lauren Olamina has an all-consuming desire for Earthseed's success that interrupts her care for those around her, including her own daughter. Monique Verdin never shies away from the complications and implications in her own Louisiana love. Thomas Jefferson Flanagan frames a profound ambivalence over the beauty and destruction evident in Providence Canyon. However, what each of these creators foregrounds is acknowledging and airing out—not walling in or covering over—the wound.

Similarly, I do not think every individual person from the West is incapable of caring about the earth. The term *settler* is not a slur, but it is a posture. It's a status of implication as well as an attitude and an orientation that when left uninterrogated confuses the desire *for* with the care *of*. This slippage appears all around the globe. One of the key things I have learned over the past few years is just how pervasive concerns about erosion are for everyone, everywhere. Each time I have spoken about this work, friends and colleagues have talked to me afterward about the places they know and love that are affected by erosion: damming projects in the Caucasus region that will affect sturgeon spawning and sediment flows; rates of sea-level rise in Thailand; mass wasting and soil instability from waterfront construction in Belgrade; dam erosion in Mosul, tackled by an international partnership of Iraq, Italy, and the United States—and the list will surely go on. Thus, just as erosion is not limited to one area of the United States, it's also not a uniquely US issue. Erosion is never exceptional. To illustrate this point, I close with a series of pulsating snapshots across time and space, zooming in on the particles of soil at stake within a global crisis. In what is less than a gesture to a "one world" global finale, I seek to follow the methodology of the "pluriverse" as articulated by Mario Blaser and Marisol de la Cadena. Working from an ethos articulated by the Ejército Zapatista de Liberación Nacional (EZLN; also known as the Zapatista Army of National Liberation) of Chiapas, Mexico, Blaser and de la Cadena argue that "to open up the possibility of a world where many worlds fit, it is not enough for the Anthropocene to disrupt the nature and culture divide that makes the world one."[7] This call suggests that a "reduction of scales" initiates a productive question: "Could the moment of the Anthropocene bring to the fore the possibility of the pluriverse? Could it offer the opportunity for a condition to emerge that, instead of destruction, thrives on the encounter of heterogeneous worldings, taking place alongside each other with their divergent here(s) and now(s)"?[8] The following snapshots attempt to engage these "here(s) and now(s)" in critical regionalisms that exceed the nation-state or large-scale divisions of homogeneous Global Norths and Souths. In order to make sense of the shifting earth, each of these local issues has been narrated in particular ways that often display the anxieties those in power have about the disappearance of their own existence and continued unquestioned relevance.

One of the main areas of the globe affected by the erosion crisis in the mid-twentieth century was apartheid South Africa. The South African government invited Hugh Hammond Bennett to tour the country's farmlands in 1944. The resulting articles that appeared in South Africa's *Daily Mail*, *Sunday Times*, and other news outlets all seemed eager to blame the problem on South Africa's "natives," noting that the soil issues were the worst on the reserves, where overgrazing had caused serious depletions of ground cover and topsoil productivity. The *Daily Mail* blamed this problem on "the native custom of buying wives with cattle."[9] While overgrazing of cattle is certainly detrimental to the soil, all of the South African press coverage of Bennett's visit blames Black South Africans for the entire erosion crisis and in paternalistic tones explains how their education will be necessary to "Save Our Soil." No mention is made of deforestation in service of the mining industry or the fact that Black South Africans had been confined to territories far too small to support their populations under the repressive homeland system of the day. In this way, South Africa's apartheid government used the erosion crisis—buttressed by the visit of an expert US soil scientist—to stand in for their own anxiety of the disappearance of their white supremacist power in the region.

In the 1990s, as apartheid leadership came under increasing scrutiny, the *New York Times* reported on the Worldwatch Institute's findings that outlined how apartheid had damaged South Africa's environment, noting very similar conditions to the ones Bennett saw in the 1940s: "Half the black population has been forced to move to the 13 percent of the country the Government has designated as tribal homelands," and they state, "The overcrowded lands have been stripped of their topsoil by overgrazing and farming, and ... forests there are rapidly dwindling as trees are felled for fuel."[10] Given his harsh words for colonial agricultural systems elsewhere, it seems likely that what Bennett saw and thought about the pervasive wind erosion in South Africa in the 1940s was not reported accurately by the South African press. As Bennett explains of South Africa in his own 1939 text, *Soil Conservation*, "As the influx of white settlers reduced the amount of land available to the natives and restricted their mobility, primitive agriculture became more destructive."[11] Although Bennett offensively deems Black agriculture in the region "primitive," he is nonetheless clear on the cause of increased erosion: the influx of white settlers and the taking of lands from Black control. Thus, it seems more likely that the mid-1940s white South African press's coverage of the erosion crisis was simply a front to justify keeping more lands out of the hands of Black people in South Africa. In this case, the facts of erosion are bent in service to the narrative that Black South Africans are incapable of protecting their own homeland resources.

It isn't only the Global South, however, that is facing erosion. The matter is just as pressing in locations such as Venice and Marghera, Italy. In a place where the water is often seen as poetic and place defining, there remains the ticking background of what sea-level rise and erosion will mean for the lagoon. Perhaps, then, it's little surprise that Lewis Baltz's work eventually found its way to this subject in his *Venezia Marghera* series in 2000. Often aiming his camera away from the famous city and toward its industrial counterpart Marghera, Baltz forces the viewer to almost squint into the frame. Two of the first three images depict the massive cruise ships built in the Marghera's Fincantieri shipyard.[12] These same cruise ships are considered by many contemporary Venetians to be a plague on the city, a hypereffect of Venice's Western iconicity. In the series Baltz juxtaposes two images: one uncaptioned, where a massive cruise liner dwarfs Piazza San Marco, and another dominated by a broad-faced view of the bow of the *Disney Wonder*, under construction at the time of the photograph. Baltz closes the caption of the second image with this sentence: "Access to the *Wonder* is restricted to authorized personnel and photographs are forbidden."[13] This information places the viewer in an odd position, complicit in the forbidden photograph and yet offered an in-the-know sensation about the entanglements facing the Veneto, where, in Baltz's words, "Marghera sells aluminum, chemicals, petroleum and ships [and] Venice sells herself."[14] As Elena Longhin explains, the industrial port projects of Marghera "require continuous dredging operations that promote a ceaseless erosion of [the lagoon's] seabed."[15] Lucio De Capitani takes this juxtaposition a necessary step farther: "Marghera becomes even more relevant for a holistic understanding of environmental crises when we think that its construction was meant as the chief tool for the removal of the lower-middle and working classes from the historical city, so that it could be turned into a centre dedicated to the service sector and representative functions."[16] Thus, a city often called one of the most romantic opted initially for an eros of the tourist over care from or for its own Venetians. These working-class Venetians were removed from their city to the industrial port to labor for those seeking to consume their city as a kind of Disneyfied theme park. In his photograph Baltz gives us Wonder, indeed.

Nonetheless, changes and responses to the dangers of sea-level rise continue in ways that can be difficult to parse as wholly negative or positive. As of August 2021, activists in Venice had successfully lobbied their country's government to ban cruise ships from the San Marco basin and canal as well as the Giudecca canal.[17] This measure's success was attributed in part to the COVID-19 lockdowns that gave the city some measure of relief from the continuous onslaught of monster ships and their thousands of day-trippers. In some cases, however, these ships have been diverted to Marghera's port, which is built for industry rather than tourism, continuing the entangled logic of the two cities that Baltz and Longhin describe. Shortly before the cruise ship ban, Venice finally completed its Modulo Sperimentale Elettromeccanico (MOSE) project to address sea-level rise. As De Capitani explains, MOSE is "a series of 78 mobile gates that are able to close off the Lagoon and are intended to shield the city from the worst floods."[18] However, as he outlines, this series of floodgates is an outmoded design based in 1980s engineering logic, representing "technocratic hubris, greed and environmental short-sightedness . . . reminiscent of the dangerous enthusiasm for geo-engineering solutions to climate change of technocratic futurists."[19] Thus, as with so many other locations, there exists an ebb and flow in Venice's response to erosion—an anxiety that enacts positive changes in the rejection of an unsustainable cruise ship economy and yet also leans into the fantasy of the city that can wall out (or in) its own disappearance, all while removing Venetians from their island homelands.

The story of erosion continues, then, in places all over the planet that are reeling from the effects of removals and walls. Not surprisingly, the removal of Palestinian people from the West Bank and the construction of the West Bank Wall have contributed to soil erosion. Moreover, the United Nations Conference on Trade and Development reports that the cutting of olive and other native trees such as carob has caused soil erosion across the Palestinian landscape. Significantly, as early as 1937, the director of the Land and Afforestation Department of the Jewish National Fund, Yosef Weitz, recommended the afforestation of Palestine as a central goal of the Jewish settlement of the region, with the planting of native trees, including carob.[20] However, as Bennett notes, the population influx of the British Mandate caused an increase of about twenty-four thousand persons per year, making any plans to curb soil exhaustion difficult. Moreover, in 1939 Bennett noted that in Palestine "the feeling prevails that there is discrimination in the distribution of land."[21] In other words, the occupation of Palestine has long been bound up with concerns over erosion, land use, and deforestation, which continue into the present and have accelerated since the 2023 start of the most recent war in Gaza. The International Committee of the Red Cross reported as recently as 2021 that the Israeli military and other settlers not only harass Palestinian olive farmers during harvest but are also known to uproot and burn olive trees.[22] Similarly, the blockade of Gaza, which began long before 2023, has damaged the large coastal wetland of the Wadi Gaza, resulting in the shrinkage of a crucial area that would surely help protect Gaza as global sea levels rise and the Gaza shore is increasingly imperiled.[23]

Edward Said notes the rhetorical and visual apparatus of this occupation in *After the Last Sky*, an illustrated memoir that at first glance may seem to have very little to do with soil erosion. However, there are traces of the questions of afforestation, deforestation, and who decides what roots will grow into what ground. As Said recounts in one poignant anecdote about eggplants and food memory, "[State of Israel] Laws 1015 and 1039 . . . stipulate that any Arab on the West Bank and Gaza who owns land must get written permissions from the military governor before planting either a new vegetable—for example, an eggplant—or fruit tree. Failure to get permission risks one the destruction of the tree or vegetable plus one year's imprisonment."[24] This story is surrounded by two images by photographer Jean Mohr—one of a makeshift scarecrow in a small kitchen garden that seems scraggly at best and the other of Palestinian refugee laborers at a farm during the eggplant harvest. The Palestinian children in the second photo smile at the camera as they pack the vegetables into produce boxes destined for some place out there, some place like the markets where Said imagines he himself shops in diaspora. The images and the critical text outlining the laws *against* Palestinian planting create an uneasy dissonance. They seem to ask not only who deserves to claim the land or use the land, but who even deserves to save the earth. *Who even deserves to plant a tree?* In the criminalization of Palestinian planting, the real stakes of erosion control come forward. It was never about the care of the earth; it was always about who can claim land, who labors on that land for whom, and who is able to establish and care for roots.[25]

Although erosion affects every area of the planet in one way or another, I hypothesize that it often accompanies sites of political anxiety as it haunts the modern nation-state with concerns over both the disappearance of physical territory and the food insecurity that might destabilize a population. One might glean from the previous sentence that what I have been talking about all along without naming it as such is the synthesis of Michel Foucault's ideas on security, territory, and population into the concept of erosion. In invoking this work, I do not mean to simply back into my argument in the final pages but rather to advance sideways like Foucault's misinvoked crayfish (it's likely he meant the crab—an animal that actually travels sideways, unlike crayfish, which often move backward—in this famous quip).[26] The anxieties of erosion and disappearance glide past one another like edges of tectonic plates existing together in ways that might be apprehended at one moment as correlation and at other times as causation. These plates snag, disrupt, float, and move in ways that give shape to the landslides that appear as Octavia Butler's deus ex machina in *Parable of the Talents* and lurk beneath every human's feet. The anxieties of erosion and disappearance get mistaken for one another time and time again, serving as twinned metaphors of crisis.

In the writing of this book, nowhere did this relationship seem more concrete to me than my own temporary location in Budapest, Hungary (location of the first international conference on soil science in 1909), where I composed most of the manuscript. I witnessed as the ruling party, Fidesz, led by Prime Minister Viktor Orbán, dealt in rhetorical claims of supposedly disappearing ethnic Magyars, and government officials railed against George Soros in blatantly anti-Semitic speeches. Perhaps (not) coincidentally, one-fourth of the entire area of Hungary is affected by water erosion via gullying, and another 16 percent of the country's landmass is affected by wind erosion.[27] However, if one conducts a Google search for "erosion in Hungary," one is likely to retrieve just as many web pages and articles discussing the erosion of democracy as there are the erosion of soil. Of course, both are happening. As I hope I have demonstrated across this book, erosion almost always comes with a set of anxieties about human existence and disappearance. These anxieties are flexible in application, speaking to both those on the right invested in anxieties of the loss of white supremacy and those on the left who sublimate erosion into their fears over the fall of liberal democracy. Lost in these debates are concerns about the physical earth that sustains every human on the planet. Even further occluded are those humans who have suffered the brunt of white supremacy, climate change, and a neoliberal occupying nation-state.

Ultimately, what I propose in this book about settler colonialism and erosion is far from novel. Bennett himself drew similar conclusions in the 1930s. Writing of Peru, he notes, "The Incas and their predecessors had developed highly efficient methods of soil conservation," and "the care used in conserving soil indicates that the ancient Peruvians had experienced soil erosion."[28] My call to return homelands to Indigenous control and stewardship has nothing to do with mysticism or romantic ideas of ecological essentialism; it's about a long history of recognizing the science and knowledge gleaned from living for thousands of years in place and with—even in Bennett's own words—care. It is also about the human labor in this stewardship. Bennett's conclusions are unequivocal about the methods of Indigenous agriculture in Peru: "It is evidently the result of definite planning [and] the result has been the most effective method of controlling erosion developed anywhere in the world though probably the most expensive so far as labor is concerned. The disintegration of much of this highly developed agriculture may be directly attributed to the devastation accompanying the Conquest and not to any intrinsic flaws in the methods employed."[29] Bennett then expands these conclusions outward, tying the local and global together to explain the erosion crises of his (and our) own time: "The current erosion crisis in North America can be definitely attributed to the exploitation of land that followed European settlement," and he goes on to say, "Although the processes of erosion operate locally, the economic effects have become international in scope."[30]

The point is that erosion has to be felt up close at the local level by those who notice the particularities of the particles of soil that touch their lives. At the same time, as I have attempted to show in these concluding remarks, we have to zoom out to see how the local is everywhere, all the time, in ways that exceed a collapsed global "we." In 2019 Eduardo Mansur of the United Nations' Food and Agriculture Organization exhorted, "We must stop soil erosion to save our future."[31] This exhortation was not metaphorical even as it appealed to collective humanity. It is entirely possible the planet will in the not-too-distant future run out of the necessary topsoil to meet the food needs of human and other animal populations.[32] Indeed, such shortages would also have negative effects on the stability of nations, but "we" do a profound disservice in collapsing these planetary concerns into debates about the best forms of settler-state government ("the erosion of democracy") or render them metaphors for the right's anxieties over the end of white America.[33]

In their own ways, authors working across temporal and geographic removes in the United States have grappled with these problems of how to make sense of a disappearing earth and their own existential investment in a land they call home. During the course of my earliest research for what would become this book, I spoke to a geologist at Providence Canyon who was quick to point out that all sediment goes somewhere; it doesn't simply disappear. It moves to places where I personally might not find it appropriate or valuable, but it doesn't simply evaporate into space. It may be rendered unfit for human agriculture. It may end up flung out into the Gulf of Mexico, causing compounding issues for humans and other animals, but it does not vanish from the planet. He jokingly said, "The Earth is a giant sphere that wants to be smooth. It's trying to reach equilibrium; we're just holding on while it does."[34] I've returned to this geological premise, expressed offhand and simplified for my neophyte consideration, and pondered what the earth wants, what the earth desires. As a companion comment, around the same time, one of my dear friends and mentors wondered aloud in a conversation we had while driving along Lake Berryessa, discussing Indigenous rights and recognition in north-central California: "How long until we all have permission to exist on this planet?" I cannot help but try to compile a narrative out of these two statements—the earth's desires and any given person's permission to exist on the planet. The underside of this is when the anthropomorphized earth's desire to smooth itself bumps up against some humans' desires for extraction. This desire insists that other humans have no right to exist on the same planet with the same access to the lands and waters necessary for survival. It is the conceptual space where particles of soil become land, where land becomes territorialized earth, and where earth becomes planet.

And thus I return to the opening mistake of this conclusion. When are we talking about erosion, and when are we talking about an all-consuming desire for the comfort of our own existence on the planet—an eros based on consumption and land title? Indigenous studies calls for a necessary awareness of the pronouns and possessive adjectives in that question. The lessons of Indigenous stewardship of homelands necessitate a step back from a narcissistic conflation of erosion and settler self-interest, a return to care not based on immediate returns. If erosion is not a metaphor, then what types of settler cessions and temporal reconsiderations will be necessary to address its material demands in the future? What narratives about the earth will yield an understanding of how one goes forward in care without falling victim to the abdication inherent in lost cause–isms? These narratives and structures of care give no credence to the settler anxieties provoked by calls for returning homelands—all of them—to Indigenous control. For settlers' own planetary survival, the only way forward is to let go.

Notes

INTRODUCTION: EROSION

1. National Resources Board, "General Distribution of Erosion" insert, 170–71; see also Sutter, *Let Us Now Praise*, 88.
2. See Handelsman, *World without Soil*, 60.
3. Jo Handelsman's recent work, *World without Soil*, stands as a productive example against this trend.
4. Nishime and Williams, "Why Racial Ecologies," 8. See also Haymes, "Africana Studies Critique."
5. DeLoughrey, *Allegories of the Anthropocene*, 34.
6. Bruyneel, *Settler Memory*, xiii.
7. Yusoff, *Billion Black Anthropocenes*, xiii.
8. E. Anderson and Taylor, "Letting the Other Story Go," 87.
9. Liboiron, *Pollution Is Colonialism*, 66.
10. Todd, "Indigenizing the Anthropocene," 243.
11. For a discussion of the recursivity of this process, see Nichols, *Theft Is Property!*
12. Blaser and de la Cadena, "Pluriverse," 3.
13. DeLoughrey, *Allegories of the Anthropocene*, 2.
14. This experience is not necessarily synonymous with *ecoanxiety*, which the American Psychological Association links to mental health concerns over

feelings of "watching the slow and seemingly irrevocable impacts of climate change unfold, and worrying about the future for oneself, children, and later generations." Ecoanxiety and mental health are often attached to post-traumatic stress issues from experiencing extreme climate events and affect communities already made vulnerable by other failures of infrastructure. See Clayton et al., "Mental Health," 27.

15 Nixon, *Slow Violence*, 62.
16 Nixon, *Slow Violence*, 62.
17 Lyons, *Vital Decomposition*, 4.
18 See L. Brooks, "Primacy of the Present"; and Caison, *Red States*, 14–27.
19 E. Anderson and Taylor, "Letting the Other Story Go," 88; and DeLoughrey, *Allegories of the Anthropocene*, 10.
20 Romine, *Real South*, 9.
21 This push and pull also works within the tension described by Mimi Sheller in *Island Futures*, where she outlines how, on the one hand, nations participate in an *islanding* of environmental degradation to the offshore spaces of empire and, on the other, nations remain aware of how these problems often "overspill these regional boundaries" (21).
22 Nichols, *Theft Is Property!*, 9.
23 Nichols, *Theft Is Property!*, 9.
24 Howe, *Choctalking on Other Realities*, 31.
25 Ursula Heise lays out a similar approach to understanding narratives of extinction in *Imagining Extinction: The Cultural Meanings of Endangered Species*.
26 Blaikie, "State of Land Management Policy," 42.
27 For example, see Caison, *Red States*; and Caison, "Removal."
28 See T. King, *Truth about Stories*.
29 Sarris, *Keeping Slug Woman Alive*, 46.
30 Rifkin, *Beyond Settler Time*, 19.
31 Whyte, "Time as Kinship," 39.
32 Whyte, "Time as Kinship," 54.
33 Handelsman, *World without Soil*, 44.
34 Tuck and Yang, "Decolonization Is Not a Metaphor," 10.
35 I want to acknowledge that methodological appropriation is something that non-Native scholars such as myself must take incredibly seriously. For starters I ask myself continually, What good does my work provide for the Indigenous community from which the knowledge derives? Who are you learning from, and how do you honor that learning? I would also like to

say, however, that I see far too many non-Native scholars reject Indigenous methodologies under the guise of not wanting to appropriate, when in reality I suspect that they don't want to do the hard work of trying to learn, to get it right, and to be accountable to the best of their ability. In this move they can maintain their privilege in the praxis of Western research methodologies and perform a certain "right-speaking" liberalness that simply means they don't have to work to learn Indigenous methodologies. So while non-Native scholars should be vigilant when working through something like an Indigenous methodology of temporal construction, they shouldn't simply reject the work because it seems hard and then paper over that rejection with the buzzword *appropriation*.

36 La Paperson, "Ghetto Land Pedagogy," 124.
37 Neimanis, *Bodies of Water*, 14.
38 Neimanis, *Bodies of Water*, 15.
39 Tuck and Yang, "Decolonization Is Not a Metaphor."
40 Tuck and Yang, "Decolonization Is Not a Metaphor," 2.
41 As Lyons explains, "Soil scientists' practices tend to take place in laboratories and depend on state research funding cycles, alliances with industrial trade associations, and soil samples transported from rural violence to relative urban safety." *Vital Decomposition*, 6.
42 Allen, *Earthworks Rising*, 26.
43 Goodman, *Planetary Lens*, 3.
44 Mirzoeff, *White Sight*, 1–23.
45 Liboiron, *Pollution Is Colonialism*, 7; and Yang, "Sustainability as Plantation Logic," 1.
46 See Moreton-Robinson, *White Possessive*; and Treuer, "Return the National Parks."
47 T. Williams, *Erosion*, xi.
48 T. Williams, *Erosion*, 4.
49 T. Williams, *Erosion*, 16–17.
50 T. Williams, *Erosion*, 20 (emphasis added).
51 La Paperson, "Ghetto Land Pedagogy," 121.
52 T. Williams, *Erosion*, 184.
53 Ghosh, *Great Derangement*, 87.
54 Ghosh, *Great Derangement*, 87.
55 See Krech, *Ecological Indian*; as well as Harkin and Lewis, *Native Americans and the Environment*.
56 E. Anderson and Taylor, "Letting the Other Story Go," 75.

57 Handelsman, *World without Soil*, 136–52.
58 Scott, *Seeing Like a State*, 331.
59 Scott, *Seeing Like a State*, 323.
60 Rankin, Barrier, and Horsley, "Evaluating Narratives of Ecocide," 369.
61 Rankin, Barrier, and Horsley, "Evaluating Narratives of Ecocide," 370 citing Lopinot and Woods, "Wood Overexploitation."
62 Rankin, Barrier, and Horsley, "Evaluating Narratives of Ecocide," 384.
63 Rankin, Barrier, and Horsley, "Evaluating Narratives of Ecocide," 384.
64 Rankin, Barrier, and Horsley, "Evaluating Narratives of Ecocide," 383 citing Mt. Pleasant, "New Paradigm"
65 Rankin, Barrier, and Horsley, "Evaluating Narratives of Ecocide," 383–84.
66 See also Handelsman's discussion of mistaken, and yet nonetheless persistent, theories of Maya city collapse as a result of soil mismanagement. *World without Soil*, 136–41.
67 *Oxford English Dictionary Online*, s.v. "erode, v., 1a" Oxford University Press, July 2023, https://doi.org/10.1093/OED/1062137317.
68 *Oxford English Dictionary Online*, s.v. "erode, v. 1a"
69 Oxford English Dictionary *Online*, s.v. "erosion (n.) 1a." Oxford University Press, July 2023, https://doi.org/10.1093/OED/4026214323.
70 Liboiron, *Pollution Is Colonialism*, 66.
71 T. L. King, *Black Shoals*, 40, quoted in Liboiron, *Pollution Is Colonialism*, 66.
72 Nixon, Slow Violence.
73 Nixon, *Slow Violence*, 62 (emphasis added).
74 McKittrick, "Plantation Futures."

CHAPTER 1: LANDSLIDES AND HORIZONS OF THE WEST

1 Xia, "California Coast Is Disappearing."
2 Xia published her reporting in a book titled *California Against the Sea* in 2023. The book does show more of an awareness of Indigenous peoples and communities in California than the initial reporting, but I would not say it *centers* a Native perspective of the Land Back movement as integral to the questions of the California coast.
3 Xia, Kannan, and Castleman, "Ocean Game."
4 Xia, Kannan, and Castleman, "Ocean Game."
5 See Whyte, "Time as Kinship."
6 See Akins and Bauer, *We Are the Land*; Bauer, *California through Native Eyes*; and Baldy, *We Are Dancing for You*.

7   Xia, "California Coast Is Disappearing."
8   See Norton, *Genocide in Northwestern California*; Lindsay, *Murder State*; and Madley, *American Genocide*.
9   Xia, "California Coast Is Disappearing" (emphasis added). Notably, the rail line to San Diego was closed in fall of 2022 due to coastal erosion. While the line had been projected to reopen in 2023, the date was pushed back, and experts have noted that the only lasting option is to move the tracks altogether. See Diehl, "Erosion Repairs."
10 Yusoff, *Billion Black Anthropocenes*, 53–54.
11 Xia, "California Coast Is Disappearing."
12 Solnit, *Infinite City*, vii.
13 Tribe, "Foreword" in Manovich, *Language of New Media*, x.
14 Barbrook and Cameron, "Californian Ideology" (emphasis added).
15 Barbrook and Cameron, "Californian Ideology," 45.
16 As early as 1776, Jefferson writes to Edmund Pendleton about the necessity of Indigenous Removal in the Southeast. Jefferson, *Writings*, 754. Also see R. Owens, *Mr. Jefferson's Hammer*.
17 Dean, *Against the Tide*, 121.
18 Dean, *Against the Tide*, 121.
19 Dean, *Against the Tide*, 121.
20 Dean, *Against the Tide*, 122.
21 Dean, *Against the Tide*, 122.
22 See Stone et al., "Sand Rights."
23 Dean, *Against the Tide*, 124–30.
24 M. Anderson, *Tending the Wild*, 2.
25 M. Anderson, *Tending the Wild*, 3.
26 M. Anderson, *Tending the Wild*, 3.
27 Reed, *Settler Cannabis*, 1, quoting Orona (in brackets) from *Stories of the River*.
28 Baltz, *Lewis Baltz: Texts*, 40–41.
29 Baltz, *Lewis Baltz: Texts*, 44.
30 Baltz, *Lewis Baltz: Texts*, 41.
31 Baltz, *Lewis Baltz: Texts*, 41.
32 Baltz scholar Chris Balaschak identifies the 1970 NASA book of photographs *The Moon as Viewed by Lunar Orbiter* by Kosofsky and El-Baz as a key influence on the decade's emergent New Topographics movement. Balaschak, "New Worlds," 39.

33. Baltz, *Lewis Baltz: Texts*, 43–44.
34. In addition to M. Anderson, *Tending the Wild*, see Manning, *Trust in the Land*; and Manning, *Upstream*.
35. A 2023 exhibition titled *Ansel Adams in Our Time*, at the de Young Museum in San Francisco, explored these exact issues, juxtaposing Adams's work with that of contemporary artists working to (re)evaluate his legacy for Western landscape art. Fine Arts Museums of San Francisco, "*Ansel Adams*."
36. Brower, "Ansel Adams at 100."
37. Scheppe, "Lewis Baltz," 91.
38. Scheppe, "Lewis Baltz," 92.
39. Scheppe, "Lewis Baltz," 92.
40. See Mirzoeff, *White Sight*.
41. Balaschak, "New Worlds," 36.
42. Sekula, "Dismantling Modernism, Reinventing Documentary," 54 quoted in Balaschak, "New Worlds," 36.
43. Interestingly, Baltz's father was a mortician.
44. Baltz, *Lewis Baltz: Texts*, 34–35.
45. In his notes to accompany eight images from the series, Baltz writes that the typical locations of the industrial parks he photographs are "previously unimproved land" and that the land criteria of the developers include "flat land requiring a minimum of grading and which poses no unusual problems in soil or foundation engineering. Typical choices are valley bottoms or the flood plains of rivers." Baltz, "Untitled," 14. The language of improvement here suggests an entirely Western view of what constitutes the appropriate use of land. Moreover, the fact that these industrial parks are built in floodplains indicates a potential instability in how these buildings will handle long-term risks, and it echoes the earlier mistakes of invading Catholic missionaries, who had to abandon their first construction of the San Gabriel Mission because they had built it in a floodplain.
46. Baltz, *Lewis Baltz: Texts*, 70. See also Baltz, "Last Interview." In the latter work, Jeff Rian asks Baltz about reactions to his work:

    RIAN What were the critical reactions to "The New Topographics" show?

    BALTZ None to speak of. It was reviewed in *Art in America*—Eastman House paid the journalist to come, put him up, etc. That was it. Then nothing, though it traveled to the Otis Art Institute, because I knew the curator, and then it went to Princeton University, in New Jersey, in the summer, when no one is there. Then six months later people were talking about the backlash against the show.

> **RIAN** Against it?
>
> **BALTZ** Well, I never saw a "fore-lash." People said the photographs were cold, anti-humanistic—emperor's new clothes. I liked it, and the show was redone again only a few years ago, which terrified me.

47 Baltz, *Lewis Baltz: Texts*, 70.
48 Baltz, *Lewis Baltz: Texts*, 70–71.
49 See Goode, *Agrotopias*.
50 Baltz, "Last Interview." Schell, *Fate of the Earth*.
51 See Climate Central, "Surging Seas."
52 Baltz, "Last Interview." Anyone familiar with *Wizard of Oz* author L. Frank Baum's writings about Native people will pause at this comparison, as Baum expressed *strong* anti-Indigenous sentiments in writings during the time of the Wounded Knee massacre. See Sutherland, "L. Frank Baum Advocated Extermination."
53 Baltz mis-paraphrases Burroughs, who writes in *Naked Lunch*, "America is not a young land: it is old and dirty and evil before the settlers, before the Indians. The evil is there waiting." Burroughs, *Naked Lunch*, 11.
54 See Ewen, "Bering Strait Theory"; and Daley, "First Humans."
55 Griswold, "There's No Place," 464. Griswold also notes that Baum himself later named his southern California estate Ozcot.
56 La Paperson, "Ghetto Land Pedagogy," 121.
57 J. King, "Rising Reality."
58 J. King, "Rising Reality."
59 J. King, "Rising Reality."
60 Balaschak, "New Worlds," 30.
61 Balaschak, "New Worlds," 30.
62 J. King, "Rising Reality."
63 See "Candlestick" and "Explore Candlestick" on the developer's website, *Candlesticksf.com*.
64 Dean, *Against the Tide*, 1–14.
65 Davenport and Robertson, "Resettling." The plans have also proven contentious for residents, as federal and state officials failed to consider diverging interests in what resettlement means for citizens of different tribal nations. See Loginova and Cassel, "Leaving the Island."
66 This phrase is repeated across both *Parable of the Sower* and *Parable of the Talents*. It first appears as Earthseed scripture at the beginning of Chapter 7 in *Sower*.

67 As Gerry Canavan notes, "Readers of the Parables books typically refer to the character as 'Lauren,' with a soft, almost parental fondness—but in Butler's personal notes she is always 'Olamina.'" Canavan, *Octavia E. Butler*, 164.

68 See Herzog, *Parables as Subversive Speech*.

69 Butler, *Parable of the Sower*, 59.

70 Butler, *Parable of the Sower*, 63.

71 Butler, *Parable of the Sower*, 63–64.

72 See Chiotakis and Bordal, "Acorns Helped LA's Indigenous People"; Alvarez and Peri, "Acorns"; and Ortiz, *It Will Live Forever*.

73 It is possible that this speculative future novel likewise has an alternate history; however, the past tense used around the loan of the California Native botany book does not indicate a radically alternative past for California Native people in the novel's universe.

74 T. L. King, *Black Shoals*, 4.

75 Rifkin, *Fictions of Land and Flesh*, 3.

76 Rifkin, *Fictions of Land and Flesh*, 5.

77 La Paperson, "Ghetto Land Pedagogy," 115.

78 La Paperson, "Ghetto Land Pedagogy," 115.

79 Butler, *Parable of the Sower*, 180.

80 As Butler describes their location, "We spread maps on the ground, studied them as we ate breakfast, and decided to turn off U.S. 101 this morning. We'll follow a smaller, no doubt emptier road inland to the little town of San Juan Bautista, then east along State Route 156. From 156 to 152 to Interstate 5. We'll use I-5 to circle around the Bay Area. For a time we'll walk up the center of the state instead of along the coast. We might have to bypass I-5 and go farther east to State 33 or 99. I like the emptiness around much of I-5." *Parable of the Sower*, 247.

81 Butler, *Parable of the Sower*, 326–27.

82 Butler, *Parable of the Sower*, 327.

83 Butler, *Parable of the Sower*, 327–28.

84 Butler, *Parable of the Sower*, 328.

85 Butler, *Parable of the Sower*, 328.

86 Butler, *Parable of the Sower*, 328.

87 Butler, *Parable of the Sower*, 313.

88 See Norton, *Genocide in Northwestern California*.

89 Butler, *Parable of the Sower*, 273.

90   Butler, *Parable of the Sower*, 273.
91   Reed, *Settler Cannabis*, 62–69. Also see Widick, *Trouble in the Forest*.
92   See Roering et al., "Beyond the Angle of Repose"; and Lisle, "Eel River, Northwestern California."
93   See Dean, *Against the Tide*.
94   Butler, *Parable of the Talents*, 21.
95   Butler, *Parable of the Talents*, 113–14.
96   Butler, *Parable of the Talents*, 114.
97   Butler, *Parable of the Talents*, 114.
98   Butler, *Parable of the Talents*, 139.
99   Butler, *Parable of the Talents*, 139–40.
100 Butler, *Parable of the Talents*, 140.
101 Butler, *Parable of the Talents*, 137.
102 Butler, *Parable of the Talents*, 184.
103 Notably, one of the many Eel River massacres was reported on September 26, 1863, in the *Marysville Appeal* and is believed to have occurred around September 12, making Lauren Olamina's account of the Acorn massacre occur exactly 170 years to the day after the reporting of a specific Eel River attack on Indigenous peoples. See State of California, Native American Heritage Commission, "Timeline of Genocide Incidents."
104 Butler, *Parable of the Talents*, 253.
105 Butler, *Parable of the Talents*, 254.
106 Butler, *Parable of the Talents*, 254.
107 Butler, *Parable of the Talents*, 406.
108 Butler, *Parable of the Talents*, 406.
109 Canavan, *Octavia E. Butler*, 178–79.
110 Canavan, *Octavia E. Butler*, 179.
111 Miranda, *Bad Indians*, 202–3.
112 Miranda, *Bad Indians*, 203.
113 Miranda, *Bad Indians*, 203.
114 Miranda, *Bad Indians*, 203.
115 Miranda, *Bad Indians*, 204.
116 See Tuck and Yang, "Decolonization Is Not a Metaphor."
117 Miranda, *Bad Indians*, 207.
118 Miranda, *Bad Indians*, 208.
119 Xia, "California Coast Is Disappearing."

CHAPTER 2: SURFACES AND ALLOTMENTS OF THE HEARTLAND

1. Meister, *Dorothea Lange, Migrant Mother*, 12.
2. See Goggans, *California on the Breadlines*.
3. Definitions of regions such as the Midwest and South often vary, and I could make an argument for Oklahoma's inclusion in either region. It is, however, speaking strictly spatially, in the middle of the country if one is heading west.
4. According to the family memoir *Migrant Mother*, by Oleta Kay Sprague Ham and Roger Sprague Sr., Thompson's parents were Mary Jane and Jack Christie. A Mary J. and Jackson Christie are listed both independently and as married on the Dawes Rolls. Mary J. Christie (Card Number 9460) is listed as twenty-three years old and IW (intermarried white). Her Cherokee enrollment was ultimately "refused" as her marriage to the enrolled Jackson did not occur before November 1, 1875. Jackson Christie (Card Number 6924) is listed as thirty-one years old and seven-eighths Cherokee. He is listed as also having married a Liza Glass on July 11, 1904, and deceased on October 7, 1904. The family memoir suggests that Mary Jane left Jackson while pregnant with Florence and established a subsequent long-term relationship with a white man named Charlie Akman. However, it also recounts that Mary Jane chose to give birth to Florence among her "Cherokee relatives," making her designated bureaucratic status as "intermarried white" possibly more complicated. Nonetheless, if Jackson Christie is Florence Christie Thompson's father, then she and her descendants would indeed likely be citizens, or would be eligible for citizenship, in the present-day Cherokee Nation. Her mother's presumed identity as not Cherokee would perhaps complicate a traditional clan belonging and kinship for her daughter Florence but could very well *not* have undermined Florence's understanding of herself as a Cherokee person at that time. This account of evidence is not meant to prove or disprove any of Thompson's or her family's claims to Cherokee identity. It should, however, give some potential clues to why Florence might not have had an allotment and how she might have navigated her own choices as a young woman in Indian Territory at the turn of the twentieth century. I offer this independent research in the spirit of transparency around the understanding of claims to Cherokee identity and belonging, which unfortunately can be fraught in the context of others' specious claims to Native identity. For a discussion of the complexities brought out by examining the Dawes Rolls in this period, see Justice, "Narrated Nationhood."
5. Dunn, "Photographic License."
6. Meister, *Dorothea Lange, Migrant Mother*, 12. Sarah Hermanson Meister summarizes Lange's account of driving home after several weeks on the

road and seeing a sign for a pea pickers' camp, which prompted her to turn around so she could photograph the conditions at the site. Her rushed time with Thompson was apparently somewhat atypical for Lange, who regularly inquired about her photographic subjects' lives; however, according to Meister, not recording the subjects' names was only moderately unusual for Lange.

7   The texts that offer the most concrete biographical context for Thompson include Bill Ganzel's *Dust Bowl Descent* (1984) and Ham and Sprague's family memoir, *Migrant Mother*. Ganzel worked with the *Modesto Bee* to establish the identity of the previously nameless woman in Lange's photograph, and Sprague is Thompson's grandson. The family memoir and Ganzel's text are both currently out of print. Texts that address or mention Thompson's identity in a more generalized Indigenous frame include Meister, *Dorothea Lange, Migrant Mother*; Gordon, *Dorothea Lange*; and Dunn, "Photographic License."

8   Gerstle, *American Crucible*, 180–81, quoted in Stein, "Passing Likeness," 354.

9   The scholars that come the closest to aligning these details include Hannah Holleman (*Dust Bowls of Empire*) and Brad Lookingbill (*Dust Bowl, USA*). Both Holleman and Lookingbill outline the relationship between colonialism and the ecological events of the period. Neither, however, centers the Indigenous contours of the conditions of the Dust Bowl on the southern plains.

10 Guthrie, "This Land Is Your Land." Notably, Guthrie wrote two versions of the song. The original 1940 lyrics were much more critically ambivalent than the postwar version of 1956.

11 Lynn Riggs, quoted in Weaver, foreword, xv.

12 See Egan, *Worst Hard Time*; and Worster, *Dust Bowl*.

13 See An Act to Provide for the Allotment of Lands in Severalty to Indians on the Various Reservations (General Allotment Act or Dawes Act), Public Laws, 49th Cong., 2nd Sess., 25 U.S.C. (1887), https://uscode.house.gov/view.xhtml?path=/prelim@title25/chapter9&edition=prelim; and Burton, *Indian Territory*.

14 See Chang, *Color of the Land*; and Justice and O'Brien, *Allotment Stories*.

15 Notably, the Curtis Act was named for Republican senator Charles Curtis, who was a member of the Kaw Nation, the first non-European-descended vice president (under Herbert Hoover), and an advocate of assimilation policies for Native people. See Unrau, *Mixed-Bloods and Tribal Dissolution*.

16 The 2020 decision in *McGirt v. Oklahoma* (140 S. Ct.), where the US Supreme Court affirmed that much of the eastern part of the state is indeed tribal lands, may very well fundamentally change how we consider Oklahoma history. See Leeds and Beard, "Wealth of Sovereign Choices."

17. An Act to Provide for an Enlarged Homestead, Pub. L. No. 245, 60th Cong., 2nd Sess., chap. 160 (1909). See also Gregg, "Imagining Opportunity."
18. Montgomery, *Dirt*, 146–47.
19. See Perdue, *Cherokee Women*.
20. See Blackhawk, *Violence over the Land*; and Hämäläinen, *Comanche Empire*.
21. See Denson, *Demanding the Cherokee Nation*.
22. Roberts, *I've Been Here*, 39–40.
23. See Hämäläinen, *Comanche Empire*; and Egan, *Worst Hard Time*.
24. Egan, *Worst Hard Time*, 24; and Worster, *Dust Bowl*, 201. Jo Handelsman draws a convincing historical line from Thomas Jefferson's development of iron plow technology to the Dust Bowl. See Handelsman, *World without Soil*, 82–83.
25. See Egan, *Worst Hard Time*; and Worster, *Dust Bowl*.
26. Egan, *Worst Hard Time*, 42–43.
27. For a discussion of Hoover's work with food relief, see M. Cox, *Hunger in War and Peace*.
28. Egan, *Worst Hard Time*, 101, 151.
29. Egan, *Worst Hard Time*, 126.
30. Hugh Hammond Bennett, quoted in Egan, *Worst Hard Time*, 125.
31. Bennett, quoted in Egan, *Worst Hard Time*, 266–67.
32. See Schloss, *In Visible Light*, 201–31.
33. Worster, *Nature's Economy*, 223.
34. Hedgpeth, "John Steinbeck," 296.
35. L. Owens, *Grapes of Wrath*, 60.
36. Steinbeck, *Grapes of Wrath*, 1.
37. Steinbeck, *Grapes of Wrath*, 33.
38. Steinbeck, *Grapes of Wrath*, 33.
39. Steinbeck, *Grapes of Wrath*, 33.
40. Steinbeck, *Grapes of Wrath*, 33.
41. Steinbeck, *Grapes of Wrath*, 34.
42. Nichols, *Theft Is Property!*, 8.
43. Nichols, *Theft Is Property!*, 9.
44. Souder, *Mad at the World*, 218–19.
45. See, for example, Han, "Defense of Steinbeck's Intercalary Chapters."
46. Steinbeck, *Grapes of Wrath*, 325–26.
47. L. Owens, *Grapes of Wrath*, 61.

48 Steinbeck, *Grapes of Wrath*, 339.
49 Steinbeck, *Grapes of Wrath*, 339.
50 L. Owens, *Grapes of Wrath*, 63.
51 Gordon, *Dorothea Lange*, 84–87.
52 Stein, "Passing Likeness," 350–51.
53 Dorothea Lange, quoted in Swensen, *Picturing Migrants*, 19.
54 Mirzoeff, *White Sight*, 4.
55 Gordon, *Dorothea Lange*, 287–300.
56 Gordon, *Dorothea Lange*, 253.
57 Taylor, *On the Ground*, 17.
58 Gordon, *Dorothea Lange*, 165.
59 Lange quoted in Gordon, *Dorothea Lange*, 165. See Lange "Oral History," 7. (emphasis added).
60 Gordon, *Dorothea Lange*, 165.
61 Gordon, *Dorothea Lange*, 167–68.
62 Gordon, *Dorothea Lange*, 226.
63 Arthur Raper, quoted in Lange and Taylor, *American Exodus*, 29.
64 Gordon, *Dorothea Lange*, 282.
65 As explained in note 4 of this chapter, Thompson's mother may or may not have been Cherokee, complicating her own clan belonging in a traditional matrilineal sense. Nonetheless, had she been born in traditional Cherokee homelands, she still would have possibly experienced some effects of a matrilineal society, where women traditionally had a larger say in land management and agricultural decisions.
66 Perdue, *Cherokee Women*, 25.
67 For an understanding of the myriad ways that the plantation enslavement economy changed Cherokee life, see Miles, *House on Diamond Hill*; and Miles, *Ties That Bind*; as well as McClinton, *Moravian Springplace Mission*.
68 See, for example, Wilkins, *Cherokee Tragedy*; and Gaul, *To Marry an Indian*.
69 See Champagne, *Social Order and Political Change*; and McLoughlin, *Cherokee Renascence*.
70 See Caison, *Red States*, 72–78.
71 Perdue, *Cherokee Women*, 195.
72 Dunn, "Photographic License."
73 Dunn, "Photographic License."
74 Braunlich, *Haunted by Home*, 21–35.
75 Braunlich, *Haunted by Home*, 21–35.

76 Braunlich, *Haunted by Home*, 21–35.

77 Examples of this can be found in correspondence with Paul Green in the Paul Green Papers in UNC's Southern Historical Collection (collection number 03693, subseries 1.1–4) as well as journals and correspondence in the Lynn Riggs Papers (collection number YCAL MSS 61, series 1–3) at Yale's Beinecke Library.

78 Lynn Riggs, quoted in Braunlich, *Haunted by Home*, 71.

79 Braunlich, *Haunted by Home*, 71; and Romine, *Real South*, 9.

80 See Allen, "When a Mound"; J. Cox, *Red Land to the South*; J. Cox and Pettit, "Fugitive Indigeneity"; Justice, *Our Fire Survives the Storm*; Weaver, *That the People Might Live*; Weaver, *Other Words*; Womack, *Red on Red*; and Womack, *Art as Performance*.

81 K. Brown, *Stoking the Fire*, xii.

82 K. Brown, *Stoking the Fire*, xv–xvi.

83 K. Brown, *Stoking the Fire*, 28.

84 Womack, *Art as Performance*, 117.

85 Weaver, *Other Words*, 105–6.

86 Weaver, *Other Words*, 105.

87 Weaver, *Other Words*, 105–6.

88 Riggs, *Green Grow the Lilacs*, 9.

89 Riggs, *Green Grow the Lilacs*, 96.

90 C. Smith, Strickland, and Smith, *Building One Fire*, 162, 170; and Zogry, *Anetso*, 121–23.

91 Womack, *Art as Performance*, 139. Also, please note that *Green Grow the Lilacs* gives the date as 1906, not 1900.

92 In his 2019 article for *Vulture*, "Oklahoma Was Never Really O.K.," Frank Rich outlines how the most recent revival of the musical reveals the more sinister undertones of the play; he argues that Curly is a white protagonist and follows Tim Carter's assertion that that Jud is modeled on the real-life "Jetar Davis (1889–1958), 'a contemporary of Riggs's who was also half-Cherokee and the town drunk,'" making the villain of the play Cherokee. Not only does this language of "half" demonstrate a shocking disregard for how Cherokee community and identity might be thought about then and now, it also forecloses the not-so-far-fetched idea that the Cherokee Riggs could have created a Cherokee protagonist or even *both* a Cherokee protagonist *and* a Cherokee villain. I contend that not seeing a Cherokee protagonist in Riggs's work says more about the viewer than it does the text.

93 Riggs, *Green Grow the Lilacs*, 70–71.

94 Riggs, *Green Grow the Lilacs*, 76–77.

95 Riggs, *Green Grow the Lilacs*, 100.

96 See Egan, *Worst Hard Time*; Handelsman, *World without Soil*; Holleman, *Dust Bowls of Empire*; Montgomery, *Dirt*; and Worster, *Dust Bowl*.

97 Riggs, *Green Grow the Lilacs*, 83.

98 These words might also take on an even more disturbing air if we consider Riggs wrote this play after the Tulsa race massacre of 1921. For an analysis of the event that considers Black and Indigenous historical contexts of the event together, see Roberts, *I've Been Here*.

99 Interestingly, sometime between 1924 and 1925, Riggs clipped and saved a news article about a white man who is suing his white wife for divorce and has included her Hopi lover as a co-respondent. The wife's mother is quoted in the piece as saying of her daughter's affair, "An Indian? That's absurd!" Riggs made no notes on the clipping, but he regularly clipped articles and made notes about character, plot, and setting while working on new plays. This clipping then is suggestive, but far from conclusive, of Riggs considering the reception of a romantic affair between a Native man and a white woman leading up to the composition of *Green Grow the Lilacs*. See "Names Hopi Guide as Co-respondent." Additionally, a relatively well-known piece of Cherokee history is that when Elias Boudinot and John Ridge married white women while students in Massachusetts, the townspeople (including some of the women's own family members) harassed and even burned the couples in effigy outside the respective ceremonies.

100 Gritz, "Behind 'Oklahoma!'"

101 See Burton, *Indian Territory*.

102 Riggs, *Green Grow the Lilacs*, 104.

103 Riggs, *Green Grow the Lilacs*, 20.

104 Riggs, *Green Grow the Lilacs*, 20.

105 Riggs, *Green Grow the Lilacs*, 4.

106 J. Cox, *Red Land to the South*, 104–5.

107 Womack, *Art as Performance*, 137.

108 See Riggs, "Journal."

109 Womack, *Art as Performance*, 116.

110 Womack, *Art as Performance*, 116.

111 Goeman, "Ongoing Storms and Struggles," 108, 123.

112 Riggs, "Journal," 36.

113 For a history of the Cherokee Female Seminary, see Mihesuah, *Cultivating the Rosebuds*.

114 Riggs, *Cream in the Well*, 158–59. All quotations are from the 1947 published version unless otherwise noted.

115 Riggs, *Cream in the Well*, 159.

116 *Oxford English Dictionary Online*, s.v. "erode, v., 1a" Oxford University Press, July 2023. https://doi.org/10.1093/OED/1062137317, https://www.oed.com/view/Entry/64061?redirectedFrom=erode&.

117 Riggs, *Cream in the Well*, 163.

118 It's never indicated if Mr. Sawters is Cherokee or not. There is a compelling reading to be made for either possibility. Mrs. Sawters refers to her children's mixedblood identity, but whether this genealogy derives from their father directly or from previous generations is never made clear.

119 Riggs, *Cream in the Well*, 193. Interestingly, in the 1940 typescript, Riggs has penciled in an edit where at the top of this passage Mr. Sawters says, "This was long before the gover'ment allotted us Indian land." Then Riggs also adds a note to that addition, stating, "I guess we understand that," suggesting his ambivalence about the edit. Ultimately, he keeps the detail about Allotment in the final published version.

120 Riggs, *Cream in the Well*, 167.

121 Riggs, *Cream in the Well*, 171.

122 Riggs, *Cream in the Well*, typescript draft 2.

123 Riggs, *Cream in the Well*, 218. Interestingly, all the typescript drafts say "corn" instead of the more generic "grain" that appears in the published version. Given the association of Cherokee women with Selu, the Corn Mother, this edit is suggestive of Riggs thinking through the transition from traditional corn to a wheat-based monocropping that will ultimately contribute to the Dust Bowl.

124 Pushkin, "Prophet," and Riggs, "Vine Theater," 280.

125 Riggs, "Vine Theater," 282.

126 Riggs, "Vine Theater," 277.

127 Riggs, *Cream in the Well*, 217.

128 Riggs, "Vine Theater," 277.

129 Holleman, *Dust Bowls of Empire*, 163.

130 Holleman, *Dust Bowls of Empire*, 163.

131 Holleman, *Dust Bowls of Empire*, 163.

CHAPTER 3: DISAPPEARING GROUNDS AND BACKGROUNDS OF THE GULF

1 The land loss estimates also compute this average across swaths of time, which yields the commonly quoted figure. Indeed, some hours or days the actual amount of land loss will vary, or land loss may even halt, and in a year

with strong storms, rates of land loss may accelerate. For various estimates and imaginative metaphors, see Restore the Mississippi River Delta, "Land Loss"; B. Anderson, "Louisiana Loses Its Boot"; Louisiana Coastal Protection and Restoration Authority, "Changing Landscape"; Snell, "Coast in Crisis"; and Swenson, "These Six Factors."

2 Solnit and Snedeker, *Unfathomable City*, 19.

3 Prud'homme-Cranford, "Summoning Swamp Songs," 96.

4 For example, Mike Tidwell's popular *Bayou Farewell* centers the experiences of Cajun people and includes his interactions with Houma communities in a few short chapters. Not only does Tidwell use dialect for many of his Indigenous interviewees in deeply troubling ways; he also writes about Indigenous people in stereotypical terms. Despite devoting an entire chapter narrating the time he spent in a Houma community, he later mentions "forgotten Indian warriors" (111), signaling his own amnesia about the Indigenous communities he had just visited. David Burley's otherwise excellent sociological study *Losing Ground* also relies most heavily on the accounts and experiences of settler descendants of the region, based on their own records of who responded to their qualitative surveys. Burley's work attempts to balance these perspectives with a short afterword by United Houma Nation Council members T. Mayheart Dardar and Thomas Dardar.

5 See Ware, *Cajun Women*.

6 See Bernard, *Cajuns*.

7 As Jennie Lightweis-Goff argues, even spaces such as New Orleans that aren't as defined by their Cajun connections are still "long imagined as an exception to both national and regional norms," suggesting a nested set of exceptionalisms where a distinct regional focus illuminates the contours of catastrophe. Lightweis-Goff, "Louisiana Lost and Found," 166–67.

8 Burley, *Losing Ground*; and Hemmerling, *Louisiana Coastal Atlas*.

9 Theriot, *American Energy, Imperiled Coast*.

10 See Dardar, "Global Climate Change," 3, in Verdin, *Return to Yakni Chitto*.

11 As Delia Byrnes draws from Terrell Scott Herring concerning the planetary turn in cultural studies, "It occludes the 'forgotten spaces' of region and microregion that remain crucial to disrupting the 'loopy binary' of the local/planetary circuit, which can too easily obscure multiple scales (the sublocal and the bioregional, for example) that overlap within the 'planetary.'" Byrnes, "Plantation Pasts," 39.

12 Nixon, *Slow Violence*, 62.

13 Nixon, *Slow Violence*, 62.

14 Solnit and Snedeker, *Unfathomable City*, 1.

15 Solnit and Snedeker, *Unfathomable City*, 1.
16 Solnit and Snedeker, *Unfathomable City*, 2.
17 See Ellwood et al., "LSU Campus Mounds."
18 Olson, "Hiding in Plain Sight."
19 Forkner, *Audubon on Louisiana*, xvii.
20 Mirzoeff, *White Sight*, 97–102.
21 Richards, "Joseph R. Mason."
22 See Olson, *Audubon's Aviary*, 26.
23 The Migratory Bird Treaty Act of 1918 called for the protection of nearly a thousand different species of birds. Passed largely in response to the plume trade's use of feathers in women's fashion, the act is directly responsible for saving the snowy egret from certain extinction. Relentlessly hunted in the swamps of the Florida Everglades, the egret is considered one of the darling success stories of the modern environmental movement. In gaining traction against the fashion trade, the act capitalized on the common knowledge that the plume industry was primarily controlled by women and Jewish merchants. Thus, the earliest "environmental" concerns must be placed fully in their eco-social context, as the legislation was motivated by a desire to control the income avenues of "others."
24 Nobles, "Myth of John James Audubon."
25 Nobles, "Myth of John James Audubon."
26 Mirzoeff, *White Sight*, 101.
27 Audubon, *Snowy Heron*.
28 Mirzoeff, *White Sight*, 103.
29 It should be noted that of all the plantation crops, rice tends to produce the least amount of soil erosion.
30 See Forkner, *Audubon on Louisiana*, 211–15.
31 Audubon, "Squatters," 245.
32 Audubon, "Squatters," 245–46.
33 Audubon, "Squatters," 246.
34 Audubon, "Squatters," 246.
35 Audubon, "Squatters," 247.
36 Audubon, "Squatters," 247–48.
37 Audubon, "Squatters," 248.
38 See Theriot, *American Energy, Imperiled Coast*; and Hemmerling, *Louisiana Coastal Atlas*.
39 Audubon, "Squatters," 248.

40   Audubon, "Squatters," 248–49.
41   For example, see Irmscher, "Controversies"; and Jiménez, "Audubon Society Keeps Name."
42   Audubon, "Improvements," 235.
43   Audubon, "Flood," 242.
44   Audubon, "Flood," 244.
45   Audubon, "Flood," 244.
46   Audubon, "Flood," 243.
47   See Parrish, *Flood Year 1927*.
48   Audubon, "Flood," 244.
49   Audubon, "To Reverend John Bachman," 285.
50   The history of the Louisiana Purchase follows the logic explicated by Robert Nichols in *Theft Is Property!*
51   See d'Oney, *Kingdom of Water*.
52   See d'Oney, *Kingdom of Water*; and Klopotek, *Recognition Odysseys*.
53   Cable, *Old Creole Days*, 121.
54   See Pesantubbee, *Choctaw Women*.
55   Cable, *Old Creole Days*, 122.
56   Cable, *Old Creole Days*, 122.
57   Kornecki and Fouss, "Sugarcane Residue Management Effects," 597.
58   Cable, *Old Creole Days*, 123.
59   Cable, *Old Creole Days*, 123.
60   Padgett, "George W. Cable's Gardens," 66.
61   For the remainder of the chapter, I omit the racial epithet of Charlie's name, as it isn't necessary to repeatedly mention it for the content of the story itself or my analysis.
62   For example, see Padgett, "George W. Cable's Gardens"; as well as Burnett, "Moving toward a 'No South'"; Castillo, "Stones in the Quarry"; and Payne, "Emergence of Alternate Masculinity."
63   Padgett, "George W. Cable's Gardens," 68.
64   Nichols, *Theft Is Property!*, 9.
65   Cable, *Old Creole Days*, 137.
66   Cable, *Old Creole Days*, 137.
67   Texas Invasive Species Institute, "Bermudagrass."
68   Cable, *Old Creole Days*, 140.
69   Cable, *Old Creole Days*, 140–41.

70   Cable, *Old Creole Days*, 142.

71   Rifkin, *Beyond Settler Time*.

72   Cable, *Old Creole Days*, 143.

73   See North Carolina Cooperative Extension, "*Gelsemium sempervirens*"; and North Carolina Cooperative Extension, "*Gelsemium rankinii*."

74   Cable, *Old Creole Days*, 144.

75   Allewaert, *Ariel's Ecology*, 35.

76   Allewaert, *Ariel's Ecology*, 35.

77   Allewaert, *Ariel's Ecology*, 35.

78   Cable, *Old Creole Days*, 144–45.

79   This image also calls up a prequel sensation to the famous Iron Eyes Cody antilittering and antipollution images, featuring an Italian actor who claimed in his real life to be Native. Cody was in fact from a New Orleans Sicilian-immigrant family.

80   Isle de Jean Charles Resettlement Program, "About"; and also see Marshall, "People of Isle de Jean Charles," for more information on the topic of resettlement.

81   See Loginova and Cassel, "Leaving the Island."

82   See Caison, *Red States*, 211–15.

83   In her interview with Kirstin Squint, Verdin recounts,

> In the making of the documentary, a lot of our advisors wanted us to give our viewers a pat ending like, "Well, they're doing this. Or, oh, this is possible." We really pushed back on that and left it very open-ended. That it was this cycle. It started with the land rights being taken away, then they came in with the oil and gas pipelines and exploration canals and shipping canals. . . . My collaborators and I thought, "We've been working on it for a couple of years, we're gonna wrap this up." They had me going home to Grandma's house, and I just kept telling them, "Guys, this is not right. We can't leave people with that. I don't want to give them the answer, because I know we don't have that, but the end is not just 'Doo-da-doo, go home.'" Squint, "Monique Verdin's Louisiana Love," 121.

84   See Howe, *Choctalking on Other Realities*.

85   See Goodman, *Planetary Lens*.

86   See Theriot, *American Energy, Imperiled Coast*; and Hemmerling, *Louisiana Coastal Atlas*.

87   See Theriot, *American Energy, Imperiled Coast*.

88   Dardar, "Global Climate Change," 3.

89   Dardar, "Global Climate Change," 3.

90   Verdin, *Return to Yakni Chitto*, 24.
91   Verdin, *Return to Yakni Chitto*, vii.
92   As Melissa Nelson outlines in "Getting Dirty,"
      Many tribal nations were equally matriarchal or women-centered. In these cases, these stories could speak to women's ability to define their own rules and protocols; to test and break taboos (in many cases without severe consequence); and to be self-sufficient, productive, and happy without a human man or with other-than-human husbands and partners. These narratives also illustrate that women have a profound connection with the natural elements—wind, water, soil—and with plant and animal species and sticks and rocks. This is not meant to imply the old, essentialist "woman as nature" trope. It is simply a comment on the diversity of relations women have in these stories. It could also speak to a unique aspect of women's psychology and fluid sexual behavior that (as noted earlier) is currently being researched by contemporary female scientists, with surprising discoveries and intriguing theories (254–55).
93   Verdin, *Return to Yakni Chitto*, 39.
94   Verdin, *Return to Yakni Chitto*, 38–39.
95   Verdin, *Return to Yakni Chitto*, 38–39.
96   For a discussion of the failure of paper *and* digital maps to account for lived spatiality, see Lightweis-Goff, "Louisiana Lost and Found."
97   See Whyte, "Time as Kinship."
98   Dardar, "Global Climate Change," 3.
99   Verdin, *Return to Yakni Chitto*, 71.
100   Verdin, *Return to Yakni Chitto*, 71.
101   Dardar, "Global Climate Change," 3.
102   Verdin, *Return to Yakni Chitto*, 107.
103   Verdin, *Return to Yakni Chitto*, 107.
104   Squint, "Monique Verdin's Louisiana Love."
105   For example, see Byrnes, "Plantation Pasts."

CHAPTER 4: GULLIES AND REMOVALS OF THE PLANTATION SOUTH

1   *Chicago Daily News*, "Soil and Sanctuary."
2   See Egan, *Worst Hard Time*; and Worster, *Dust Bowl*.
3   Soil Conservation Act, Pub. L. No. 74–76, 74th Cong., 1st Sess. (April 27, 1935); and Soil Conservation and Domestic Allotment Act, Pub. L. No. 74–461, 74th Cong., 2nd Sess. (February 29, 1936).

4   Lambert, "Making of *Gone with the Wind*."
5   See essays in Crank, *New Approaches*; and Pauly, "Hollywood Histories of the Depression."
6   Rifkin, *Beyond Settler Time*, xvii–xviii, 1, 4.
7   Mirzoeff, *White Sight*, 14; and McKittrick, "Plantation Futures," 2.
8   While I composed this line without outside influence, I later learned in the Southern Historical Collection at UNC's Wilson Library that I am not, unfortunately, the first to make this clever association. In the April 4, 1937, *Atlanta Constitution*, an unnamed cartoonist depicts giant dust storms hovering over gullied land where a distressed Uncle Sam is standing on one remaining pillar of soil reminiscent of Providence Canyon's topography. The caption along the top reads "Gone With the Wind—" and the bottom concludes "And Going With the Water!" *Atlanta Constitution*, "Gone."
9   Mitchell, "Interview."
10 See Bennett, *Soil Conservation*.
11 Bennett, *Soil Conservation*, 4.
12 Pauly, "Hollywood Histories of the Depression," 172.
13 Mirzoeff, *White Sight*, 94.
14 McKittrick, "Plantation Futures," 11.
15 Yang, "Sustainability as Plantation Logic," 2.
16 Oxford English Dictionary *Online*, s.v. "erosion (n.) 1a." Oxford University Press, July 2023, https://doi.org/10.1093/OED/4026214323, https://www.oed.com/view/Entry/64074?redirectedFrom=erosion.
17 Lyell, *Second Visit*, v. 2, 28–30.
18 Sutter, *Let Us Now Praise*, 27–30.
19 Lyell, *Second Visit*, v. 2, 35.
20 Lyell, *Second Visit*, v. 2, 35.
21 For an extended analysis of Lyell's rhetoric, see Gould, *Time's Arrow, Time's Cycle*.
22 See T. L. King, *Black Shoals*; and Rifkin, *Fictions of Land and Flesh*.
23 Bennett, *Soil Conservation*, 4.
24 See Sutter, *Let Us Now Praise*.
25 Wilmeth, "Georgia's Great Canyon."
26 Wilmeth, "Georgia's Great Canyon."
27 Sutter, *Let Us Now Praise*, 1–10.
28 Kreyling, *Inventing Southern Literature*.
29 Twelve Southerners, *I'll Take My Stand*.

30   Wilson, *Baptized in Blood*; and Jackson, "Southern Disaster Complex."
31   Wilson, *Baptized in Blood*, 13.
32   Jackson, "Southern Disaster Complex," 559.
33   Watson, "Other Matter of the South."
34   Jackson, "Southern Disaster Complex," 569.
35   For a discussion of the plantationocene, see Haraway, "Anthropocene, Capitalocene, Plantationocene, Chthulucene"; as well as Aikens et al., "South to the Plantationocene."
36   See Clukey, "Plantation Modernity."
37   Yeager, *Dirt and Desire*, 166.
38   Yeager, *Dirt and Desire*, 108.
39   Mitchell, *Gone with the Wind*, 26.
40   Mitchell, *Gone with the Wind*, 29.
41   One might go so far as to point out that Mitchell names her protagonist Scarlett—itself a word indicating "red"—but such an argument would be somewhat complicated by the fact that Mitchell had originally named her protagonist Pansy and was asked to change the name by her press.
42   Beckert, *Empire of Cotton*, 91.
43   Bennett quoted in Cook, "Hugh Hammond Bennett."
44   Mitchell, *Gone with the Wind*, 29.
45   Mitchell, *Gone with the Wind*, 97.
46   Turner, "Georgia State Symbols."
47   Turner, "Georgia State Symbols."
48   Mitchell, *Gone with the Wind*, 41, 47, 90, 106, 150–51, 167, 333, 346, 376, 405, 408, 460, 522, 531.
49   Mitchell, *Gone with the Wind*, 657.
50   Margaret Mitchell, quoted in Gray, foreword, xiv.
51   Bennett, *Soil Conservation*, 880.
52   Mitchell, *Gone with the Wind*, 407.
53   Mitchell, *Gone with the Wind*, 413.
54   Mitchell, *Gone with the Wind*, 413.
55   Nichols, *Theft Is Property!*, 9.
56   Mitchell, *Gone with the Wind*, 463.
57   Mitchell, *Gone with the Wind*, 468. I have omitted Mitchell's racial slur, as it provides no substantive content to the selected quote or my analysis and thus does not bear repeating in my own text.
58   Mitchell, *Gone with the Wind*, 657.

59 Bennett, *Soil Conservation*, 3.

60 In a provocative but ultimately unfollowable lead, Bennett's pocket notebook from his 1933 visit to Georgia includes a note regarding a Mrs. Marsh (Margaret Mitchell's everyday name was Peggy Marsh) who resides in a boardinghouse. If Bennett indeed met Peggy Marsh on his visit to survey soils in the state, this meeting has possibly been lost to time. See Bennett, "Notebook."

61 Goode, *Agrotopias*, 3.

62 Goode, *Agrotopias*, 16.

63 Mitchell, *Gone with the Wind*, 928–29.

64 Mitchell, *Gone with the Wind*, 956.

65 Mitchell, *Gone with the Wind*, 400.

66 Mitchell, *Gone with the Wind*, 63.

67 See Hudson, *Creek Paths and Federal Roads*, for a discussion of infrastructure and fraudulent treaties to gain control of that infrastructure in the Southeast during this time.

68 Mitchell, *Gone with the Wind*, 63–64.

69 Mitchell, *Gone with the Wind*, 64.

70 Mitchell, *Gone with the Wind*, 149.

71 Mitchell, *Gone with the Wind*, 405.

72 Mitchell, *Gone with the Wind*, 395.

73 See Haraway, "Anthropocene, Capitalocene, Plantationocene, Chthulucene"; and Aikens et al., "South to the Plantationocene."

74 Coates, *Trail of Tears*.

75 Sutter, *Let Us Now Praise*, 124.

76 Sutter, *Let Us Now Praise*, 124.

77 Sutter, *Let Us Now Praise*, 16.

78 Sutter, *Let Us Now Praise*, 20.

79 Sutter, *Let Us Now Praise*, 92. See also L. Williams, "Providence Canyon"; Bush, "Canyon Began as a Path"; and T. Ham, "Wasteful, Beautiful Providence Canyons."

80 In a somewhat surprising twist, creationists have also used Providence Canyon to support their belief in a biblical time. As Rebecca Gibson writes in the magazine *Creation*,

> So, it does not take millions of years for huge canyons to form—it just takes the right conditions. If it had not been seen to happen, hardly anyone would have believed it. Erosion after the global Flood would have been especially rapid through the still soft, freshly laid sediments.

> In fact, it has been documented in this magazine that erosion overall is happening so fast that the continents cannot be millions of years old or they would have all eroded away. Providence Canyon beautifully illustrates how the geology of the earth is consistent with the short timescale of the Bible, provided we understand the conditions properly. Gibson, "Canyon Creation."

This confluence of geology, settler agriculture, and creationism surely requires a bit of pseudoscientific gymnastics.

81 Sutter, *Let Us Now Praise*, 2–3.
82 Sutter, *Let Us Now Praise*, 4.
83 Arthur Raper, quoted in Webb, *For the Mud Holds*.
84 Carlsson, catalog description; and Flaherty, *Land*.
85 T. L. King, *Black Shoals*, 4. See Gilroy, *Black Atlantic*.
86 T. L. King, *Black Shoals*, 13.
87 Webb, *For the Mud Holds*.
88 Webb, *For the Mud Holds*.
89 Webb, *For the Mud Holds*.
90 Sutter, *Let Us Now Praise*, 177–81.
91 Flanagan, *Canyons*, 4.
92 Flanagan, *Canyons*, 4.
93 Flanagan, *Canyons*, 5–6.
94 Yang, "Sustainability as Plantation Logic," 3.
95 Sutter, *Let Us Now Praise*, 78.
96 Flanagan, *Canyons*, 10.
97 Flanagan, *Canyons*, 9.
98 Flanagan, *Canyons*, 6.
99 See Allen, *Earthworks Rising*; Howe and Wilson, "Life"; and E. Anderson, "Earthworks."

CHAPTER 5: LITTORAL CELLS AND LITERAL SELLS OF THE ATLANTIC

1 Orrin Pilkey has devoted his career to outlining these issues, with much of it focused on North Carolina. See, for example, Kaufman and Pilkey, *Beaches Are Moving*; and Pilkey and Pilkey, *Sea Level Rise*.
2 See Caison, "Teaching Region."
3 Lipman, *Saltwater Frontier*, 9.
4 Kaufman and Pilkey, *Beaches Are Moving*, 191.
5 See Rountree, *Manteo's World*.

6   Wood, *Markings on Earth*, 21.
7   Wood, *Markings on Earth*, 21.
8   Wood, *Markings on Earth*, 21.
9   Wood, *Markings on Earth*, 22.
10 Wood, *Markings on Earth*, 22.
11 Wood, *Markings on Earth*, 23.
12 Wood, *Markings on Earth*, 24.
13 Wood, *Markings on Earth*, 21.
14 Herbert Verdin, quoted in Verdin, *Return to Yakni Chitto*, 24.
15 Wood, *Markings on Earth*, 34.
16 Wood, *Markings on Earth*, 34.
17 This phenomenon is not unlike the argument that Monique Allewaert forwards about William Bartram's accounts of the Georgia sea islands in *Ariel's Ecology*.
18 Rountree, *Manteo's World*, 7.
19 Lipman, *Saltwater Frontier*, 8.
20 See Zucchino, *Wilmington's Lie*.
21 See Kahrl, *Land Was Ours*.
22 See "New Hanover County."
23 See Kahrl, *Land Was Ours*, 158–59.
24 Kahrl, *Land Was Ours*, 159.
25 *Wilmington News*, quoted in Kahrl, *Land Was Ours*, 159.
26 Kahrl, *Land Was Ours*, 159.
27 Marc Farinella, quoted in Boyd, "Forgotten History."
28 See Boyd, "Forgotten History"; Kahrl, *Land Was Ours*; and Maurer, "White Sand, Shell Island," for varying accounts of the events.
29 Marc Farinella, quoted in Maurer, "White Sand, Shell Island."
30 "To Nelson MacRae."
31 *Wilmington Morning Star*, "Hot Weather."
32 *Wilmington Morning Star*, "Hot Weather."
33 Bache, "Shell Island," 16. I first encountered the story at the Wilson Library Special Collections at UNC Chapel Hill when researching Shell Island as a place. It was also published in Bache's collection *The Value of Kindness* in 1993. Unless otherwise noted, I refer to final the manuscript version at the Wilson Library.
34 Bache, "Shell Island," 16.

35 Bache, "Shell Island," 16.
36 Bache, "Shell Island," 6.
37 Bache, "Shell Island," 3.
38 Bache, "Shell Island," 3.
39 Bache, "Shell Island," 5.
40 Kahrl, *Land Was Ours*, 58.
41 Bache, "Shell Island," 8.
42 Bache, "Shell Island," 8.
43 Bache, "Shell Island," 8.
44 Bache, "Shell Island," 26.
45 Bache, "Shell Island," 26.
46 Bache, "Shell Island," 26.
47 Bache, "Shell Island," 17.
48 Fletcher, "Saving Hatteras Light."
49 William Harris, quoted in Fletcher, "Saving Hatteras Light."
50 See Xia, Kannan, and Castleman, "Ocean Game."
51 See B. C. Brooks, "John Lawson's Indian Town."
52 See Rountree, *Manteo's World*, 123–24.
53 See Dolin, *Brilliant Beacons*.
54 While by all accounts Morton's politics were certainly *far better* than those of his grandfather MacRae (not a high bar to clear), he was still part of a particular North Carolina elite legacy that valued a certain view of issues such as historical preservation and environmental awareness.
55 See Snider, *Helms and Hunt*.
56 Hilliard, "Battle for the Beacon."
57 Hugh Morton, quoted in Hilliard, "Battle for the Beacon."
58 US Geological Survey, USGS Coastal Change Hazards Portal.
59 See Nestor, *Indian Placenames in America*, 80. I am not entirely persuaded by the accuracy of this source, but the translation is provocative.
60 See Young, "Archaeology Saves the Bay"; Strickland et al., *Rappahannock Indigenous Cultural Landscape*; and Hatfield, *Atlantic Virginia*.
61 J. Smith, *Writings*, 335.
62 J. Smith, *Writings*, 336.
63 In some theological accounts of the Limbo of the Fathers, notably that of Albertus Magnus, Limbo might very well be uninhabited because Christ harrowed all the virtuous dead to heaven, not simply those who were part

of a Jewish monotheistic tradition. In most versions, however, so-called pagan figures such as Virgil remain in Limbo. See Vanhoutte, *Limbo Reapplied*, 62.

64 I am speaking specifically here of the Limbo of the Fathers rather than the limbo space for unbaptized infants. Notably, Pope Benedict XVI removed Limbo from Catholic theology altogether in 2007.

65 Vanhoutte, *Limbo Reapplied*, 16.

66 *Oxford English Dictionary*, s.v. "harrow (v.2), sense a," last updated September 2023, https://doi.org/10.1093/OED/5974150879.

67 *Oxford English Dictionary*, s.v. "harrow (v.1), sense 1.a," last updated September 2023, https://doi.org/10.1093/OED/5175935279.

68 *Oxford English Dictionary*, s.v. "harrow (n.1), sense 1.a," last updated September 2023, https://doi.org/10.1093/OED/7603335447.

69 See Handelsman, *World without Soil*, 27, 74–77, 82–83.

70 There is also perhaps a conceptual link here to what Ralph Bauer describes via Carl Schmitt as the state of New World exceptionalism whereby a kind of "global linear thinking" began to divide the earth into temporal and spatial spaces for colonial control. See R. Bauer, *Alchemy of Conquest*, 19–21.

71 Vanhoutte, *Limbo Reapplied*, 19, 27.

72 Vanhoutte, *Limbo Reapplied*, 74.

73 Vanhoutte, *Limbo Reapplied*, 31.

74 Dante, *Divine Comedy*, ll. 42, 98.

75 Vanhoutte, *Limbo Reapplied*, 93.

76 Vanhoutte, *Limbo Reapplied*, 93. John Milton would revise Limbo in *Paradise Lost* some sixty years following Smith's account.

77 Vanhoutte, *Limbo Reapplied*, 109.

78 Vanhoutte, *Limbo Reapplied*, 109.

79 Swift, *Chesapeake Requiem*, 6.

80 Residents quoted in Swift, *Chesapeake Requiem*, 19.

81 Swift, *Chesapeake Requiem*, 67.

82 Swift, *Chesapeake Requiem*, 19.

83 Carol Moore, quoted in Swift, *Chesapeake Requiem*, 19.

84 Swift, *Chesapeake Requiem*, 22.

85 Swift, *Chesapeake Requiem*, 216.

86 Pilkey and Pilkey, *Sea Level Rise*, 4.

87 Kaufman and Pilkey, *Beaches Are Moving*, 191, 195.

88 Swift, *Chesapeake Requiem*, 218.

89   Swift, *Chesapeake Requiem*, 218.
90   Swift, *Chesapeake Requiem*, 321.
91   Residents quoted in Swift, *Chesapeake Requiem*, 367.
92   Mayor quoted in Swift, *Chesapeake Requiem*, 370.
93   W. Brown, *Walled States, Waning Sovereignty*, 24.
94   Wood, *Markings on Earth*, 71.
95   Wood, *Markings on Earth*, 71.
96   Wood, *Markings on Earth*, 71.
97   See Nixon, *Slow Violence*; Howe, *Choctalking*; and Whyte, "Time as Kinship."

CONCLUSION: WHAT WE TALK ABOUT WHEN WE TALK ABOUT EROSION

1   See Export, "*Eros/ion*."
2   Nichols, *Theft Is Property!*, 13.
3   Miranda, *Bad Indians*, 202.
4   Lyons, *Vital Decomposition*, 5.
5   Miranda, *Bad Indians*, 208.
6   Riggs, "Vine Theater," 282.
7   Blaser and de la Cadena, "Pluriverse," 15.
8   Blaser and de la Cadena, "Pluriverse," 16.
9   *Daily Mail*, "Natives Being Taught."
10   *New York Times*, "Apartheid Hurts the Environment."
11   Bennett, *Soil Conservation*, 922.
12   Baltz, *Lewis Baltz*, 297–99.
13   Baltz, *Lewis Baltz*, 299.
14   Baltz, *Lewis Baltz*, 297.
15   Longhin, "Barene and Petrochemicals," 52.
16   De Capitani, introduction, 15.
17   Pianigiani and Bubola, "Italy's Government to Ban."
18   De Capitani, introduction, 14–15.
19   De Capitani, introduction, 15.
20   Bennett, *Soil Conservation*, 25. Notably, Bennett quotes Weitz's claim blaming the erosion of the Palestinian landscape on the expulsion of Jewish people from the land. This account may be more politically than ecologically informed. See the documentary *Blue Box* (2021), by Michal Weits, Weitz's

great-granddaughter, who explores the way that land acquisition, afforestation, and the erasure of Palestinian communities often worked hand in hand.

21. Bennett, *Soil Conservation*, 905.
22. International Committee of the Red Cross, "2021 Olive Harvest Season."
23. As reported by Reuters, "In Central Gaza Strip, Radwan al-Shantaf, from Al-Zahra city municipality, said the authorities had used large quantities of the rubble of houses destroyed in the May 2021 Israeli bombardment to barricade beaches," demonstrating a tragic irony almost beyond comprehension. Al-Mughrabi, "Palestinians Strive."
24. Said, *After the Last Sky*, 28.
25. In addition to Weits's documentary, see Tal, *Pollution in a Promised Land*, for a further discussion of Weitz's work.
26. Foucault, *Birth of Biopolitics*, 78.
27. See Kertész and Gergely, "Gully Erosion in Hungary."
28. Bennett, *Soil Conservation*, 48–49.
29. Bennett, *Soil Conservation*, 50.
30. Bennett, *Soil Conservation*, 54.
31. United Nations, "Soil Erosion Must Be Stopped."
32. See Handelsman, *World without Soil*; as well as Tato and Hurni, *Soil Conservation*.
33. For example, see the online event hosted by the *Atlantic*, Disinformation and the Erosion of Democracy; and Hsu, "End of White America?"
34. Although these comments happened in our unrecorded background conversations, for a version of my conversation with this geologist, see Poole, "Providence and the Anthropocene."

Bibliography

Adams, Robert. *The New West: Landscapes Along the Colorado Front Range*. Denver: Colorado Associated University Press, 1974.
Agee, James, and Walker Evans. *Let Us Now Praise Famous Men*. Boston: Houghton Mifflin, 1941.
Aikens, Natalie, Amy Clukey, Amy K. King, and Isadora Wagner. "South to the Plantationocene." *asap/J* (blog), October 17, 2019. https://asapjournal.com/south-to-the-plantationocene-natalie-aikens-amy-clukey-amy-k-king-and-isadora-wagner/.
Akins, Damon B., and William J. Bauer Jr. *We Are the Land: A History of Native California*. Berkeley: University of California Press, 2021.
Allen, Chadwick. *Earthworks Rising: Mound Building in Native Literature and Arts*. Minneapolis: University of Minnesota Press, 2022.
Allen, Chadwick. "When a Mound Isn't a Mound, but Is: Figuring (and Fissuring) Earthworks in Lynn Riggs's *The Cherokee Night*." In *The Routledge Handbook of North American Indigenous Modernisms*, edited by Kirby Brown, Stephen Ross, and Alana Sayers, 17–28. London: Routledge, 2022.
Allewaert, Monique. *Ariel's Ecology: Plantations, Personhood, and Colonialism in the American Tropics*. Minneapolis: University of Minnesota Press, 2013.
Al-Mughrabi, Nidal. "Palestinians Strive to Stop Gaza Shore Erosion with Concrete and Rubble." Reuters, August 1, 2011. https://www.reuters.com/world/middle-east/palestinians-strive-stop-gaza-shore-erosion-with-concrete-rubble-2022-07-27/.
Alvarez, Susan H., and David W. Peri. "Acorns: The Staff of Life." *News from Native California* 1, no. 4 (Summer 1987): 10–14.

Anderson, Brett. "Louisiana Loses Its Boot." *Matter* (blog), March 11, 2015. https://medium.com/matter/louisiana-loses-its-boot-b55b3bd52d1e.

Anderson, Eric Gary. "Earthworks and Contemporary Indigenous American Literature: Foundations and Futures." *Native South* 9 (2016): 1–26.

Anderson, Eric Gary, and Melanie Benson Taylor. "Letting the Other Story Go: The Native South in and beyond the Anthropocene." *Native South* 12 (2019): 74–98.

Anderson, M. Kat. *Tending the Wild: Native American Knowledge and the Management of California's Natural Resources*. Berkeley: University of California Press, 2013.

Atlanta Constitution. "Gone with the Wind—and Going with the Water!" Uncredited cartoon, April 4, 1937. Collection no. 03527, folder 46. Hugh H. Bennett Papers, 1906–1966, 1984. Southern Historical Collection. University of North Carolina at Chapel Hill.

Audubon, John James. "A Flood." In Forkner, *Audubon on Louisiana*, 240–44.

Audubon, John James. "Improvements in the Navigation of the Mississippi." In Forkner, *Audubon on Louisiana*, 235–39.

Audubon, John James. *Snowy Heron, or White Egret*. Birds of America, Audubon, plate 242. Accessed November 25, 2014. https://www.audubon.org/birds-of-america/snowy-heron-or-white-egret.

Audubon, John James. "The Squatters of the Mississippi." In Forkner, *Audubon on Louisiana*, 245–49.

Audubon, John James. "To Reverend John Bachman February 24, 1837." In Forkner, *Audubon on Louisiana*, 285–87.

Bache, Ellyn. "Shell Island." Collection no. 04980, Series 1, Box 3. Ellyn Bache Papers, 1974–1998. Southern Historical Collection. University of North Carolina at Chapel Hill.

Bache, Ellyn. "Shell Island." In *The Value of Kindness: Stories*, 72–90. Kansas City, MO: Helicon Nine Editions, 1993.

Balaschak, Chris. "New Worlds: Lewis Baltz and a Geography of Aesthetic Decisions." In Pfeifer, *Reconsidering the New Industrial Parks*, 25–55.

Baldy, Cutcha Risling. *We Are Dancing for You: Native Feminisms and the Revitalization of Women's Coming-of-Age Ceremonies*. Seattle: University of Washington Press, 2018.

Baltz, Lewis. *Candlestick Point*. Göttingen, Germany: Steidl, 2011.

Baltz, Lewis. "Last Interview of Lewis Baltz with Jeff Rian." Interview by Jeff Rian. *Eye of Photography Magazine*, January 1, 2015. https://loeildelaphotographie.com/en/last-interview-of-lewis-baltz-with-jeff-rian/.

Baltz, Lewis. *Lewis Baltz*. Göttingen, Germany: Steidl, 2017.

Baltz, Lewis. *Lewis Baltz: Texts*. Göttingen, Germany: Steidl, 2012.

Baltz, Lewis. "Untitled." In Pfeifer, *Reconsidering the New Industrial Parks*, 13–16.

Barbrook, Richard, and Andy Cameron. "The Californian Ideology." *Science as Culture* 6, no. 1 (January 1996): 44–72.

Bauer, Ralph. *The Alchemy of Conquest: Science, Religion, and the Secrets of the New World*. Charlottesville, VA: University of Virginia Press, 2019.

Bauer, William J., Jr. *California through Native Eyes: Reclaiming History*. Seattle: University of Washington Press, 2016.

Beckert, Sven. *Empire of Cotton: A Global History*. New York: Vintage, 2015.

Bennett, Hugh Hammond. "Notebook." Collection no. 03527, Box 17. Hugh H. Bennett Papers, 1906–1966, 1984. Southern Historical Collection. University of North Carolina at Chapel Hill.

Bennett, Hugh Hammond. *Soil Conservation*. New York: McGraw-Hill, 1939.

Bernard, Shane K. *The Cajuns: Americanization of a People*. Jackson: University Press of Mississippi, 2003.

Blackhawk, Ned. *Violence over the Land: Indians and Empires in the Early American West*. Cambridge, MA: Harvard University Press, 2008.

Blaikie, Piers M. "The State of Land Management Policy, Present and Future." In *Soil Conservation for Survival*, edited by Kebede Tato and Hans Hurni, 29–44. Ankeny, IA: Soil and Water Conservation Society, 1992.

Blaser, Mario, and Marisol de la Cadena. "Pluriverse: Proposals for a World of Many Worlds." In *A World of Many Worlds*, edited by Mario Blaser and Marisol de la Cadena, 1–23. Durham, NC: Duke University Press, 2018.

Boyd, Rachel. "A Forgotten History: The Puzzle of Shell Island." Spectrum News 1, October 4, 2021. https://spectrumlocalnews.com/nc/charlotte/news/2021/10/04/shell-island-resort.

Braunlich, Phyllis Cole. *Haunted by Home: The Life and Letters of Lynn Riggs*. Norman: University of Oklahoma Press, 1988.

Brooks, Baylus C. "John Lawson's Indian Town on Hatteras Island, North Carolina." *North Carolina Historical Review* 91, no.2 (April 2014): 171–207.

Brooks, Lisa. "The Primacy of the Present, the Primacy of Place: Navigating the Spiral of History in the Digital World." PMLA 127, no. 2 (March 2012): 308–16.

Brower, Kenneth. "Ansel Adams at 100." *Atlantic*, July 1, 2002. https://www.theatlantic.com/magazine/archive/2002/07/ansel-adams-at-100/302533/.

Brown, Kirby. *Stoking the Fire: Nationhood in Cherokee Writing, 1907–1970*. Norman: University of Oklahoma Press, 2018.

Brown, Wendy. *Walled States, Waning Sovereignty*. Cambridge, MA: Zone Books, 2017.

Bruyneel, Kevin. *Settler Memory: The Disavowal of Indigeneity and the Politics of Race in the United States*. Chapel Hill: University of North Carolina Press, 2021.

Burley, David M. *Losing Ground: Identity and Land Loss in Coastal Louisiana*. Jackson: University Press of Mississippi, 2010.

Burnett, Katharine A. "Moving toward a 'No South': George Washington Cable's Global Vision in *The Grandissimes*." *Southern Literary Journal* 45, no. 1 (Fall 2012): 21–38.

Burroughs, William S. *Naked Lunch: The Restored Text*. Edited by James Grauerholz and Barry Miles. New York: Grove, 2013.

Burton, Jeffrey. *Indian Territory and the United States, 1866–1906: Courts, Government, and the Movement for Oklahoma Statehood*. Norman: University of Oklahoma Press, 1995.

Bush, Luthene. "Canyon Began as a Path." *Albany Georgia Herald*, August 5, 1992.

Butler, Octavia E. *Parable of the Sower*. New York: Warner, 1993.

Butler, Octavia E. *Parable of the Talents*. New York: Grand Central, 1998.

Byrnes, Delia. "Plantation Pasts and the Petrochemical Present: Energy Culture, the Gulf Coast, and *Petrochemical America*." In *Ecocriticism and the Future of Southern Studies*, edited by Zackary Vernon, 38–52. Baton Rouge: Louisiana State University Press, 2019.

Cable, George Washington. *Old Creole Days*. New York: Charles Scribner's Sons, 1883.

Caison, Gina. *Red States: Indigeneity, Settler Colonialism, and Southern Studies*. Athens: University of Georgia Press, 2018.

Caison, Gina. "Removal." In *American Literature in Transition*, vol. 2, *1820–1860*, edited by Justine S. Murison, 104–18. Cambridge: Cambridge University Press, 2022.

Caison, Gina. "Teaching Region: Native American Studies." In *Appalachia in Regional Context: Place Matters*, edited by Dwight B. Billings and Ann E. Kingsolver, 236–41. Louisville: University Press of Kentucky, 2018.

Caldwell, Erskine. *God's Little Acre*. New York: Viking, 1933.

Caldwell, Erskine. *Tobacco Road*. New York: Scribner's, 1932.

Caldwell, Erskine, and Margaret Bourke-White. *You Have Seen Their Faces*. New York: Viking, 1937.

Canavan, Gerry. *Octavia E. Butler*. Urbana: University of Illinois Press, 2016.

Candlesticksf.com, "Candlestick." San Francisco: FivePoint, accessed April 2, 2024. https://www.candlesticksf.com.

Candlesticksf.com, "Explore Candlestick." San Francisco: FivePoint, accessed April 2, 2024. https://www.candlesticksf.com/explore-candlestick/.

Carlsson, Chris. Catalog description for *The Land*, by Robert Flaherty. Internet Archive, uploaded May 28, 2015. https://archive.org/details/TheLand_201505.

Castillo, Susan. "Stones in the Quarry: George Cable's Strange True Stories of Louisiana." *Southern Literary Journal* 31, no. 2 (Spring 1999): 19–34.

Champagne, Duane. *Social Order and Political Change: Constitutional Governments among the Cherokee, the Choctaw, the Chickasaw, and the Creek*. Stanford, CA: Stanford University Press, 1992.

Chang, David. *The Color of the Land: Race, Nation, and the Politics of Land Ownership in Oklahoma*. Chapel Hill: University of North Carolina Press, 2010.

Chicago Daily News. "Soil and Sanctuary." July 30, 1940. Collection no. 03527, folder 54. Hugh H. Bennett Papers, 1906–1966, 1984. Southern Historical Collection. University of North Carolina at Chapel Hill.

Chiotakis, Steve, and Christian Bordal. "Acorns Helped LA's Indigenous People through Winter." KCRW, December 22, 2021. https://www.kcrw.com/news/shows/greater-la/native-americans-la-acorns-forage-indigenous.

Clayton, Susan, Christie Manning, Kirra Krygsman, and Meighen Speiser. *Mental Health and Our Changing Climate: Impacts, Implications, and Guidance*. Washington, DC: American Psychological Association and ecoAmerica, 2017.

Climate Central. "Surging Seas: Risk Zone Map." Accessed March 5, 2021. https://ss2.climatecentral.org.

Clukey, Amy. "Plantation Modernity: *Gone with the Wind* and Irish-Southern Culture." *American Literature* 85, no. 3 (September 2013): 505–30.

Coates, Julia. *Trail of Tears*. Santa Barbara, CA: Greenwood, 2014.

Cook, Maurice. "Hugh Hammond Bennett: The Father of Soil Conservation." Department of Soil Science, College of Agriculture and Life Sciences, North Carolina State University, 2005. http://www.soil.ncsu.edu/about/century/hugh.html.

Cox, James H. *The Red Land to the South: American Indian Writers and Indigenous Mexico*. Minneapolis: University of Minnesota Press, 2012.

Cox, James H., and Alexander Pettit. "Fugitive Indigeneity in Paul Green's *The Last of the Lowries* and Lynn Riggs's *The Cherokee Night*." In *The Routledge Handbook of North American Indigenous Modernisms*, edited by Kirby Brown, Stephen Ross, and Alana Sayers, 182–90. London: Routledge, 2022.

Cox, Mary E. *Hunger in War and Peace: Women and Children in Germany, 1914–1924*. Oxford: Oxford University Press, 2019.

Crank, James A., ed. *New Approaches to "Gone with the Wind."* Baton Rouge: Louisiana State University Press, 2015.

Daily Mail. "Natives Being Taught to 'Save Our Soil.'" November [day illegible], 1944. Collection no. 03527, Box 17. Hugh H. Bennett Papers, 1906–1966, 1984. Southern Historical Collection. University of North Carolina at Chapel Hill.

Daley, Jason. "First Humans Entered the Americas along the Coast, Not through the Ice." *Smithsonian Magazine*, August 11, 2016. https://www.smithsonianmag.com/smart-news/humans-colonized-americas-along-coast-not-through-ice-180960103/.

Dante Alighieri. *The Divine Comedy*. Vol. 1, *Inferno*. Translated by Mark Musa. New York: Penguin Classics, 1971.

Dardar, T. Mayheart. "Global Climate Change: A Houma Perspective." In *Return to Yakni Chitto: Houma Migrations*, by Monique Verdin, 1–3. New Orleans: University of New Orleans Press, 2019.

Davenport, Coral, and Campbell Robertson. "Resettling the First American 'Climate Refugees.'" *New York Times*, May 3, 2016. https://www.nytimes.com/2016/05/03/us/resettling-the-first-american-climate-refugees.html.

Dean, Cornelia. *Against the Tide: The Battle for America's Beaches*. New York: Columbia University Press, 1999.

De Capitani, Lucio. Introduction to *Venice and the Anthropocene: An Ecocritical Guide*, edited by Cristina Baldacci, Shaul Bassi, Lucio De Capitani, and Pietro Daniel Omodeo, 9–24. Venice: Wetlands Books, 2022.

DeLoughrey, Elizabeth M. *Allegories of the Anthropocene*. Durham, NC: Duke University Press, 2019.

Denson, Andrew. *Demanding the Cherokee Nation: Indian Autonomy and American Culture, 1830–1900*. Lincoln: University of Nebraska Press, 2004.

Diehl, Phil. "Erosion Repairs Could Halt Passenger Rail Service to San Diego through Year's End." *Los Angeles Times*, October 19, 2022. https://www.latimes.com/california/story/2022-10-19/san-clemente-rail-repairs-will-take-longer-than-expected.

Disinformation and the Erosion of Democracy. Event, hosted by the University of Chicago Institute of Politics and the *Atlantic*, April 6–8, 2022. https://www.theatlantic.com/live/disinformation-democracy-uchicago-conference-2022/.

Dolin, Eric Jay. *Brilliant Beacons: A History of the American Lighthouse*. New York: Liveright, 2017.

d'Oney, Daniel J. *A Kingdom of Water: Adaptation and Survival in the Houma Nation*. Lincoln: University of Nebraska Press, 2020.

Dowson, Ernest. "Non Sum Qualis Eram Bonae Sub Regno Cynarae." 1894. In *Poems of Ernest Dowson*, edited by Mark Longaker, 58. Berlin: De Gruyter, 1962.

Dunn, Geoffrey. "Photographic License." *New Times: San Luis Obispo*, January 17, 2002. Published initially in *Santa Clara Metro*, January 19–25, 1995. https://www.newtimesslo.com/archive/2003-12-03/archives/cov_stories_2002/cov_01172002.html.

Egan, Timothy. *The Worst Hard Time: The Untold Story of Those Who Survived the Great American Dust Bowl*. Boston: Houghton Mifflin, 2006.

Ellwood, Brooks B., Sophie Warny, Rebecca A. Hackworth, Suzanne H. Ellwood, Jonathan H. Tomkin, Samuel J. Bentley, Dewitt H. Braud, and Geoffrey C. Clayton. "The LSU Campus Mounds, with Construction Beginning at ~11,000 BP, Are the Oldest Known Extant Man-Made Structures in the Americas." *American Journal of Science* 322, no. 6 (June 2022): 795–827.

Ewen, Alexander. "The Death of the Bering Strait Theory." ICT News, August 12, 2016; updated September 13, 2018. https://indiancountrytoday.com/archive/the-death-of-the-bering-strait-theory.

Export, Valie. "*Eros/ion*." Media Art Net, accessed January 2, 2023. http://www.medienkunstnetz.de/works/eros-ion/images/2/.

Faulkner, William. *Absalom! Absalom!* 1936. New York: Vintage, 1990.

Faulkner, William. *The Wild Palms (If I Forget Thee, Jerusalem)*. 1939. New York: Vintage, 1995.

Fine Arts Museums of San Francisco. "*Ansel Adams in Our Time*." https://www.famsf.org/exhibitions/ansel-adams-in-our-time.

Flaherty, Robert, dir. *The Land*. Documentary. Washington, DC: US Department of Agriculture, 1942. Documentary. Internet Archive, uploaded May 28, 2015. Accessed July 11, 2023. https://archive.org/details/TheLand_201505.

Flanagan, Thomas Jefferson. *The Canyons at Providence (The Lay of the Clay Minstrel)*. Atlanta: Morris Brown College Press, 1940. Collection no. 990565409. Atlanta University Center Special Collections.

Fleming, Victor, dir. *Gone with the Wind*. Los Angeles: Metro-Goldwyn-Mayer, 1939.

Fletcher, Stephen. "Saving Hatteras Light." A View to Hugh, University Libraries, University of North Carolina at Chapel Hill, September 8, 2021. https://blogs.lib.unc.edu/morton/2021/09/08/saving-hatteras-light/.

Forkner, Ben, ed. *Audubon on Louisiana: Selected Writings of John James Audubon*. Baton Rouge: Louisiana State University Press, 2018.

Foucault, Michel. *The Birth of Biopolitics: Lectures at the College de France, 1978–1979*. Translated by Graham Burchell. New York: Palgrave Macmillan, 2008.

Ganzel, Bill. *Dust Bowl Descent*. Lincoln: University of Nebraska Press, 1984.

Gaul, Theresa Strouth. *To Marry an Indian: The Marriage of Harriett Gold and Elias Boudinot in Letters, 1823–1839*. Chapel Hill: University of North Carolina Press, 2005.

Gentry, Curt. *The Last Days of the Late, Great State of California*. New York: G. P. Putnam's Sons, 1968.

Gerstle, Gary. *American Crucible: Race and Nation in the Twentieth Century*. Princeton, NJ: Princeton University Press, 2001.

Ghosh, Amitav. *The Great Derangement: Climate Change and the Unthinkable*. Chicago: University of Chicago Press, 2016.

Gibson, Rebecca. "Canyon Creation." *Creation* 4, no. 22 (September 2000). https://creation.com/canyon-creation.

Gilroy, Paul. *The Black Atlantic: Modernity and Double Consciousness*. London: Vintage, 1993.

Goeman, Mishuana R. "Ongoing Storms and Struggles: Gendered Violence and Resource Exploitation." In *Critically Sovereign: Indigenous Gender, Sexuality, and Feminist Studies*, edited by Joanne Barker, 99–126. Durham, NC: Duke University Press, 2018.

Goggans, Jan. *California on the Breadlines: Dorothea Lange, Paul Taylor, and the Making of a New Deal Narrative*. Berkeley: University of California Press, 2010.

Goldsmith, Oliver. *A History of the Earth and Animated Nature*. 8 vols. London: J. Nourse, 1774.

Goode, Abby L. *Agrotopias: An American Literary History of Sustainability*. Chapel Hill: University of North Carolina Press, 2022.

Goodman, Audrey. *A Planetary Lens: The Photo-Poetics of Western Women's Writing*. Lincoln: University of Nebraska Press, 2021.

Gordon, Linda. *Dorothea Lange: A Life beyond the Limits*. New York: W. W. Norton, 2009.

Gould, Stephen Jay. *Time's Arrow, Time's Cycle: Myth and Metaphor in the Discovery of Geological Time*. Cambridge, MA: Harvard University Press, 1987.

Gray, Richard. Foreword to *South to a New Place: Region, Literature and Culture*, edited by Suzanne W. Jones and Sharon Monteith, xiii–xxiii. Baton Rouge: Louisiana State University Press, 2022.

Greeson, Jennifer Rae. *Our South: Geographic Fantasy and the Rise of National Literature*. Cambridge, MA: Harvard University Press, 2010.

Gregg, Sarah M. "Imagining Opportunity: The 1909 Enlarged Homestead Act and the Promise of the Public Domain." *Western Historical Quarterly* 50, no. 3 (Autumn 2019): 257–79.

Griswold, Jerry. "There's No Place but Home: *The Wizard of Oz*." *Antioch Review* 45, no. 4 (Autumn 1987): 462–75.

Gritz, Jennie Rothenberg. "Behind 'Oklahoma!' Lies the Remarkable Story of a Gay Cherokee Playwright." *Smithsonian Magazine*, March 30, 2023. https://www.smithsonianmag.com/history/behind-oklahoma-lies-remarkable-story-gay-cherokee-playwright-lynn-riggs-180981895/.

Guthrie, Woody. "This Land Is Your Land." Mt. Kisco, NY: Woody Guthrie Publications and New York: TRO-Ludlow Music, 1956. https://www.woodyguthrie.org/Lyrics/This_Land.htm.

Ham, Oleta Kay Sprague, and Roger Sprague Sr. *Migrant Mother: The Untold Story*. Mustang, OK: Tate, 2013.

Ham, Tom. "Wasteful, Beautiful Providence Canyons: Where Cows Going to Water Made Georgia a Wonderland." *Atlanta Journal*, November 25, 1941. Collection no. 03527, folder 46. Hugh H. Bennett Papers, 1906–1966, 1984. Southern Historical Collection. University of North Carolina at Chapel Hill.

Hämäläinen, Pekka. *The Comanche Empire*. New Haven, CT: Yale University Press, 2009.

Han, John J. "A Defense of Steinbeck's Intercalary Chapters in *The Grapes of Wrath*." In *"The Grapes of Wrath": A Re-consideration*, edited by Michael J. Meyer, 603–17. Leiden: Brill, 2009.

Handelsman, Jo. *A World without Soil: The Past, Present, and Precarious Future of the Earth beneath Our Feet*. New Haven, CT: Yale University Press, 2022.

Haraway, Donna. "Anthropocene, Capitalocene, Plantationocene, Chthulucene: Making Kin." *Environmental Humanities* 1, no. 6 (May 2015): 159–65.

Harkin, Michael E., and David Rich Lewis, eds. *Native Americans and the Environment: Perspectives on the Ecological Indian*. Lincoln: University of Nebraska Press, 2007.

Hatfield, April Lee. *Atlantic Virginia: Intercolonial Relations in the Seventeenth Century*. Philadelphia: University of Pennsylvania Press, 2004.

Haymes, Stephen Nathan. "An Africana Studies Critique of Environmental Ethics." In *Racial Ecologies*, edited by Leilani Nishime and Kim D. Hester Williams, 34–49. Seattle: University of Washington Press, 2018.

Hedgpeth, Joel. "John Steinbeck: Late-Blooming Environmentalist." In *Steinbeck and the Environment: Interdisciplinary Approaches*, edited by Susan F. Beegel, Susan Shillinglaw, and Wesley N. Tiffany Jr., 293–309. Tuscaloosa: University of Alabama Press, 1997.

Heise, Ursula K. *Imagining Extinction: The Cultural Meanings of Endangered Species*. Chicago: University of Chicago Press, 2016.

Hemmerling, Scott A. *A Louisiana Coastal Atlas: Resources, Economies, and Demographics*. Baton Rouge: Louisiana State University Press, 2017.

Herzog, William R., II. *Parables as Subversive Speech: Jesus as Pedagogue of the Oppressed*. Louisville: Westminster John Knox, 1994.

Hilliard, Jack. "Battle for the Beacon." A View to Hugh, University Libraries, University of North Carolina at Chapel Hill, July 9, 2009. https://blogs.lib.unc.edu/morton/2009/07/09/battle-for-the-beacon/.

Holleman, Hannah. *Dust Bowls of Empire: Imperialism, Environmental Politics, and the Injustice of "Green" Capitalism.* New Haven, CT: Yale University Press, 2018.

Hong, Sharon Linezo, dir. *My Louisiana Love.* Vison Maker Media, 2012. Documentary.

Howe, LeAnne. *Choctalking on Other Realities.* San Francisco: Aunt Lute Books, 2013.

Howe, LeAnne, and Jim Wilson. "Life in a 21st Century Mound City." In *The World of Indigenous North America*, edited by Robert Warrior, 3–26. New York: Routledge, 2014.

Hsu, Hua. "The End of White America?" *Atlantic*, January 1, 2009. https://www.theatlantic.com/magazine/archive/2009/01/the-end-of-white-america/307208/.

Hudson, Angela Pulley. *Creek Paths and Federal Roads: Indians, Settlers, and Slaves and the Making of the American South.* Chapel Hill: University of North Carolina Press, 2010.

International Committee of the Red Cross. "2021 Olive Harvest Season in the West Bank amidst a Triple Challenge." News release, October 12, 2021. https://www.icrc.org/en/document/2021-olive-harvest-season-west-bank-amidst-triple-challenge.

Irmscher, Christoph. "Controversies Remind Us of How Complex John James Audubon Always Was." Library of America, September 24, 2022. https://www.loa.org/news-and-views/1979-christoph-irmscher-controversies-remind-us-of-how-complex-john-james-audubon-always-was.

Isle de Jean Charles Resettlement Program. "About the Isle de Jean Charles Resettlement." Isle de Jean Charles Resettlement, July 25, 2015. https://isledejeancharles.la.gov/about-isle-de-jean-charles-resettlement.

Jackson, Robert. "The Southern Disaster Complex." *Mississippi Quarterly* 63, nos. 3–4 (Summer and Fall 2010): 555–70.

Jefferson, Thomas. *Writings.* Edited by Merrill D. Peterson. New York: Library of America, 1984.

Jiménez, Jesus. "Audubon Society Keeps Name despite Slavery Ties, Dividing Birders." *New York Times*, March 15, 2023. https://www.nytimes.com/2023/03/15/science/audubon-society-name-change.html.

Justice, Daniel Heath. "Narrated Nationhood and Imagined Belonging: Fanciful Family Stories and Kinship Legacies of Allotment." In *Allotment Stories: Indigenous Land Relations under Settler Siege*, edited by Daniel Heath Justice and Jean M. O'Brien, 17–34. Minneapolis: University of Minnesota Press, 2022.

Justice, Daniel Heath. *Our Fire Survives the Storm: A Cherokee Literary History.* Minneapolis: University of Minnesota Press, 2006.

Justice, Daniel Heath, and Jean M. O'Brien, eds. *Allotment Stories: Indigenous Land Relations under Settler Siege.* Minneapolis: University of Minnesota Press, 2022.

Kahrl, Andrew W. *The Land Was Ours: How Black Beaches Became White Wealth in the Coastal South.* Chapel Hill: University of North Carolina Press, 2016.

Kaufman, Wallace, and Orrin H. Pilkey Jr. *The Beaches Are Moving: The Drowning of America's Shoreline.* Durham, NC: Duke University Press, 1983.

Kertész, Ádám, and Jakab Gergely. "Gully Erosion in Hungary, Review and Case Study." In "The 2nd International Geography Symposium-Mediterranean Environment 2010," edited by Recep Efe and Munir Ozturk. Special issue, *Procedia—Social and Behavioral Sciences* 19 (January 2011): 693–701.

King, John. "Rising Reality: Defying the Tides." *San Francisco Chronicle*, September 2016. https://projects.sfchronicle.com/2016/sea-level-rise/part3/.

King, Thomas. *The Truth about Stories: A Native Narrative*. Minneapolis: University of Minnesota Press, 2008.

King, Tiffany Lethabo. *The Black Shoals: Offshore Formations of Black and Indigenous Studies*. Durham, NC: Duke University Press, 2019.

Klopotek, Brian. *Recognition Odysseys: Indigeneity, Race, and Federal Tribal Recognition Policy in Three Louisiana Indian Communities*. Durham, NC: Duke University Press, 2011.

Kornecki, Ted S., and James L. Fouss. "Sugarcane Residue Management Effects in Reducing Soil Erosion from Quarter-Drains in Southern Louisiana." *Applied Engineering in Agriculture* 27, no. 4 (2011): 597–603.

Kosofsky, L. J., and Farouk El-Baz. *The Moon as Viewed by Lunar Orbiter*. Washington, DC: Office of Technology Utilization National Aeronautics and Space Administration, 1970.

Krech, Shepard, III. *The Ecological Indian: Myth and History*. New York: W. W. Norton, 2000.

Kreyling, Michael. *Inventing Southern Literature*. Jackson: University Press of Mississippi, 1998.

Lambert, Gavin. "The Making of *Gone with the Wind*." *Atlantic*, February 1973. https://www.theatlantic.com/magazine/archive/1973/02/the-making-of-gone-with-the-wind-part-i/306455/.

Lange, Dorothea, "Oral History Interview." Interview by Richard K. Doud. Washington, DC: Archives of American Art, Smithsonian Institution, 1964.

Lange, Dorothea, and Paul Taylor. *An American Exodus: A Record of Human Erosion*. New York: Reynal and Hitchcock, 1939.

La Paperson. "A Ghetto Land Pedagogy: An Antidote for Settler Environmentalism." *Environmental Education Research* 20, no. 1 (February 2014): 15–130.

Leeds, Stacey, and Lonnie Beard. "A Wealth of Sovereign Choices: Tax Implications of *McGirt v. Oklahoma* and the Promise of Tribal Economic Development." *Tulsa Law Review* 56 no. 3 (Spring 2021). https://digitalcommons.law.utulsa.edu/tlr/vol56/iss3/9.

Liboiron, Max. *Pollution Is Colonialism*. Durham, NC: Duke University Press, 2021.

Lightweis-Goff, Jennie. "Louisiana Lost and Found." In *Remediating Region: New Media and the U.S. South*, edited by Gina Caison, Lisa Hinrichsen, and Stephanie Rountree, 162–79. Baton Rouge: Louisiana State University Press, 2021.

Lindsay, Brendan C. *Murder State: California's Native American Genocide, 1846–1873*. Lincoln: University of Nebraska Press, 2015.

Lipman, Andrew. *The Saltwater Frontier: Indians and the Contest for the American Coast*. New Haven, CT: Yale University Press, 2015.

Lisle, Thomas E. "The Eel River, Northwestern California: High Sediment Yields from a Dynamic Landscape." In *The Geology of North America*, vols. 0–1, *Surface Water Hydrology*, edited by M. G. Wolman and H. C. Riggs, 311–14. Boulder, CO: Geological Society of America, 1990. https://www.fs.usda.gov/research/treesearch/7853.

Loginova, Olga, and Zak Cassel. "Leaving the Island: The Messy, Contentious Reality of Climate Relocation." Center for Public Integrity, August 17, 2022. https://publicintegrity.org/environment/harms-way/leaving-isle-de-jean-charles-climate-relocation/.

Longhin, Elena. "Barene and Petrochemicals: The Landscapes of Porto Marghera." In *Venice and the Anthropocene: An Ecocritical Guide*, edited by Cristina Baldacci, Shaul Bassi, Lucio De Capitani, and Pietro Daniel Omodeo, 51–53. Venice: Wetlands Books, 2022.

Lookingbill, Brad D. *Dust Bowl, USA: Depression America and the Ecological Imagination, 1929–1941*. Athens: Ohio University Press, 2001.

Lopinot, N. H., and W. I. Woods. "Wood Overexploitation and the Collapse of Cahokia." In *Foraging and Farming in the Eastern Woodlands*, edited by C. M. Scarry, 206–31. Gainesville: University Press of Florida, 1993.

Louisiana Coastal Protection and Restoration Authority. "A Changing Landscape." Accessed June 13, 2020. https://coastal.la.gov/whats-at-stake/a-changing-landscape/.

Lyell, Charles. *Principles of Geology*. 3 vols. London: John Murray, 1830–33.

Lyell, Charles. *A Second Visit to the United States of North America*. New York: Harper and Brothers, 1849.

Lyons, Kristina M. *Vital Decomposition: Soil Practitioners and Life Politics*. Durham, NC: Duke University Press, 2020.

MacLeish, Archibald. *Land of the Free*. New York: Harcourt Brace, 1938.

Madley, Benjamin. *An American Genocide: The United States and the California Indian Catastrophe, 1846–1873*. New Haven, CT: Yale University Press, 2016.

Manning, Beth Rose Middleton. *Trust in the Land: New Directions in Tribal Conservation*. Tucson: University of Arizona Press, 2011.

Manning, Beth Rose Middleton. *Upstream: Trust Lands and Power on the Feather River*. Tucson: University of Arizona Press, 2018.

Manovich, Lev. *The Language of New Media*. Cambridge, MA: MIT Press, 2001.

Marshall, Bob. "The People of Isle de Jean Charles Aren't the Country's First Climate Refugees." *Lens* (blog), December 6, 2016. https://thelensnola.org/2016/12/06/the-people-of-isle-de-jean-charles-arent-the-countrys-first-climate-refugees/.

Maurer, Kevin. "White Sand, Shell Island: The 20th Century 'Negro Atlantic City' That Never Was." *Port City Daily*, February 10, 2022. https://portcitydaily.com

/local-news/2022/02/10/white-sand-shell-island-the-20th-century-negro-atlantic-city-that-never-was/.

McClinton, Rowena. *The Moravian Springplace Mission to the Cherokees*. 2 vols. Lincoln: University of Nebraska Press, 2007.

McKittrick, Katherine. "Plantation Futures." *Small Axe* 17, no. 3 (2013): 1–15.

McLoughlin, William G. *Cherokee Renascence in the New Republic*. Princeton, NJ: Princeton University Press, 1986.

Meister, Sarah Hermanson. *Dorothea Lange, Migrant Mother*. New York: Museum of Modern Art, 2018.

Mihesuah, Devon. *Cultivating the Rosebuds: The Education of Women at the Cherokee Female Seminary, 1851–1909*. Urbana: University of Illinois Press, 1997.

Miles, Tiya. *The House on Diamond Hill: A Cherokee Plantation Story*. Chapel Hill: University of North Carolina Press, 2010.

Miles, Tiya. *Ties That Bind: The Story of an Afro-Cherokee Family in Slavery and Freedom*. Berkeley: University of California Press, 2005.

Miranda, Deborah A. *Bad Indians: A Tribal Memoir*. Berkeley, CA: Heyday Books, 2013.

Mirzoeff, Nicholas. *White Sight: Visual Politics and Practices of Whiteness*. Cambridge, MA: MIT Press, 2023.

Mitchell, Margaret. *Gone with the Wind*. 1936. New York: Scribner, 2011.

Mitchell, Margaret. "Interview with Margaret Mitchell from 1936." Interview by Medora Perkerson. First broadcast July 3, 1936, on WSB, Atlanta, Georgia. Rereleased on *American Masters*, season 26, episode 3, "Margaret Mitchell: American Rebel," aired March 12, 2012, on PBS. https://www.pbs.org/wnet/americanmasters/margaret-mitchell-american-rebel-interview-with-margaret-mitchell-from-1936/2011/.

Montgomery, David R. *Dirt: The Erosion of Civilizations*. Berkeley: University of California Press, 2012.

Moreton-Robinson, Aileen. *The White Possessive: Property, Power, and Indigenous Sovereignty*. Minneapolis: University of Minnesota Press, 2015.

Mt. Pleasant, Jane. "A New Paradigm for Pre-Columbian Agriculture in North America." *Early American Studies* 13, no. 2 (Spring 2015): 374–412.

"Names Hopi Guide as Co-respondent." In "Santa Fe Notebook." Collection no. YCAL MSS 61, Series II, Box 8, Folder 162. Lynn Riggs Papers, 1899–1954. Beinecke Rare Book and Manuscript Library, Yale University.

National Resources Board. "General Distribution of Erosion." In *A Report on National Planning and Public Works in Relation to Natural Resources and Including Land Use and Water Resources, with Findings and Recommendations*, 171–72. Washington, DC: US Government Printing Office, 1934.

Neimanis, Astrida. *Bodies of Water: Posthuman Feminist Phenomenology*. London: Bloomsbury Academic, 2019.

Nelson, Melissa K. "Getting Dirty: The Eco-Eroticism of Women in Indigenous Oral Literatures." In *Critically Sovereign: Indigenous Gender, Sexuality, and Feminist*

Studies, edited by Joanne Barker, 229–60. Durham, NC: Duke University Press, 2017.

Nestor, Sandy. *Indian Placenames in America*. New York: McFarland, 2012.

"New Hanover County, State of North Carolina." Collection no. 05656, Box 21. Hugh MacRae Papers, 1887–1970s, 2006. Southern Historical Collection. University of North Carolina at Chapel Hill.

New York Times. "Study Says Apartheid Hurts the Environment." May 13, 1990. https://www.nytimes.com/1990/05/13/world/study-says-apartheid-hurts-the-environment.html.

Nichols, Robert. *Theft Is Property! Dispossession and Critical Theory*. Durham, NC: Duke University Press, 2019.

Nishime, Leilani, and Kim D. Hester Williams. "Why Racial Ecologies." In *Racial Ecologies*, edited by Leilani Nishime and Kim D. Hester Williams, 3–14. Seattle: University of Washington Press, 2018.

Nixon, Rob. *Slow Violence and the Environmentalism of the Poor*. Cambridge, MA: Harvard University Press, 2013.

Nobles, Gregory. "The Myth of John James Audubon." Audubon, July 31, 2020. https://www.audubon.org/news/the-myth-john-james-audubon.

North Carolina Cooperative Extension. "*Gelsemium rankinii* (Swamp Jasmine, Swamp Yellow Jessamine, Yellow Jessamine)." North Carolina Extension Gardener Plant Toolbox. Accessed July 14, 2020. https://plants.ces.ncsu.edu/plants/gelsemium-rankinii/.

North Carolina Cooperative Extension. "*Gelsemium sempervirens* (Carolina Jasmine, Carolina Jessamine, Carolina Yellow Jessamine, Yellow Jessamine)." North Carolina Extension Gardener Plant Toolbox. Accessed July 14, 2020. https://plants.ces.ncsu.edu/plants/gelsemium-sempervirens/.

Norton, Jack. *Genocide in Northwestern California: When Our Worlds Cried*. San Francisco: Indian History Press, 1979.

Olson, Roberta J. M. *Audubon's Aviary: The Original Watercolors for "The Birds of America."* New York: New-York Historical Society and Rizzoli Electa, 2013.

Olson, Roberta J. M. "Hiding in Plain Sight: New Evidence about the Birth, Identity, and Strategic Pseudonyms of John James Audubon." *Bulletin of the Museum of Comparative Zoology* 163, no. 4 (September 2021): 129–50.

Orona, Brittani, curator. *Stories of the River, Stories of the People: Memory on the Klamath River Basin*. Accessed August 1, 2023. https://www.nativewomenscollective.org/storiesoftheriver.html.

Ortiz, Bev. *It Will Live Forever: Traditional Yosemite Indian Acorn Preparation*. Berkeley, CA: Heyday Books, 1991.

Owens, Louis. *The Grapes of Wrath: Trouble in the Promised Land*. Boston: Twayne, 1989.

Owens, Robert M. *Mr. Jefferson's Hammer: William Henry Harrison and the Origins of American Indian Policy*. Norman: University of Oklahoma Press, 2011.

Padgett, Ieva. "George W. Cable's Gardens: Planting the Creole South and Uprooting the Nation." *Southern Literary Journal* 47, no. 2 (Spring 2015): 55–72.

Parrish, Susan Scott. *The Flood Year 1927: A Cultural History*. Princeton, NJ: Princeton University Press, 2016.

Pauly, Thomas H. "*Gone with the Wind* and *Grapes of Wrath* as Hollywood Histories of the Depression." In *Movies as Artifacts: Cultural Criticism of Popular Film*, edited by Michael T. Marsden, John G. Nachbar, and Samm L. Grogg, 164–76. Chicago: Nelson-Hall, 1982.

Payne, James Robert. "Emergence of Alternate Masculinity in George Washington Cable's 'Sieur George' and 'Belles Demoiselles Plantation.'" *American Literary Realism* 32, no. 3 (Spring 2000): 244–55.

Perdue, Theda. *Cherokee Women: Gender and Culture Change, 1700–1835*. Lincoln: University of Nebraska Press, 1998.

Pesantubbee, Michelene E. *Choctaw Women in a Chaotic World: The Clash of Cultures in the Colonial Southeast*. Albuquerque: University of New Mexico Press, 2005.

Pfeifer, Mario. *Reconsidering the New Industrial Parks near Irvine, California by Lewis Baltz, 1974*. New York: Sternberg, 2009.

Pianigiani, Gaia, and Emma Bubola. "Italy's Government to Ban Cruise Ships from Venice." *New York Times*, July 13, 2021. https://www.nytimes.com/2021/07/13/world/europe/venice-italy-cruise-ship-ban.html.

Pilkey, Orrin H., and Keith C. Pilkey. *Sea Level Rise: A Slow Tsunami on America's Shores*. Durham, NC: Duke University Press, 2019.

Pilkey, Orrin H., and Rob Young. *The Rising Sea*. Washington, D.C.: Island Press, 2009.

Poole, Josh. "Providence and the Anthropocene." *About South* (podcast hosted by Gina Caison), season 2, episode 14, November 2, 2017. https://ivy.fm/podcast/about-south-438943.

Prud'homme-Cranford, Rain. "Summoning Swamp Songs: Decolonizing Creole-Indigenous Textual Tributaries." In *Swamp Souths: Literary and Cultural Ecologies*, edited by Kirstin L. Squint, Eric Gary Anderson, Taylor Hagood, and Anthony Wilson, 81–98. Baton Rouge: Louisiana State University Press, 2020.

Purdy, Jedediah. *This Land Is Our Land: The Struggle for a New Commonwealth*. Princeton: Princeton University Press, 2015.

Pushkin, Alexander. "The Prophet." In *From the Ends to the Beginning: A Bilingual Anthology of Russian Poetry Online*. Edited by Ilya Kutik and Andrew Wachtel. https://max.mmlc.northwestern.edu/mdenner/Demo/texts/prophet.htm.

Rankin, Caitlin G., Casey R. Barrier, and Timothy J. Horsley. "Evaluating Narratives of Ecocide with the Stratigraphic Record at Cahokia Mounds State Historic Site, Illinois, USA." *Geoarchaeology* 36, no. 3 (May 2021): 369–87.

Reed, Kaitlin. *Settler Cannabis: From Gold Rush to Green Rush in Indigenous Northern California*. Seattle: University of Washington Press, 2023.

Restore the Mississippi River Delta. "Land Loss." Accessed April 4, 2024. https://mississippiriverdelta.org/our-coastal-crisis/land-loss/.

Rich, Frank. "Oklahoma Was Never Really O.K." *Vulture*, April 2, 2019. https://www.vulture.com/2019/04/frank-rich-oklahoma.html.

Richards, Irving T. "Joseph R. Mason and John Neal." *American Literature* 6, no. 2 (May 1934): 122–40.

Rifkin, Mark. *Beyond Settler Time: Temporal Sovereignty and Indigenous Self-Determination*. Durham, NC: Duke University Press, 2017.

Rifkin, Mark. *Fictions of Land and Flesh: Blackness, Indigeneity, Speculation*. Durham, NC: Duke University Press, 2019.

Riggs, Lynn. *The Cherokee Night*. 1932. In *The Cherokee Night and Other Plays*, edited by Jace Weaver, 106–211. Norman: University of Oklahoma Press, 2003.

Riggs, Lynn. *The Cream in the Well*. 1940. In *Four Plays*, 155–222. New York: Samuel French, 1947.

Riggs, Lynn. *The Cream in the Well*. Draft 2 typescript. Collection 1940, no. YCAL MSS 61, Series II, Box 13, Folder 216. Lynn Riggs Papers, 1899–1954. Beinecke Rare Book and Manuscript Library, Yale University.

Riggs, Lynn. *Green Grow the Lilacs*. 1930. In *The Cherokee Night and Other Plays*, edited by Jace Weaver, 3–105. Norman: University of Oklahoma Press, 2003.

Riggs, Lynn. "Journal, Holograph (with Clippings), 1940 Feb–Aug." Collection no. YCAL MSS 61, Series II, Box 8, Folder 163. Lynn Riggs Papers, 1899–1954. Beinecke Rare Book and Manuscript Library, Yale University.

Riggs, Lynn. "The Vine Theater." *Texas Studies in Literature and Language* 59, no. 3 (Fall 2017): 274–86.

Roberts, Alaina E. *I've Been Here All the While: Black Freedom on Native Land*. Philadelphia: University of Pennsylvania Press, 2021.

Roering, Joshua J., Benjamin H. Mackey, Alexander L. Handwerger, Adam M. Booth, David A. Schmidt, Georgina L. Bennett, and Corina Cerovski-Darriau. "Beyond the Angle of Repose: A Review and Synthesis of Landslide Processes in Response to Rapid Uplift, Eel River, Northern California." *Geomorphology* 236 (February 2015): 109–31.

Romine, Scott. *The Real South: Southern Narrative in the Age of Cultural Reproduction*. Baton Rouge: Louisiana State University Press, 2008.

Rountree, Helen C. *Manteo's World: Native American Life in Carolina's Sound Country before and after the Lost Colony*. Chapel Hill: University of North Carolina Press, 2021.

Said, Edward W. *After the Last Sky: Palestinian Lives*. Photographs by Jean Mohr. New York: Columbia University Press, 1999.

Sarris, Greg. *Keeping Slug Woman Alive: A Holistic Approach to American Indian Texts*. Berkeley: University of California Press, 1993.

Schell, Jonathan. *The Fate of the Earth*. New York: Knopf, 1982.

Scheppe, Wolfgang. "Lewis Baltz and the Garden of False Reality." In *Candlestick Point*, by Lewis Baltz, 83–104. Göttingen, Germany: Steidl, 2011.

Schloss, Carol. *In Visible Light: Photography and the American Writer, 1840–1940*. New York: Oxford University Press, 1987.

Scott, James C. *Seeing Like a State: How Certain Schemes to Improve the Human Condition Have Failed*. New Haven, CT: Yale University Press, 1998.

Scranton, Roy. *Learning to Die in the Anthropocene: Reflections on the End of a Civilization*. San Francisco, City Lights, 2015.

Sekula, Allan. "Dismantling Modernism, Reinventing Documentary (Notes on the Politics of Representation)." *Massachusetts Review* 19, no. 4 (1978): 859–83.

Sheller, Mimi. *Island Futures: Caribbean Survival in the Anthropocene*. Durham, NC: Duke University Press, 2020.

Silko, Leslie Marmon. *Storyteller*. New York: Seaver, 1981.

Smith, Chadwick Corntassel, Rennard Strickland, and Benny Smith. *Building One Fire: Art and World View in Cherokee Life*. Tahlequah, OK: Cherokee Nation, 2010.

Smith, John. *Captain John Smith: Writings with Other Narratives of Roanoke, Jamestown, and the First English Settlement of America*. Edited by James Horn. New York: Library of America, 2007.

Snell, John. "Coast in Crisis: Louisiana Suffers the Fastest Rate of Land Loss in North America." FOX 8, December 24, 2021. https://www.fox8live.com/2021/12/24/coast-crisis-louisiana-suffers-fastest-rate-land-loss-north-america/.

Snider, William D. *Helms and Hunt: The North Carolina Senate Race, 1984*. Chapel Hill: University of North Carolina Press, 1985.

Solnit, Rebecca. *Infinite City: A San Francisco Atlas*. Berkeley: University of California Press, 2010.

Solnit, Rebecca, and Rebecca Snedeker. *Unfathomable City: A New Orleans Atlas*. Berkeley: University of California Press, 2013.

Souder, William. *Mad at the World: A Life of John Steinbeck*. New York: W. W. Norton, 2020.

Squint, Kirstin L. "Monique Verdin's Louisiana Love: An Interview." *American Indian Quarterly* 42, no. 1 (Winter 2018): 117–33.

State of California, Native American Heritage Commission. "Timeline of Genocide Incidents in the Greater Humboldt Region." Accessed November 22, 2023. https://nahc.ca.gov/cp/timelines/humboldt/.

Stein, Sally. "Passing Likeness: Dorothea Lange's 'Migrant Mother' and the Paradox of Iconicity." In *Only Skin Deep: Changing Visions of the American Self*, edited by Coco Fusco and Brian Wallis, 345–55. New York: International Center of Photography and Harry N. Abrams, 2003.

Steinbeck, John. *The Grapes of Wrath*. 1939. New York: Penguin, 2006.

Stone, Katherine, Orville Magoon, Billy Edge, and Lesley Ewing. "Sand Rights." In *Encyclopedia of Coastal Science*, edited by Maurice L. Schwartz, 820–21. Dordrecht, Netherlands: Springer, 2015. 26, 2022. https://www.kpcnews.com/outdoors/article_611af078-19af-543e-8988-b2fb66685a53.html.

Strickland, Scott M., Julia A. King, G. Anne Richardson, Martha McCartney, and Virginia Busby. *Defining the Rappahannock Indigenous Cultural Landscape*. St. Mary's City: St. Mary's College of Maryland, 2016.

Sutherland, J. J. "L. Frank Baum Advocated Extermination of Native Americans." NPR, October 27, 2010. https://www.npr.org/sections/thetwo-way/2010/10/27/130862391/l-frank-baum-advocated-extermination-of-native-americans.

Sutter, Paul S. *Let Us Now Praise Famous Gullies: Providence Canyon and the Soils of the South*. Athens: University of Georgia Press, 2015.

Swensen, James R. *Picturing Migrants: "The Grapes of Wrath" and New Deal Documentary Photography*. Norman: University of Oklahoma Press, 2015.

Swenson, Dan. "These Six Factors Explain Why Louisiana Is Rapidly Losing Land." NOLA.com, May 31, 2021. https://www.nola.com/news/these-six-factors-explain-why-louisiana-is-rapidly-losing-land-see-graphics/article_59675b8c-bfbe-11eb-9602-47cf4c0429dc.html.

Swift, Earl. *Chesapeake Requiem: A Year with the Watermen of Vanishing Tangier Island*. New York: Dey Street Books, 2018.

Tal, Alon. *Pollution in a Promised Land: An Environmental History of Israel*. Berkeley: University of California Press, 2002.

Tato, Kebede, and Hans Hurni, eds. *Soil Conservation for Survival*. Ankeny, IA: Soil and Water Conservation Society, 1992.

Taylor, Paul. *On the Ground in the Thirties*. Salt Lake City: Gibbs M. Smith, 1983.

Texas Invasive Species Institute. "Bermudagrass." Texas State University System. Accessed July 14, 2020. http://tsusinvasives.org/home/database/cynodon-dactylon.

Theriot, Jason P. *American Energy, Imperiled Coast: Oil and Gas Development in Louisiana's Wetlands*. Baton Rouge: Louisiana State University Press, 2014.

Tidwell, Mike. *Bayou Farewell: The Rich Life and Tragic Death of Louisiana's Cajun Coast*. New York: Vintage, 2003.

Todd, Zoe. "Indigenizing the Anthropocene." In *Art in the Anthropocene: Encounters among Aesthetics, Politics, Environments and Epistemologies*, edited by Heather Davis and Etienne Turpin, 241–54. London: Open Humanities Press, 2015.

"To Nelson MacRae." July 12, 1926. Collection no. 05656, Box 21. Hugh MacRae Papers, 1887–1970s, 2006. Southern Historical Collection. University of North Carolina at Chapel Hill.

Treuer, David. "Return the National Parks to the Tribes." *Atlantic*, May 2021. https://www.theatlantic.com/magazine/archive/2021/05/return-the-national-parks-to-the-tribes/618395/.

Tribe, Mark. Foreword to *The Language of New Media*, by Lev Manovich, x–xiii. Cambridge, MA: MIT Press, 2001.

Tuck, Eve, and K. Wayne Yang. "Decolonization Is Not a Metaphor." *Decolonization: Indigeneity, Education and Society* 1, no. 1 (September 2012): 1–40.

Turner, Colbi. "Georgia State Symbols: Flowers and Wildflowers." Lincoln County UGA Extension, September 17, 2021. https://site.extension.uga.edu/lincoln/2021/09/georgia-state-symbols-flowers-and-wildflowers/.

Twelve Southerners. *I'll Take My Stand: The South and the Agrarian Tradition*. 1930. Edited by Susan V. Donaldson. Baton Rouge: Louisiana State University Press, 2006.

United Nations. "Soil Erosion Must Be Stopped 'to Save Our Future', Says UN Agriculture Agency." UN News, December 5, 2019. https://news.un.org/en/story/2019/12/1052831.

Unrau, William. *Mixed-Bloods and Tribal Dissolution: Charles Curtis and the Quest for Indian Identity*. Lawrence: University Press of Kansas, 1989.

US Geological Survey. USGS Coastal Change Hazards Portal. Accessed December 22, 2022. https://marine.usgs.gov/coastalchangehazardsportal/.

Vanhoutte, Kristof K. P. *Limbo Reapplied: On Living in Perennial Crisis and the Immanent Afterlife*. London: Palgrave Macmillan, 2018.

Verdin, Monique. "A City in Time: La Nouvelle-Orleans over 300 Years." In *Unfathomable City: A New Orleans Atlas*, by Rebecca Solnit and Rebecca Snedeker, 19–24. Berkeley: University of California Press, 2013.

Verdin, Monique. *Return to Yakni Chitto: Houma Migrations*. New Orleans: University of New Orleans Press, 2019.

Ware, Carolyn E. *Cajun Women and Mardi Gras: Reading the Rules Backward*. Urbana: University of Illinois Press, 2007.

Watson, Jay. "The Other Matter of the South." PMLA 131, no. 1 (January 2016): 157–61.

Weaver, Jace. Foreword to *The Cherokee Night and Other Plays*, by Lynn Riggs, ix–xv. Norman: University of Oklahoma Press, 2003.

Weaver, Jace. *Other Words: American Indian Literature, Law, and Culture*. Norman: University of Oklahoma Press, 2001.

Weaver, Jace. *That the People Might Live: Native American Literatures and Native American Community*. New York: Oxford University Press, 1997.

Webb, Elizabeth M. *For the Mud Holds What History Refuses (Providence in Four Parts)*. Elizabeth M. Webb's website, 2019. http://www.elizabethmwebb.com/portfolio/providence-in-four-parts/.

Weits, Michal, dir. *Blue Box*. Tel Aviv: Norma Productions, 2021. 82 min.

Whyte, Kyle Powys. "Time as Kinship." In *The Cambridge Companion to Environmental Humanities*, edited by Jeffrey Jerome Cohen and Stephanie Foote, 39–55. Cambridge: Cambridge University Press, 2021.

Widick, Richard. *Trouble in the Forest: California's Redwood Timber Wars*. Minneapolis: University of Minnesota Press, 2009.

Wilkins, Thurman. *Cherokee Tragedy: The Ridge Family and the Decimation of a People*. Boston: Macmillan, 1970.

Williams, Lloyd. "Providence Canyon: Georgia's Big Beautiful Gully." *Georgia Life*, Autumn 1975. Collection no. 03527, folder 46. Hugh H. Bennett Papers, 1906–1966, 1984. Southern Historical Collection. University of North Carolina at Chapel Hill.

Williams, Terry Tempest. *Erosion: Essays of Undoing*. New York: Picador Paper, 2019.

Wilmeth, Dudley. "Georgia's Great Canyon." *Atlanta Journal*, May 21, 1939. Collection no. 03527, Folder 54. Hugh H. Bennett Papers, 1906–1966, 1984. Southern Historical Collection. University of North Carolina at Chapel Hill.

Wilmington Morning Star. "Hot Weather Has Turned Fishing Season Upside Down." November 22, 1985. Collection no. 04980, Series 1, Box 3. Ellyn Bache Papers, 1974–1998. Southern Historical Collection. University of North Carolina at Chapel Hill.

Wilson, Charles Reagan. *Baptized in Blood: The Religion of the Lost Cause, 1865–1920.* Athens: University of Georgia Press, 1980.

Womack, Craig. *Art as Performance, Story as Criticism: Reflections on Native Literary Aesthetics.* Norman: University of Oklahoma Press, 2009.

Womack, Craig. *Red on Red: Native American Literary Separatism.* Minneapolis: University of Minnesota Press, 1999.

Wood, Karenne. *Markings on Earth.* Tucson: University of Arizona Press, 2001.

Worster, Donald. *Dust Bowl: The Southern Plains in the 1930s.* Oxford: Oxford University Press, 2004.

Worster, Donald. *Nature's Economy: A History of Ecological Ideas.* 2nd ed. Cambridge: Cambridge University Press, 1994.

Xia, Rosanna. *California Against the Sea: Visions for Our Vanishing Coastline.* Berkeley, CA: Heyday, 2023.

Xia, Rosanna. "The California Coast Is Disappearing under the Rising Sea. Our Choices Are Grim." *Los Angeles Times*, July 7, 2019. https://www.latimes.com/projects/la-me-sea-level-rise-california-coast/.

Xia, Rosanna, Swetha Kannan, and Terry Castleman. "The Ocean Game: The Sea Is Rising. Can You Save Your Town?" *Los Angeles Times*, July 7, 2019. https://www.latimes.com/projects/la-me-climate-change-ocean-game/.

Yang, K. Wayne. "Sustainability as Plantation Logic, or, Who Plots an Architecture of Freedom?" *e-flux Architecture*, February 28, 2021. http://worker01.e-flux.com/pdf/article_353587.pdf.

Yeager, Patricia. *Dirt and Desire: Reconstructing Southern Women's Writing, 1930–1990.* Chicago: University of Chicago Press, 2000.

Young, Mary. "Archaeology Saves the Bay: The Sustainability of the Chesapeake Bay Oyster Fishery." Undergraduate honors thesis, Williamsburg, College of William and Mary, 2021.

Yusoff, Kathryn. *A Billion Black Anthropocenes or None.* Minneapolis: University of Minnesota Press, 2018.

Zogry, Michael J. *Anetso, the Cherokee Ball Game: At the Center of Ceremony and Identity.* Chapel Hill: University of North Carolina Press, 2010.

Zucchino, David. *Wilmington's Lie: The Murderous Coup of 1898 and the Rise of White Supremacy.* New York: Grove, 2021.

Index

Note: Page numbers in *italics* indicate illustrations. *Gone with the Wind* is abbreviated as GWTW.

Adams, Ansel, 39, 47, 226n35
Adams, Robert (photographer), 43
afforestation, 215–16
African Americans: Black Native people, 164–65; distinguished as nonsettlers, 53; history of land and labor dispossession of, 57–58; intermarriage, lynchings and, 93–94, 235nn98–99; as subjects in *American Exodus*, 84. *See also* Butler, Octavia; enslavement; Flanagan, Thomas Jefferson; racism; Wrightsville Beach, North Carolina
Agamben, Giorgio, 202
Agee, James (and Walker Evans): *Let Us Now Praise Famous Men*, 84
agrarianism, 152, 161. *See also* Confederate Lost Cause
agriculture of settler colonialism, 39, 44, 182–83, 215–16, 217. *See also* plantation enslavement economy —as cause of erosion: in California, 32, 44, 53, 65, 67; and the Dust Bowl, 74, 77–78, 232n24, 236n123; prevention methods known to settlers, 123, 154, 155, 158, 160; soil science of nineteenth century onward, 158, 160; soil scientists' view on gullying, 150, 166; soil scientists' view on rampant topsoil destruction, 78, 149–50, 158, 161, 213, 217; Steinbeck on, 79–81; wheat-growing craze, 77–78, 236n123. *See also* plantation enslavement economy—agriculture as cause of erosion

Akman, Charlie, 230n4
Alabama, 149, 151
Alaska, 31
Alexander of Hales, 201
Algonquian people, 179, 182–83, 187
Allen, Chadwick, 15, 88–89
Allewaert, Monique, 128, 246n17
Allotment policies, 74–75, 87, 230n4, 231n15; Dawes Rolls and, 230n4; and erosion of women's power, 96–97, 98–102, 103, 104; mixedblood Cherokees eligible for allotments, 81; Riggs's family allotment, 87; Steinbeck's misrepresentation of, 81–82

—erosion as precipitated by: 1934 map, 1–3, *2*; the Dust Bowl and, 74–76, 92–93. *See also* Removal—erosion as precipitated by

American exceptionalism, 13–14, 17–18, 32–34, 84–85, 167–68, 180–81, 205, 225n16

Anderson, Eric Gary, 4, 6–7, 20

Anderson, M. Kat, 36–37

Anthropocene discourse: capitalism and, 19; naturalization of, 144–45, 167, 210; the novelty of crisis as focus, 5, 45; and the southern disaster complex, 152–53; turn to, 4; the unequal effects of colonialism, 19, 28, 138–39, 153; as unsettling but not decolonizing, 4–5

anti-Semitism, 216

Apache people, 76

Arapaho people, 76

Atakapa people, 121

Atlantic Seaboard. *See* Eastern Seaboard

Audubon, John James, 25; background of, as self-invention, 112–13, 114; on clear-cutting cypress swamp forests, 117–18, 120; as destroying what he purports to love (eros/ion), 115–16, 210–11; and enslavement, as apologist for and participant in, 113, 114–20; on erosion of land and replenishment of soil by the Mississippi, 118–20; as foregrounding the degradation of bird habitat, 113, 118; on the future, 118, 120; and Indigenous peoples, presence of, 120–21, 122; and pathos of loss in the plantation economy, 119–20, 129; Removal witnessed by, 120–21; and settler anxiety of disappearing white supremacy, 109, 113, 114–15, 117, 140; on soil exhaustion, 116–17; Verdin's significant historical differences from, 131–32. Works: *Birds of America*, 25, 113–15, 120; "A Flood," 119; "Improvements in the Navigation of the Mississippi," 118–19; letter to John Bachman (1837), 120–21; *Ornithological Biography*, 116; *Snowy Heron, or White Egret*, 114–15, *115*, 238n23; "The Squatters of the Mississippi," 116–18, 139

Augustine, 200

Bache, Ellyn: "Shell Island" (1993), 27, 180, 190–93, 202, 203

Bachman, John, 120–21

background and foreground: and composition of visual works, 109, 111, 112; cultural representation of Louisiana and, 109, 111–12, 123–24, 139–40, 237n7; erosion in southern Louisiana, 140; narrative construction and, 111–12; spatiality and temporality as linked in, 110–12; Verdin's backgrounds as adjusting and confounding perspective, 131, *132*, *133*–39, *134*, *136*, *138*

Balaschak, Chris, 41–42, 49, 225n32

Baltz, Lewis, 23–24, 33, 37–38; background of, 34, 38, 45, 226n43; the horizon line as formal construct, 23–24, 38–39, 40, 42–43; and horizon line limiting understanding of settler colonial history, 40, 42, 43–44, 45–46, 226n45; humanity as bound to be destructive, 37, 45–46, 227n53; and industrial land abuse, 38–43; on landscape art, 39–40, 45; and the mortuarial, 41, 42–43, 44–45; and optimism, 38, 40, 46–47, 64; and quotidian backgrounds, 109; and space as "final frontier," 39, 50; and the surface, engagement with, 41–43, 45; *Wizard of Oz* moment of switch to color, 45–48, 64, 227n52. Works: *Candlestick Point* series, 40, 44–49, 46–47, 64; *New Industrial Parks* series, 40–44, *42*, 49, 226n45; *Park City* series, 38–39; review essay on photographer Robert Adams, 43; *San Quentin Point* series, 44–45; *South Corner, Riccard America Company, 3184 Pullman, Costa Mesa*, 41; *South Wall, PlastX, 350 Lear, Costa Mesa*, 41, *42*;

South Wall, Unoccupied Industrial Structure, 16812 Milliken, Irvine, 41; *Unnamed Image 23,* 46; *Unnamed Image 24,* 47; *Venezia Marghera* series, 213–14. *See also* New Topographics photography movement
Barbrook, Richard, 33–34
Bartram, William, 128, 246n17
Baton Rouge, Louisiana, 112
Bauer, Ralph, 248n70
Baum, L. Frank: *Wizard of Oz,* 45, 47–48, 64, 227nn52–53
beaches: attempts to slow/reverse erosion, 198; damming/controlling rivers and starvation of, 35, 41, 58; dredging and increasing of erosive wave action, 36; erosion of, as theft, 35–36; littoral cell system, 178–80, 181, 186, 198, 205; natural processes of nourishment for, 35; rubble of houses used as barricades for (Gaza), 250n23; sand and gravel mining, 35, 36; sand rights (*jus publicum*), 35–36; sea walls, 35, 203; structural nourishment plans for remediation of loss, 197–98
Beckert, Sven, 155
Beck, Ulrich, 111
Bennett, Hugh Hammond: 15, 26; as chair of the US Soil Erosion Service, 78; on government policies causing the Dust Bowl, 78; possible contact with Margaret Mitchell, 244n60
—*Soil Conservation* (1939): on ancient Peruvian methods of soil conservation, 217; on the global effects of local erosion, 217; on link between enslaved labor and soil destruction, 159; on nineteenth-century soil science recommendations, 158; on Palestine, 215, 249–50n20; Providence Canyon featured in, 151; on settler colonialism as devastating Indigenous knowledge, 217; on soil exhaustion and sheet erosion, 155; on South Africa, 212, 213; types of water runoff erosion, 144; on white settlers' record of topsoil destruction, 161, 213, 217
Bering Strait theory, 45
Bermuda grass, 126, 128
biblical narratives: Limbo, 200; parable of the sower, 51–52, 56; parable of the talents, 51, 59
Black Native people, 164–65
Black studies, and intersections with Indigenous studies, 53–54, 64, 150, 169–70
Blaikie, Piers, 9–10
Blaser, Mario, 5, 27, 212
Boise City, Oklahoma, 70
Boudinot, Elias, 235n99
Bourke-White, Margaret (and Erskine Caldwell): *You Have Seen Their Faces,* 84
Brower, Kenneth, 39–40
Brown, Kirby, 89–90, 105
Brown, Wendy, 205
Bruyneel, Kevin, 4, 5, 23–24
buffalo grass, 74, 77, 232n24
Burley, David, 237n4
Burroughs, William: *Naked Lunch,* 45, 227n53
Butler, Octavia, 23–24, 33; background and death of, 34, 66
—Earthseed series: the acorn as food and symbol in, 52, 53, 54–55, 58; Acorn community in, 56, 57–58, 59, 61–62, 63; agriculture and soil science in, 51–52; as airing out the wound of eros/ion, 211; antinomy in, 63; biblical parables and, 51–52, 56, 59; California exceptionalism and, 53, 54–55; caution against Black futurism taking the form of settler colonialism, 53–58, 62–64; Earthseed religion, 56, 59, 61; and erasure of Indigenous history and presence, 23–24, 34, 50–51, 52–53, 54, 63, 64, 66, 228n73; erosion events as deus ex machina, 51, 59–60, 61, 216; horizon line limiting possibility in, 51, 61; humanity as bound to be destructive, 37; Indigenous knowledges utilized in alternate

Butler—Earthseed series (continued) understanding of landscape, 34, 52–53, 54, 56, 63; managed retreat in, 60; optimism and, 55–57, 58, 64; *Parable of the Sower* (1993), 34, 51, 52–53, 54–58, 63, 228n80; *Parable of the Talents* (1998), 34, 50–52, 58–64, 216; private land ownership and, 55, 57, 63; and settler anxiety of disappearance, 60–61, 216; space as "final frontier" in, 50–51, 62–63, 64, 227n66; unrealized third book ("Parable of the Trickster"), 51, 63; walled communities as insulation from social and ecological collapse, 52, 65

Byrnes, Delia, 237n11

Cable, George Washington: backgrounding of enslavement by, 123–24; Indigenous characters of, 122; Verdin quoting, on "trembling prairie" of Terrebonne Parish, 132; Verdin's significant historical differences from, 131–32

—"Belles Demoiselles Plantation" (1883), 25, 109; critical focus on race relations and kinship affiliations in, 122, 124–25; and erasure of Indigenous homeland attachment, 125–29, 240n79; erosive processes of the Mississippi River, 123, 126–27, 129; foregrounding of plantation house and backgrounding of enslavement in, 123–24; and settler anxiety of disappearing white supremacy, 109, 140; surface erosion and, 124; vine metaphor of collectivization, 127–28, 129; white pathos of loss in, 122, 124–25, 127, 129

Caddo people, 76

Cahokia, 20–21

Cajun culture, 108, 109, 130, 140, 237n4, 237n7

Caldwell, Erskine: *God's Little Acre*, 144; *Tobacco Road*, 144

Caldwell, Erskine (and Margaret Bourke-White): *You Have Seen Their Faces*, 84

California: and the Dust Bowl migration, 70, 78, 86–87; flood of 1938, 35; floodplains, building in, 226n45; Gold Rush, 32, 53; industrialized agriculture in, 32, 44, 53, 65, 67; as interrelated with all regions of the US, 70; Mexican ranchero labor system, 53; nonrecognition of Indigenous land management, 31, 36–37, 39; petroculture, 53; "pristine wilderness" myth of, 36–37, 39–40, 47, 226n35; Spanish Catholic mission system, 32, 53, 54, 56, 61, 226n45; water supply for populace, 35, 53. *See also* California exceptionalism

—and erosion: beaches, sand starvation of, 35–36, 41; dismantling of Indigenous land management strategies, 36; drought and, 32, 35; mudslides and landslides, 35, 50, 58; natural processes of beach nourishment, 35; and nature, attempts to control, 35–36; "The Ocean Game" (*Los Angeles Times*), 29–31, 45, 66, 67, 194, 198; rail line to San Diego, 32, 225n9; sea-level rise, managed retreat and, 45, 48–50; settler colonialism as cause of, 32, 36; timber extraction, 58; wildfire and, 32, 35. *See also* environmentalism and the settler colonial future

California exceptionalism, 13–14, 23–24, 30–34, 38, 40, 48–50, 52–56, 63–66

Cameron, Andy, 33–34

Canavan, Gerry, 63, 228n67

Candlestick Point (San Francisco): Baltz photo series (*Candlestick Point*), 40, 44–49, 46–47, 64; development project, 48–50

Cape Hatteras Lighthouse (North Carolina), 27, 180, 193–99, *196*

capitalism: Anthropocene discourse and, 19; conscious capitalism, 65–66; entanglement of Indigenous people in, 8, 139; immediate returns demanded by, 189; and "invisible hand," color of, 188; and protection of the egret, 238n23; static land as

requirement of, 27, 181, 186, 205; static racial and Indigenous identity as requirement of, 27, 181, 186
Carolina jessamine, 128
Carter, Tim, 234n92
Catholic religion, 32, 53, 54, 56, 61, 200–202, 226n45, 247–48nn63–64
Caucasus region, 211
Chambers, Juliette Scrimsher (stepmother of Lynn Riggs), 87, 98
Chatahoochie River, 150, 163
Cherokee Female Seminary, 97, 99
Cherokee Nation: Allotment policy and, 74, 89; eligibility of Florence Thompson for citizenship in, 230n4; as *Grapes of Wrath* starting place, 24, 78–79, 81–82, 87, 103–4; male political authority established, 85–86; Oklahoma statehood and loss of sovereignty, 89, 96; Removal to Indian Territory (Trail of Tears), 74, 86, 156–57; Removal within the southeast, 163–64; Tahlequah as capital, 70, 86. *See also* Cherokee people; Five Southeastern Tribes
Cherokee people: Booger Dance, 91; children sent to missionary schools, 85; continuation in the post-Allotment world, 89–90; and white settler plantation enslavement culture, emulation of, 75, 85, 163; women as controllers of land management and agricultural practices, 75, 85–86, 96–103, 104, 122, 133, 236n123, 241n92. *See also* Riggs, Lynn; Thompson, Florence
—Cherokee identity: Dawes Rolls and, 87, 230n6; matrilineal clan system, 85, 87, 122, 230n4, 233n65; of Lynn Riggs, 87; specious claims to, 230n4; Steinbeck's errors in representing, 81–82; of Florence Thompson, 70–71, 86–87, 230n4, 231n7, 233n65
Cherokee rose, 127, 156–57
Chesapeake Bay, 199–200. *See also* Tangier Island
Chesnutt, Charles: *The Marrow of Tradition*, 180

Cheyenne people, 76
Chiapas, Mexico, 212
Chicago Daily News: "Soil and Sanctuary," 141–42
Chickasaw people, 74. *See also* Five Southeastern Tribes
Chitimacha people, 121
Choctaw people, 74, 75, 121, 122, 124. *See also* Cable, George Washington; Five Southeastern Tribes; Howe, LeAnne
Christie, Jackson (father of Florence Thompson), 230n4
Christie, Mary Jane (mother of Florence Thompson), 230
Cimarron County, Oklahoma, 70
climate change: California exceptionalism claiming to take seriously, 49–50; ocean warming, 190–91; oil and gas industry and, 137; as perennial crisis akin to limbo, 181, 201–3, 205–6; rejection as "heresy," 49, 50; and state protection of symbols, 199; Tangier Island dwellers as skeptical of, 27, 180, 202; as too big to fight, 202; and Wrightsville Beach, 190–91. *See also* global scale of erosion; sea-level rise
Clinton, Hillary, 204
Clukey, Amy, 153
Cody, Iron Eyes, 240n79
Colbert, Stephen, 204
Comanche people, 76
Confederate Lost Cause pathos, 31, 140, 150, 152–53, 161, 165, 175; GWTW and naturalization of, 145, 156–57, 158, 161. *See also* lost-cause narratives; plantation futures; white supremacy
conservation movement. *See* environmentalism and the settler colonial future
Cooper, Anderson, 205
cotton, 79–80, 84, 154–55, 157, 158–59, 160, 162, 165–66
Coushatta Caddo people, 121
COVID-19 lockdowns, 214
Cox, James, 88–89, 96, 98

275

Creek people. *See* Muscogee Creek people
Creole culture, 108, 109, 130. *See also* Cable, George Washington
critical Indigenous studies: intersections with Black studies, 53–54, 64, 150, 169–70; methodology of, 9–12, 14–16, 22–23
critical regional studies, methodology of, 9–10, 12–14
Croatoan, 194
Curtis Act (1898), 74, 87, 231n15. *See also* Allotment policies
Curtis, Charles, 231n15
cypress swamp forests, clear-cutting, 117–18, 120

Dahlonega (Georgia) gold rush, 156, 165
Dakota Access Pipeline protests, 139–40
damming projects: as global crisis, 211; reversal of, 67; sediment retention and coastal erosion due to, 35, 41, 58
Dante: *The Divine Comedy*, 200–202
Dardar, Thomas, 237n4
Dardar, T. Mayheart, 237n4; "Part I: Global Climate Change—A Houma Perspective" (in *Return to Yakni Chitto*, by Monique Verdin), 132, 135, 137, 138–39, 140
Davis, Jetar, 234n92
Dawes Act (General Allotment Act of 1887), 74, 87, 230n4; amended to include the Five Southeastern Tribes (Curtis Act), 74, 87, 231n15. *See also* Allotment policies
Dean, Cornelia, 35, 36
De Capitani, Lucio, 214
decolonization: Anthropocene narratives not producing, 4–5; deconstructing coloniality not the same as, 12; as distinct project, 14–15; neoliberal multiculturalism efforts and, 66; settler memory and, 24. *See also* return of homelands to Indigenous control
deforestation and erosion: clear-cutting cypress swamp forests, 117–18, 120, 124, 139; cutting of trees in Palestine, 215–16; other timber extraction, 58, 149, 150, 155, 157, 161–62; salt-water intrusion and death of live oaks, 133–34. *See also* roots and erosion
de la Cadena, Marisol, 5, 27, 212
DeLoughrey, Elizabeth, 4, 5, 7
democracy, anxiety about the erosion of: as American exceptionalism, 13, 17–18, 205; concern for the physical earth as lost in debates on, 216–17, 218; and continuance of the settler-state government, 205, 218; loss of Indigenous homelands subsumed in, 17–18
democracy, Jeffersonian, 33–34, 225n16
Diamond Shoals, 195
disaster complex, southern, 152–53
Dixon, Maynard, 82
Doud, Richard, 83
Douglass, Frederick, 56
Dowson, Ernest: "Non Sum Qualis Eram Bonae Sub Regno Cynarae," 142
Dust Bowl, 72, 73–74, 76–78, 232n24, 236n123; Allotment policy as leading to, 74–76, 92–93; amount of topsoil losses, 77–78; and the Cherokee diaspora, in Steinbeck, 86–87; erasure of Indigenous peoples in narratives of, 75–76, 86–87, 103–4; *GWTW* in context of, 26, 141–43, 242n8; regional exceptionalism and, 84–85; Riggs as prophetic of, 90, 92, 95, 96; scholarly mischaracterizations of, 75–76; and settler anxiety of disappearance, 24, 72, 80, 81, 82–84; Soil Conservation and Domestic Allotment Act (1936), 142; soil exhaustion and, 103; soil scientists' knowledge of settler colonialism as cause, 78; *The Wizard of Oz* as allegory of, 47–48

Earth Day, 75
earthworks, Indigenous, 20–21, 112, 175–76

Eastern Seaboard: belying regional exceptionalism, 14; capitalism and, 27, 181, 186, 205; dawn as orienting life along, 207; Diamond Shoals and dangers to shipping trade, 195; erasure of Indigenous homelands and presence, 200; histories of English colonial invasion as subtending, 179; littoral cell system of, 178–80, 181, 186, 198, 205; settler anxiety about change, 187; settler exploitation of tobacco, 165–66, 182–84; soil exhaustion and migration of settlers, 116–17. *See also* Cape Hatteras Lighthouse; Tangier Island, Virginia; Wrightsville Beach, North Carolina
—Indigenous peoples of: in the Algonquian linguistic continuum, 179; assistance to settlers, 187, 194; colorism among Native peoples, 185–86; entanglement of Indigenous peoples in tobacco exploitation, 182–84; homelands and homewaters, dynamic management of, 26–27, 186–87, 206–7; homelands on Hatteras Island, 194–95; recognition of Indigeneity among Native peoples, 185–86; resistance to colonial invasion, 183; settler moves toward innocence, 183–84; settler nonrecognition of Indigenous identity, 181, 184–85

ecoanxiety, 221–22n14

ecocide models, as narrative, 20–22, 224n66

ecocriticism, 129, 140, 144

Eel River and basin: massacres of Native people by US troops, 57, 58, 61, 229n103; soil erosion in landslides, 58

Egan, Timothy, 77

Ejército Zapatista de Liberación Nacional (EZLN), 27, 212

Elks sisters (Hatteras Island land sale), 194–95

Elks, Thomas, 194

Emancipation, 164–65

Enlarged Homestead Act (1909), 75

enslavement: American exceptionalism as belied by, 13; Bermuda grass introduction via slave trade, 126; Black survivance, 148, 173, 175; Cable's Creole writings and elision of, 124; California ideology as reproducing the legacy of, 33–34; California Mexican ranchero system, 53; codification of Black and Indigenous identities, 27, 164, 181, 186, 190; Emancipation, and Black Native character (*GWTW*), 164–65; erasure of, 32, 150, 161, 165, 175; and Indigenous homelands theft, direct relationship of, 26, 150, 157, 159–60, 169–71, 173–74, 175; lighthouses and protection of slave trade, 195; soil scientists on link to soil destruction, 150, 159; as theft of Black life and labor converted to property, 170. *See also* plantation enslavement economy

environmentalism and the settler colonial future: appropriation of Indigenous knowledge and (settler environmentalism), 18–19; conscious capitalism, 65–66; ecocide narratives and, 20–22, 224n66; horizon line limiting the possibility of return of Indigenous homelands, 23–24; Indigenous knowledges as subsumed in, 17; myths of the "ecological Indian" and, 19–20; preservation of lands and, 5, 17, 65–66; public lands as settler birthright, 17, 19; settler anxiety of disappearing white supremacy as sublimated into, 65; sustainability, 17; technoutopianism and, 33–34; as unquestioned "good" politics, 18–19

environmental movement: protection of the egret, settler capitalism and, 238n23; public-interest advertising, 240n79; and the southern disaster complex, 152–53; and white pathos of loss, 140, 152–53. *See also* ecocriticism; environmentalism and the settler colonial future erasure of material history of Indigenous loss, 3, 5, 11, 13–14, 17–18, 24, 32, 71, 150, 161; Black enslavement and land abuse, direct relationship of, 26, 150, 157, 159–60, 169–71, 173–74, 175; in Butler's Earthseed series, 23–24, 34, 50, 52–53, 54, 63, 64, 66, 228n73; in California exceptionalist fantasies, 13–14, 23–24, 31–34, 43, 44; in ecocritical pleas for the southern Louisiana wetlands, 129–30, 140; of Indigenous presence and claims to homelands, 23–24, 31–34, 44, 111–12, 122, 125–29, 200, 240n79; white pathos subsuming Indigenous loss, 3, 156–57, 164. *See also* enslavement: erasure of

eros and eros/ion: as destroying what it purports to love, 209–11, 218–19

erosion: 1934 map (US National Resources Board), 1–3, *2*; etymology of term, 22, 99–100, 139; geological usage of term, 22, 73, 148; as hyperquotidian ubiquity, 140; as natural process of beach nourishment, 35; as natural process of land-building, 115–20, 123, 126–27, 178; not a metaphor, 15; temporal frame of, 22

Evans, Walker (and James Agee): *Let Us Now Praise Famous Men*, 84

exceptionalist paradigms, 6–7. *See also* regional exceptionalism

Export, Valie, *Eros/ion*, 210

extraction. *See* mining and erosion; oil and gas industry

Farinella, Marc, 189

Farm Security Administration, 68, 82–84

Faulkner, William, 133, 152; *Absalom! Absalom!*, 152; *The Wild Palms*, 152

Fazio, Tom, 65

Filipinx Americans, as subjects in *American Exodus*, 84

Fish, Daniel, 94

Five Southeastern Tribes (Choctaw, Cherokee, Chickasaw, Creek, Seminole): non-Native men and marriage, 122; as place-based agricultural-cultivation nations, 75, 85; and plantation enslavement economy, 75, 85–86, 163; played as "buffer" against western tribes, 76; Removal treaty broken by the Curtis Act, 74, 231n15. *See also* Allotment policies; Indian Territory; matrilineal structures of southeastern Indigenous peoples; *individual tribes*

Flaherty, Robert: *The Land* (documentary film, 1942), 168, 169

Flanagan, Thomas Jefferson: as African American poet and scholar, 145, 171; historian Sutter's reading of, 171, 174; oral history recording by, incorporated into Webb's multimedia exhibit, 168, 169; on segregated schools, 168, 169

— *The Canyons at Providence, The (The Lay of the Clay Minstrel)* (1940), 26, 145; as airing out the wound of eros/ion, 211; ambivalent appreciation of the canyon, 174, 175, 211; and Black survivance, 148, 173, 175; the canyon as built by God, 173, 174; on the canyon as unerasable marker of Black enslavement, 174, 175–76; elision of Indigenous homeland thefts, 26, 173–74, 175; enslaved Black laborers mourned in, 173; form and meaning in, 171, 172–73; on racial hierarchy as leveled by the canyon, 173, 174–75, 176; Removal (Trail of Tears) as precipitating the erosion, 172; Webb's multimedia exhibit incorporating pages from, 145, 171. Poems: "The Canyons at Providence," 171–72; "The Erosion March," 172–75

Fletcher, Stephen, 193–94

floods and flooding: 1927 Mississippi River flood, 129, 152; 1938 California flood, 35, 78; and levee building, 119, 120, 129–30; as natural land-building process, 35, 119–20

Florida, 31, 36, 113, 114, 238n23

Forbes, Jack, 22

foreground. *See* background and foreground
Forkner, Ben, 113
Foucault, Michel, 216
Fouss, James, 123
France, Louisiana Purchase and, 121
French creoles. *See* Creole culture

Galveston, Texas: elevation of, 50
Ganzel, Bill, 231n7
Gaza, Palestinian territory, 27, 215–16, 250n23
General Allotment Act. *See* Allotment policies; Dawes Act (General Allotment Act of 1887)
genocide: Allotment policy as attempt at, 74; bison destruction as intended to produce, 76; continued land theft via erosion as attempt at, 76; and the Dust Bowl, 80, 81; Eel River Basin massacres (US military) as, 57, 58, 61, 229n103; gold rushes and, 32, 53, 83; Removal as, 163
Gentry, Curt: *The Last Days of the Late, Great State of California*, 13–14
geological and earth science, 3; calls for interdisciplinary exploration of land degradation problems, 9–10; *erosion* and *erode* as terms used in, 22, 73, 248; sediment equilibrium, 218; uniformitarianism, 148–49. *See also* Lyell, Charles
— popular works in: ecocide models as justifying settler colonialism, 20–22, 224n66; regional exceptionalism as foregrounded by, 14; white pathos as evoked by, 14
— soil science: on ancient Peruvian soil conservation, 217; on burning of monocrop residues vs. mulching the land, 123; as colonial science, 15, 223n41; enslaved labor linked (or not) to soil destruction, 150, 159; on global effects of local erosion, 217; land loss in Louisiana, 107–8, 236–37n1; manure applications, 158; nineteenth-century awareness of soil degradation, 158, 160; Removal linked to gullying, 150, 166; topsoil lost in Dust Bowl, 77–78; on US policy as cause of Dust Bowl, 78; on white settler agriculture as cause of topsoil destruction, 78, 149–50, 158, 161, 213, 217. *See also* Bennett, Hugh Hammond

Georgia, 155, 156, 163–64, 165, 246n17. *See also* Mitchell, Margaret—*Gone with the Wind*; Muscogee Creek Removal; Providence Canyon
Gerstle, Gary, 71
Ghosh, Amitav, 19
Gibson, Rebecca, 244–45n80
Gillis, Rose Ella Duncan (mother of Lynn Riggs), 87
global scale of erosion, 27–28, 105–6, 110, 202, 215–18, 237n11; and eros as desire vs. philia or agape as forms of care, 211–12, 218–19; Indigenous peoples as unequally affected by, 19, 28, 138–39, 153. *See also* climate change
Goeman, Mishuana, 54, 98
gold mining, erosion and, 58, 65
gold rushes: and theft of Indigenous homelands, 32, 53, 156, 165
Goldsmith, Oliver: *A History of the Earth and Animated Nature*, 22, 148
Goode, Abby, 151
Goodman, Audrey, 16
Gordon, Linda, 83–84, 85
Gore, Al, 205
Gray, Jennifer, 204
Gray, Richard, 157–58
Great Depression: wheat-growing craze and, 77–78. *See also* Dust Bowl
Great Plains, southern, 73–74, 76–77. *See also* Dust Bowl
Greeson, Jennifer Rae, 13
Griffith, Andy, 194, 198
Griggs, Gary, 35
Griswold, Jerry, 48, 227n55
Gulf Coast: elevation of Galveston, 50. *See also* Louisiana—and land loss
Gulf of Mexico, sediments and damage to, 110, 218

279

Guthrie, Woody, "This Land Is Your Land," 72, 231n10

Haiti, 113, 114
Hamilton, Alexander, 195
Hammerstein, Oscar, II: "Old Man River," 133. See also *Oklahoma!*
Ham, Oleta Kay Sprague (and Roger Sprague Sr.): *Migrant Mother*, 230n4, 231n7
Handelsman, Jo, 11, 20, 224n66, 232n24
hardscapes. *See* seawalls and other hardscapes to prevent erosion
Harjo, Sterlin, *Reservation Dogs*, 7
Harris, William, 194, 198
Hatteras Island, North Carolina, 193–95, 198, 199. *See also* Cape Hatteras Lighthouse
Hawthorne, Nathaniel, 156
Haymes, Stephen Nathan, 4
Hedgpeth, Joel, 79
Heise, Ursula, 222n25
Helms, Jesse, 194, 197, 198, 199
Herring, Terrell Scott, 237n11
Hilliard, Jack, 198
Historic New Orleans Collection, 139
Hogan, Linda, 98
Holleman, Hannah, 75, 105, 106, 231n9
Hollywood images, 139–40
homelands: defined as Indigenous worldview, 8–9; settler longing for, 48, 64. *See also* Indigenous homelands
Homestead Act (1862), 75
homestead policy, 74–75, 78
Hong, Sharon Linezo (and Monique Verdin): *My Louisiana Love*, 130, 240n83
Hoover, Herbert, 77, 231n15
horizon line, as formal feature of photography, 23, 38–39, 40, 42–43
—as a limit of understanding: Baltz's photography and, 40, 42, 43–44, 45–46, 226n45; in Butler's Earthseed series, 51, 61; definition of, 23; environmental and land-preservation movements and, 23–24; return of Indigenous homelands as lying beyond, 23–24

Houma Nation, United: economic dependence on oil and gas, 132, 135, 137–38, 139; erasure from erosion narratives, 237n4; "informal" removals and difficulty obtaining federal recognition, 121–22. *See also* Verdin, Monique
Howe, LeAnne: tribalography, 8–9, 11, 16, 128, 131, 207
human rights, 14–15, 157. *See also* enslavement; theft of Indigenous homelands
Hungary, erosion in, 27, 216–17
Hunt, Jim, 194, 197, 198, 199
Hurricane Katrina, 109, 134

imperialism: and effects of the Anthropocene, 19, 28, 138–39, 153; as ending worlds, 4, 5; islanding of environmental degradation, 222n21; New World exceptionalism and global linear thinking, 248n70
Inca people, erosion control methods, 217
Indian Removal Act (1830), 121, 148
Indian Territory: agricultural practices in, 75, 76; Curtis Act (1898), 74, 231n15; eastern tribes played as "buffer" against western tribes, 76; location in present eastern Oklahoma, 74; *McGirt v. Oklahoma* (2020), 231n16; Isaac Parker, 95; promised by treaty to be in perpetuity (Indian Intercourse Act of 1834), 74, 75; scholarly mischaracterizations of, 75–76. *See also* Allotment policies; Cherokee Nation; Oklahoma statehood
Indigenous agriculture: acorns, 52, 53; among western Oklahoma peoples, 76; California Native peoples, 36–37, 52, 53; Cherokee women as controlling land management and agricultural practices, 75, 85–86, 96–103, 104, 122, 133, 236n123, 241n92; destruction of by settler colonialism, 217; in eastern Oklahoma (Indian Territory), 76; place-based,

among the Five Southeastern Tribes, 75, 85
Indigenous epistemological and methodological practices, 11–12, 15–16, 222–23n35
Indigenous homelands: 1934 map deeming erosion as "unimportant" on, 1–3, 2; all parts of the continent consist of, 70, 112, 121, 186; of the Eastern Seaboard, 26–27, 186–87, 194–95, 206–7; *homelands* defined as Indigenous worldview, 8–9; loss due to irreparable damage of erosion, 7, 65, 138–39; return to Indigenous control, 6, 7, 23–24, 65, 66, 67, 231n16; of Tangier Island and vicinity, 199–200, 201, 205; Verdin as recontextualizing the realities of claims to, 133. *See also* erasure of material history of Indigenous loss; Indigenous land management strategies; theft of Indigenous homelands
Indigenous identity: codification of, 27, 164, 181, 186, 190; collapse into generic "Native American" vs. specific tribal nation, 71; colorism and, 185–86; dynamic, call to, 184–85; Eastern Seaboard, 181, 184–85; Louisiana peoples and federal recognition, 121–22; phenotypic markers as fallacies of understanding, 184–86; specious claims to, 230n4, 240n79; white nonrecognition of mixed-blood people, 70, 82, 87, 184–85, 230–31n6. *See also* Cherokee people—Cherokee identity; mixed-blood people
Indigenous knowledges: Audubon's dependence on, 114, 120; Butler's Earthseed series as utilizing, 34, 52–53, 54, 56, 63; care beyond immediate returns as lesson of, 217, 218–19; longevity in place as basis of, 20, 23, 36, 37, 53, 217; myth of static knowledge, 20; myth of the "ecological Indian," 20, 21–22, 37; as science vs. mystical claim, 7–8,

19–20, 37; settler colonialism and destruction of, 217; settlers and need for access to, 187; subsumed in liberal environmentalism, 17; tribalography, 8–9, 11, 16, 128, 131, 207. *See also* kinship time
Indigenous land management strategies, 67; in California, 36–37, 67; Cherokee women as controlling land management and agricultural practices, 75, 85–86, 96–103, 104, 122, 133, 236n123, 241n92; Eastern Seaboard wetlands and accommodation of change, 26–27, 186–87, 206–7; ecocide narratives as justifying settler colonialism, 20–22, 224n66; GWTW on Muscogee Creek practices, 164; as human-labor intensive, 217; and the Mississippi River's changes in course, 118–19; Peruvian methods, 217; reciprocal change and growth with the people, 206–7; to reverse erosion, 67; underestimated or ignored, 21, 36–37, 39
Indigenous peoples: and eros/ion, airing out the wound of, 211; and tobacco, 182–83; and the unequal effects of the Anthropocene, 19, 28, 138–39, 153
Indigenous women: bodies of, as site of colonial contest, 98; multivocal multimedia storytelling by, 16, 131. *See also* matrilineal structures of southeastern Indigenous peoples
industrialization, 38–39, 40, 41–44, 226n45. *See also* Baltz, Lewis
International Soil Conservation Organization, 9–10
invasive species: Bermuda grass, 126, 128; Cherokee rose, 127, 156
Iraq, 211
islands: American exceptionalism as quarantining, 180–81; as barrier of safety, 191, 192. *See also* Eastern Seaboard
Isle de Jean Charles Indigenous community, 50, 130, 227n65

Israel, and Palestinian territories, 27, 215–16, 249–50n20, 250n23
Italy, 211, 213–15

Jackson, Andrew, 148
Jackson, Robert, 152, 153
Jamestown, Virginia. *See* Wood, Karenne
Jefferson, Thomas, 33–34, 121, 225n16, 232n24
jetties, 203
Jim Crow, school segregation, 168, 169
Justice, Daniel Heath, 88–89

Kahrl, Andrew, 188, 191
Kaufman, Wallace, 181
Kaw people, 231n15
Kichai people, 76
King, John, 48–50, 66, 211
King, Thomas (Tom), 66
King, Tiffany, 22, 53, 64, 150, 169–70
kinship time, 11, 49, 135; reciprocal change and growth, 206–7. *See also* Whyte, Kyle Powys
Kiowa people, 76
Kornecki, Ted, 123
Krasnoff, Mark, 130
Kreyling, Michael, 152

#LandBack, 7
land building as natural process, 115–20, 123, 126–27, 178
landscape art, 39–40, 43–44, 45. *See also* Baltz, Lewis; New Topographics photography movement
landslides and mudslides, 35, 50, 58
Lange, Dorothea: portraits of Southwest Indigenous people, 82; and quotidian backgrounds, 109; Verdin's reuse of image from, 132–33; white sight of, 82
—Farm Security Administration work with Paul Taylor: documentation of the US South, 84, 132–33; and settler anxiety of disappearing white supremacy, 83–84; and settler colonialism in erosion metaphors, 83, 132–33; whiteness as focus of, 82–84
—*Migrant Mother*, 69; California as location of photograph, 70, 86; Cherokee identity of subject not recognized by Lange, 70, 82, 87, 230–31n6; corrective narratives of, as dependent on white supremacy, 71; corrective narratives of, collapsing tribal identity into generic "Native American," 71; as iconic image of impoverished white migrant farmers of the Dust Bowl, 68, 70–71; name of subject not known by Lange, 70, 230–31n6; not published in *American Exodus*, 85, 86–87; reputation of Lange as cemented by, 70. *See also* Thompson, Florence (née Christie, subject of Lange's *Migrant Mother*)
Lange, Dorothea (and Paul Taylor): *An American Exodus*, 73, 82, 84–85, 86–87
La Paperson. *See* Yang, K. Wayne
Latham, Harold, 142
Lehman, George, 113, 114–15, 118
Leonardo da Vinci: *Mona Lisa*, 132
levees. *See* Mississippi River
Liboiron, Max, 4, 6, 17, 22
Lighthouse Act, 195
Lightweis-Goff, Jennie, 237n7
Limbo, 181, 200–203, 205–6, 207, 247–48nn63–64, 248n76
Lipman, Andrew, 180, 187
literary regionalism: agrarian fantasies and Confederate Lost Cause pathos, 152, 161; challenges to the universalizing "we," 6–7; historical patterns, 6; knowing where one is standing, 6; the particular and diverse vs. the exceptional and totalizing, 88; retrospective as well as prospective thinking, 5–6; spiralic time, 6
littoral cell system, 178–80, 181, 186, 198, 205
Longhin, Elena, 214
Lookingbill, Brad, 231n9
Lopinot, Neal, 21

Los Angeles River, 35
Los Angeles Times: article on coastal erosion (Xia), 29, 31–32, 35, 43, 66, 224n2; "The Ocean Game" (coastal erosion), 29–31, 45, 66, 67, 194
Lost Cause narratives: as abdicating necessity of care, 219; as born of white supremacy, 9; California exceptionalism and, 31, 55–56; erasure of Black enslavement in, 150; erasure of theft of Indigenous homelands in, 150; futility of recovery from industrial land abuses, 38–39; and New Topographics photography movement, 34; optimism of the past as, 55–56. *See also* Confederate Lost Cause pathos
Louisiana: arts and culture sponsorships by oil and gas companies, 139; Audubon's self-invention of origins and focus on, 112–13, 114, 115; colonial history of, 111–12, 122, 127; cultural representation of, foreground and background, 109, 111–12, 123–24, 139–40, 237n7; New Orleans, 109, 111–12, 113, 117, 139, 237n7
—and land loss: amount of ongoing land loss, 107–8, 140, 236–37n1; overview of causes, 110; erasure of Indigenous peoples' experience of, 108–9, 237n4; geographic and temporal scale and, 110, 237n11; Hurricane Katrina and, 109, 134; Isle de Jean Charles Indigenous community, 50, 130, 227n65; lack of sediment deposits and, 110; land subsidence and fragmentation, 110, 131; sea-level rise and, 110, 133–34; and settler anxiety of disappearing white supremacy, 109–10, 113, 114–15, 117, 140; as theft of Indigenous homelands, 117–18, 129–30, 140; white Cajun (and Creole) losses as standard of discourse, 108, 109, 130, 140, 237n4, 237n7. *See also* Mississippi River; oil and gas industry; plantation enslavement economy

—Indigenous peoples of: ancient earthworks found, 112; Audubon recording presence of, 120–21, 122; colonial history narratives and erasure of the deep history of homelands and presence, 111–12, 122, 127; economic dependence on oil and gas industry, 132, 135, 137–38, 139; illness due to oil and gas industry, 130; "informal" removals and unique difficulties in obtaining federal recognition, 121–22; invisibilization of, 122, 139–40; and liminality (*sfumato*), 131, 132, 140; as pushed into the wetlands farthest south, 121, 129. *See also* Verdin, Monique; *specific peoples*
Louisiana Purchase (1803), 116, 121, 239n50
love: eros as desire vs. philia or agape as forms of care, 209–11, 218–19
Lyell, Charles, 15, 26, 155; as apologist for enslavement, 150, 175; *erode* as term credited to, 22, 148; on Removal and gullying, 150, 166; on settler agriculture as gullying cause, 149–50; uniformitarianism, 148–49. Works: *Principles of Geology*, 22, 148–49; *A Second Visit to the United States of North America*, 149–50
Lyell Gully, 149
Lyons, Kristina, 6, 211, 223n41

McGirt v. Oklahoma (2020), 231n16
McKay, Mabel, 10
McKittrick, Katherine, 25, 143, 145, 148
MacLeish, Archibald: *Land of the Free*, 84
MacRae, Hugh: as developer and owner of Wrightsville Beach, 187, 188; and development of Shell Island, 188; fire destroying Shell Island, 189; Hugh Morton as grandson of, 194, 247n54; and the Wilmington massacre (1898), 180, 187, 188, 189
MacRae, Nelson, 189
Magnus, Albertus, 247–48n63
Magyar people, 216

managed retreat: in Butler's Earthseed series, 60; California projects of, 45, 48–50, 225n9; defined through "The Ocean Game," 30–31; elevation of communities claimed to be, 48–50; elevation of communities shown to be unsuccessful, 50; relocation of Cape Hatteras Lighthouse, 194, 198–99; sea-level rise and necessity of, 198; Tangier Island dwellers as rejecting, 203

Manifest Destiny, 166

Manovich, Lev, 33

Mansur, Eduardo, 217

Manteo, 194

Marghera, Italy, 213–15

Maryland, sand rights, 36

Mason, Joseph, 113

Massachusetts, 32–33, 235n99

matrilineal structures of southeastern Indigenous peoples: clan systems, 85, 87, 122, 230n4, 233n65; and conceptions of the feminine in nature, 133, 241n92; land management and agricultural practices controlled by women, 75, 85–86, 96–103, 104, 122, 133, 236n123, 241n92

Mattole people, 57

Maya city collapse: myth of ecocide, 224n66

Meister, Sarah Hermanson, 230–31n6

methodology, 9–10; appropriation of Indigenous methodologies, avoidance of, 35, 222–23n35; appropriation of Indigenous methodologies, used as excuse to avoid the work of engagement, 222–23n35; and the archive, listening to, 16; arrangement of chapters from west to east, 22, 31, 70; critical Indigenous studies, 9–12, 14–16, 22–23; critical regional studies, 9–10, 12–14; engagement, 208; erosion not a metaphor, 15; Indigenous epistemological and methodological practices, 11–12, 15–16, 222–23n35; and the limitations of periodization, 10; materiality and material agency, 15–16; omissions of racial slurs, 239n61, 243n57; organization within chapters as avoiding Indigenous history viewed as prequel, 23; positionality of the researcher, 10, 12, 177–78, 207–8, 209–11, 216, 222–23n35; temporality, adjustment of, 10–11, 22. *See also* visual archive

Mexican Americans: as subjects in *American Exodus* (Lange and Taylor), 84

Mexican rancheros and labor system of enslavement, 53

Mexico: Chiapas, 212; settler colonialism as arriving via, 31

Midwest: as region, 70, 230n4

Migratory Bird Treaty Act (1918), 113, 238n23

Milledgeville, Georgia, 149

Milton, John: *Paradise Lost*, 248n76

mining and erosion, 21, 35, 36, 38–39, 58, 65

Miranda, Deborah: *Bad Indians*, 24, 34, 64; finding the location of her family's land grant (El Potrero), 64–66; as multigenre memoir, 16, 64, 131; on the wound of loss of homeland as intergenerational trauma, 64, 66, 211

Mirzoeff, Nicholas, 16, 40, 82, 113, 114, 115–16, 143, 145, 148

Mississippi delta, 152

Mississippi River: flood of 1927, 129, 152; Indigenous peoples and, 118–19, 133, 241n92; levee construction, 110, 119, 120, 129–30; natural erosive processes of, 110, 115–20, 123, 126–27

Mitchell, Margaret: Bennett's possible contact with, 244n60

—*Gone with the Wind* (GWTW): 25–26, 141–45; anxiety of disappearance in, 153–54, 155–57; destroying what she purports to love (eros/ion), 210–11; film adaptation (David Selznick, dir.), 142, 153, 158; Indigenous Removal as theft of homelands, and Tara, 145, 149, 162–65, 166; and myth of agrarianism, 161, 175; as naturalizing the settler plantation landscape, 143, 144, 152; as naturalizing white Confederate Lost Cause pathos, 145, 156–57, 158, 161; and plantation futures, 142–43, 152, 158–61, 164–65; romanticization of racism in, 144, 164–65; romanticization of rape in, 144, 162; and roots, white nativist desire for, 161–62; Scarlett's name, 243n41; and settler paradox of their own destruction, 159–60, 161; and the southern grotesque, 153; southern literature as overdetermined by popular images from, 175; the surface and meaning of, 154, 155, 161–62; title taken from Ernest Dowson poem, 142–43; white supremacist, anxiety of disappearing, 158, 159–60; the work of settler memory embedded in, 143, 145, 161

—GWTW and erosion: awareness of erosion prevention, 154, 155, 158, 160; descriptions of soil exhaustion and erosion, 153–55, 156–58, 157, 243n41; Dust Bowl as context of, 26, 141–43, 242n8; enslavement in, as linked to land abuse, 150, 157, 159–60; and gullying, 155, 157; "I'll never go hungry again" scene, 153–54, 158; Indigenous Removal connections to, 153, 154, 156–57; lumber business and timber extraction, 157, 161–62; romanticization of soil exhaustion, 144, 149, 152, 155, 161; title as erosion metaphor, 26, 142

Miwok people, 37

mixedblood people: Cherokees eligible for allotments, 81; white nonrecognition of Indigenous identity of, 70, 82, 87, 184–85, 230–31n6

Mohr, Jean, 215–16

Monacan people. *See* Wood, Karenne

Montgomery, David: *Dirt*, 14, 75–76, 151

Moore, Carol, 202

Morton, Hugh, 27; *Cape Hatteras Lighthouse* (aerial view), 195, *196*, 197; Hugh MacRae as grandfather of, 194, 247n54; as photographer and activist, 194, 195, 247n54; and the "Save the Cape Hatteras Lighthouse" campaign, 194, 197, 198, 199

Mosul, Iraq, 211

mounds and earthworks, 20–21, 112, 175–76

Mt. Pleasant, Jane, 21

Muir, John: and myth of pristine wilderness, 36–37, 39

Muscogee Creek people: Allotment policy amended to include, 74; GWTW on land stewardship by, 164; plantation enslavement economy, participation in, 163. *See also* Five Southeastern Tribes; *McGirt v. Oklahoma* (2020); Muscogee Creek Removal

Muscogee Creek Removal: Audubon as witness to, 120–21; in GWTW, 145, 149, 153, 154, 156–57, 162–65, 166; plantation enslavement economy and, 150–51, 162–64, 165–66, 169–70; Providence Canyon formation precipitated by, 26, 145, 149, 150–51, 166, 169–70, 172, 175; Treaty of Indian Springs (fraudulent), 162

Naya, Ramon, 104–6, 211

Neimanis, Astrida, 12

Nelson, Melissa, 241n92

new materialism, 15–16

New Orleans, 109, 111–12, 113, 117, 139, 237n7

New Topographics photography movement: alienation of audiences by, 43, 226–27n46; and erasure of Indigenous homelands claims, 34, 43–44; lost-cause narrative undergirding, 34; as pushing against a California ideology, 34; as reaction to corporate boosterism, 49; space as "final frontier" influence on, 50, 225n32

New York, 32–33

Nichols, Robert, 8, 80, 125, 145, 159, 210, 239n50

Nishime, LeiLani, 4

Nixon, Rob, 5–6, 8, 22, 23–24, 111, 112, 131, 207
Nobles, Gregory, 114
Nongatl people, 57
North Carolina, 178n1. *See also* Cape Hatteras Lighthouse; Wrightsville Beach; Wrightsville Beach—and Shell Island
Northrop, Robert H., 187, 189

oaks: acorns, 52, 53, 54–55, 58; as metaphor of rootedness, 105; saltwater intrusion as killing, 133–34
Obama, Barack, 204
"Ocean Game, The," 29–31, 45, 66, 67, 194, 198
ocean warming, 190–91. *See also* sea-level rise
"off the grid" imaginary, 48
Ohlone/Costanoan-Esselen people, 64–66
oil and gas industry: automobile travel, 151; avarice of, 135, 137, 138–39; global impact of, 135, 137; land loss in Louisiana wetlands due to, 53, 110, 131–34; Native dependence on economic opportunities, 132, 135, 137–38, 139; pipeline protests, 139–40; pollution due to, 130, 134, 140, 240n83
Oklahoma, 70, 230n3. *See also* Dust Bowl
—statehood: Allotment policy as paving the way to, 74; effect on the Cherokee Nation, 89, 96; and erosion of Cherokee women's power, 96–97, 98–102, 103, 104; erosion of topsoil as following settler agriculture, 2, 2; in Riggs, 90, 93, 95, 96–97, 100–101, 104, 236n119
Oklahoma Territory: incorporation of (1890), 75; "land runs" by white settlers, 74–75; location in present western Oklahoma, 75; scholarly mischaracterizations of, 75–76. *See also* Indian Territory
Oklahoma! (Rodgers and Hammerstein), 24, 72, 87, 88, 90–91, 94, 95.

See also Riggs, Lynn—*Green Grow the Lilacs*
Orbán, Viktor, 216
Orona, Brittani, 37
Oroville, California, 86
Osage people, 76
Outer Banks. *See* Eastern Seaboard; North Carolina
Owens, Cleo, 86
Owens, Louis, 79, 81, 82

Padgett, Ieva, 124, 125
Palestinian territories, 27, 215–16, 249–50n20, 250n23
Parker, Isaac, 95
Parmele, C. B., 187–88, 189
Parmele, Edgar, 187
Pendleton, Edmund, 225n16
Perdue, Theda, 85, 86
Perkerson, Medora, 143
Peru, ancient Indigenous erosion control methods, 217
petroculture. *See* oil and gas industry
Pettit, Alexander, 88–89
photography: aerial, 195, *196*, 197; horizon line, 23–24, 38–39, 40, 42–43; and the myth of the pristine wilderness, 39–40; portraits as determined by the photographer, 82. *See also* Lange, Dorothea; New Topographics photography movement; Verdin, Monique
Pilkey, Keith, 203
Pilkey, Orrin, 14, 49–50, 181, 203, 245n1
plantation enslavement economy: Audubon as apologist for, 114–20; Audubon as enslaver, 113, 114; and Black survivance on the land, 148, 173, 175; collapse of, 84; entanglement of southeastern Indigenous peoples in, 75, 85–86, 163; GWTW as naturalizing landscape of, 143, 144, 152; as harbinger of erosion, 116; and levee construction on the Mississippi River, 110, 119, 120, 129–30; plantationocene, 153; and Removal, 150–51, 153–54, 162–66. *See also* agriculture of settler colonialism;

enslavement; plantation futures; white supremacy
—agriculture as cause of erosion: agrarianism and romanticization of ("agrotopias"), 144, 151–52, 161, 175; burning crop residue vs. mulching the land, 123; cotton, 79–80, 84, 154–55, 157, 158–59, 160, 162, 165–66; gullying erosion, 117, 144, 149–50, 166, 168, 170; in *American Exodus* (Lange and Taylor), 84–85, 132–33; Lyell on, 150, 166; migration of settlers due to, 113–15; natural erosive processes of the Mississippi River, 116–20, 123, 126–27; Providence Canyon (gully) formation, 144, 149–50, 155, 165–66, 175; rice, 238n29; sheet erosion, 155; sugarcane, 123, 132–33; tobacco exploitation, 165–66, 182–84; water runoff, 123, 144, 170; white pathos of loss and, 119–20, 122, 124–25, 127, 129. *See also* Mitchell, Margaret—*Gone with the Wind*—and erosion; Removal—erosion as precipitated by; soil exhaustion
plantation futures: definition of, 143; in GWTW, 142–43, 152, 158–61, 164–65; Providence Canyon as unerasable marker of Black enslavement, 174, 175–76; and settlers' paradox of their own destruction, 159–60, 161; southern disaster complex and, 152–53; and theft as generating property, 159, 160–61. *See also* McKittrick, Katherine
plantationocene, 153
plows and plowing, 74, 77, 154, 155, 200–201, 232n24
plum-leaf azalea, 170–71
pluriverse, the, 27, 212
Pocomoke people, 199–200, 201
Pomo people, 10
Pope Benedict XVI, 202, 248n64
property: African American property acquisition, 57–58; individual ownership of, 5, 57–58; private, class politics and priority over public interest, 178; rendering of Indigenous homelands into, 8; speculation, 57–58; timber extraction, 58, 117–18, 124, 149, 150, 155, 157, 161–62. *See also* real estate development; theft as generating property
Providence Canyon, 146–47; creationist theory of, 244–45n80; the geological record revealed by, 144, 148–49, 150, 151–52; as Georgia state park, 150; as geotext, 152, 175–76; as gully, 25–26, 144; as indelible marker of Black enslavement and Indigenous dispossession, 174, 175–76, 210; as "Little Grand Canyon," 144; as monument to plantation economy, 167; Muscogee Creek Removal as precipitating, 26, 145, 149, 150–51, 166, 169–70, 172, 175; named for Providence Church, 144, 166; plantation enslavement agriculture as cause of, 144, 149–50, 155, 165–66, 175; settler theory of cause blaming Indigenous walking trails, 166–67; sharecropping economy and, 175; and southern exceptionalism, 167–68; timescale of emergence, 144, 148–49, 166; tourist boosterism as "natural wonder," 26, 145, 151–52, 166, 167; as US Government poster child for land abuse, 151; and white supremacy and settler agriculture, naturalization of, 167; the work of settler memory and desire for absolution, 167. *See also* Flanagan, Thomas Jefferson; Webb, Elizabeth
Prud'homme-Cranford, Rain, 25, 108
public interest law, sand rights, 35–36
Purdy, Jedediah, 14
Pushkin, Alexander: "The Prophet," 104

race, codification of: as buttressing white supremacy, 164, 190; capitalist enslavers' demand for, 27, 181, 186 racism: anxiety of white liberal masculine futurity, 190–93; and the Black resort on Shell Island, 188, 189; California ideology and reproduction of, 33–34; colorism among Native peoples, 185–86; Flanagan on Providence Canyon as leveling racial hierarchies, 173, 174–75, 176; GWTW as romanticizing, 144, 164–65; Wrightsville Beach bridge "pass" system for Black visitors, 191. See also segregation; white supremacy

Rankin, Caitlin, 20–21

Raper, Arthur, 84, 168, 169

Reaganism, 41, 45

real estate development, 48–50, 57–58, 76–77, 181, 186, 187, 189. See also Wrightsville Beach, North Carolina

Reed, Kaitlin, 37, 58

regional exceptionalism, 13–14, 34, 84–85, 139, 180, 237n7. See also American exceptionalism; California exceptionalism; southern exceptionalism

regions: colonial teleology of east-to-west conquest, 22, 31, 70; interrelation of, 70; North/South regional boundary, debates on, 179; varying definitions of, 230n3

Removal: Audubon's narrative of witnessing, 120–21; "Cherokee rose" as romanticized symbol of, 156–57; Dahlonega gold rush and, 156, 165; fraudulent treaties, 162, 163; Indian Removal Act (1830), 121, 148; Thomas Jefferson and, 34, 121, 225n16; Louisiana peoples, 121–22; of original Indigenous peoples in southern Great Plains, 76; plantation economy and, 150–51, 153–54, 162–66; series of, and presence of Cherokee diaspora in California, 86–87; as Trail of Tears, 156–57, 172. See also Allotment policies; Indian Territory; Muscogee Creek Removal

—erosion as precipitated by: 1934 map, 1–3, 2; geologist (Lyell) on, 150, 166; in GWTW, 153, 154, 156–57; at Providence Canyon, 26, 145, 149, 150–51, 166, 169–70, 172, 175. See also Allotment policies—erosion as precipitated by

return of homelands to Indigenous control, 6, 7, 23–24, 65, 66, 67, 231n16

Rian, Jeff, 45, 226–27n46

Rich, Frank, 234n92

Ridge, John, 235n99

Rifkin, Mark, 10, 53–54, 127, 143, 150

Riggs, Lynn, 72–73; on "a kind of truth" of the particular and diverse, 88; birthplace and time of, 87, 88; Cherokee identity of, 87; Cherokee mother of, 87; the Cherokee Nation as setting of plays, 103–4; on creation vs. destruction, 105; as ethnographer, 72–73; as exploring alternative modes of recognition and belonging in the post-Allotment world, 89–90; as gay man, 87, 94, 104; importance as playwright, 72, 88–89; stepmother of, 87, 98–99; white-settler father of, 87. Works: *The Cherokee Night* (1936), 90; "The Vine Theater" (with Ramon Naya), 104–6, 211

—*Green Grow the Lilacs* (1931), 24–25, 87–88, 90; Curly as Cherokee character in, 90–95, 234nn91–92; "Green Grow the Lilacs" as closing song, 95; Jeeter's attack as attempted lynching of an interracial couple, 93–94, 235nn98–99; *Oklahoma!* musical based on, 24, 72, 87, 88, 90–91, 94, 95; Oklahoma statehood and Allotment as looming presence in, 90, 93, 95, 104; as prophetic of the Dust Bowl, 90, 92, 95, 96; settler agriculture referenced in, 90, 92–93, 96, 104; uncertainty of ending of, 94–95

—*The Cream in the Well* (1940), 24–25, 87–88, 90; as airing out the wound of eros/ion, 211; Cherokee identity of the Sawters family, 96, 236n118; Cherokees as both agent and victim of exploitative agriculture,

102–3, 104; and erosion of Cherokee women's power, 96–97, 98–102, 103, 104; gay sex work in, 97, 98; gnawed power in, 99–100, 102; murder in, 97; Oklahoma statehood and Allotment as looming presence in, 96–97, 100–101, 236n119; settler colonial agriculture referenced in, 100–101, 102–3, 104, 236n123; sibling incest in, 97, 102, 211; suicides of women in, 96, 97–98, 101–2, 104; titular metaphor, 101–2, 236n119; "the way to fight destruction is not with destruction," 105

rill erosion, 144

rivers: damming and control projects, 35, 41, 58, 211; dams, reversal of, 67; erosion and sediment transport as natural process, 58, 110; fish kills, 58, 150; soil erosion and landslides, 58; timber extraction and ecological problems in, 58, 150. *See also* Mississippi River

Roberts, Alaina, 76

Rodgers, Richard. *See Oklahoma!*

Rodriguez, Allison (cousin of Monique Verdin), 135, *136*

Romine, Scott, 7, 88

Roosevelt, Franklin D., 142

rootedness, as metaphor, 104–6

roots of native plants, erosion prevention, 74, 77, 126, 128, 129, 170–71, 232n24. *See also* deforestation

roots: shallowness of settlers', 80, 83, 161–62

Rountree, Helen, 187

Russia, settler colonialism coming from, 31

Rutherford Falls (television series), 7

Said, Edward: *After the Last Sky*, 27, 215–16

sand and gravel mining, 35, 36

sand rights, 35–36

San Francisco: San Quentin Point, 44–45, 48. *See also* Candlestick Point

San Francisco Chronicle (Candlestick Point article), 48–50, 66, 211

Santa Lucia Preserve, 65–66

Sargent, Bradley, 64, 65

Sarris, Greg, 10

Schell, Jonathan: *The Fate of the Earth*, 44–45

Scheppe, Wolfgang, 40

Schmitt, Carl, 248n70

Scott, James, 20

Scranton, Roy, 14

sea-level rise: and Candlestick Point development, 48–50; elevation of communities as "managed retreat," 48–50; elevation of communities as unsuccessful strategy, 50; erosion as accelerating with, 198, 202; glacier melt, 204; as global crisis, 211; and Hatteras Island, 193–94, 198, 199; and Louisiana land loss, 110, 133–34; oil and gas industry and, 110, 133–34; salt-water intrusion, 133–34; and structural plans to slow beach erosion, 198; and Tangier Island, 202, 205; Tangier Island residents on the politics of denying, 202; worst-case prediction for, 48

seawalls and other hardscapes to prevent erosion: damage deferred elsewhere, 129–30, 178, 203; jetties, 203; as prohibited by law in North Carolina, exceptions made for, 178; riprap walls, 203; sea cliff erosion in California, 35; and Shell Island Resort, 178; Tangier Island, 203–5

sediment: and damage to the Gulf of Mexico, 110, 218; damming of inland waterways and starvation of beaches, 35, 41, 58; equilibrium of, 218; levee control of the Mississippi River and starvation of wetlands, 110; sand and gravel mining, 35, 36

segregation: of beaches, 187, 188, 189; of schools, 168, 169, 192

Sekula, Allan, 41–42

Selu (Corn Mother), 85, 236n123

Selznick, David O. (dir.): *Gone with the Wind*, 142, 153, 158

Seminole people, 74. *See also* Five Southeastern Tribes

Serbia, 211

settler anxiety of disappearance: the American gothic trope of threatening haunted forest and, 156; Baltz's photographs and, 45; in Butler's Earthseed series, 60–61, 216; California exceptionalism and, 45; the Dust Bowl and, 24, 72, 80, 81, 82–84; ecoanxiety distinguished from, 221–22n14; and erosion anxiety, as mistaken for one another, 216; and Lange's photography, 83–84; settler time and, 10–11, 30–31, 207; in Steinbeck, 24, 80, 81; as subtending conversations about erosion, 5; Tangier Island and perennial crisis akin to limbo, 181, 201–3, 205–6. *See also* Confederate Lost Cause; democracy, anxiety about the erosion of; environmentalism and the settler colonial future; lost-cause narratives; white supremacy—anxiety of loss of

settler colonialism: appearance in multiple places in multiple times, 31; Butler's caution against Black futurism taking the form of, 53–58, 62–64; conquest as slow consumption vs. discrete event, 22; conscious capitalism as, 65–66; continual recursivity of, 145; criminalization of growing food, 215–16; denial of right of other humans to exist on the planet, 218; as destroying what it purports to love (eros/ion), 209–11, 218–19; destruction of Indigenous agriculture, 217; east-to-west progression of, as myth, 22, 31, 70; inevitability narrative of, 45–46; lighthouses as symbol of protection of, 27, 195, 199; and naturalized narratives of place, 4, 145, 148; relativisms seeking to justify, 21; Steinbeck alluding to, 79–81; and the unequal effects of the Anthropocene, 19, 28, 138–39, 153. *See also* agriculture of settler colonialism; theft as generating property; theft of Indigenous homelands; virgin untouched land, as myth

settler environmentalism, 18–19. *See also* environmentalism and the settler colonial future

settler moves toward innocence, 11, 167; as seeking reconciliation ahead of accountability, 18, 32, 183–84. *See also* work of settler memory

settlers: descendants of African American enslaved people distinguished as nonsettlers, 53; as fetishizing the moment colonial feet touched dirt, 187; longing for homelands, 48, 64; as mistaking the desire *for* with the care *of*, 211; and paradox of their own destruction, 159–60, 161; as posture vs. slur, 211; shallowness of roots of, 80, 83, 161–62

settler time, 10, 48–49, 111–12, 122, 127, 143, 160–61; and settler anxieties, 10–11, 30–31, 207. *See also* plantation futures; Rifkin, Mark

sfumato, 132, 140

sheet erosion, 155

Sheller, Mimi, 222n21

Shell Island Resort, 177–78, 179, 193, 207–8

Shell Island. *See* Wrightsville Beach—and Shell Island

shoals: as metaphor of connection for Black and Native Studies, 169–70

Sierra Club, 36–37

Silko, Leslie Marmon: *Storyteller*, 131

Sinkyone people, 57

Siouan people, 182–83

small ecologies, 3–4

Smith Island, Virginia, 199–200

Smith, John, 199–200, 205; and Limbo, 181, 200–203, 205–6, 207, 247–48nn63–64, 248n76

Snedeker, Rebecca, 108, 111–12, 127

sodbusters, 77

Soil Conservation Act (1935), 142

Soil Conservation and Domestic Allotment Act (1936), 142

soil exhaustion: cotton and, 79–80, 84, 154–55, 157, 158–59, 160, 162,

165–66; the Dust Bowl and, 103; and migration of settlers from the Eastern Seaboard, 116–17; plantation enslavement economy and, 113–15, 151, 153–55, 156–58, 157, 164, 243n41; romanticization of, 144, 149, 151–52, 155, 161, 175; soil scientists on, 155; sugarcane and, 123; as theft of Indigenous homelands, 166; tobacco exploitation, 183–84; US government response to, 142, 151
soil science. *See* geological and earth science
Solnit, Rebecca, 33, 50, 108, 111–12, 127
Soros, George, 216
Sorsby, N. T., 158
South Africa, 27, 212–13
South Carolina, 113, 114–15, *116*
southeastern Indigenous peoples. See Five Southeastern Tribes; Louisiana—Indigenous peoples of; *specific peoples*
Southern Agrarians, 152, 161; *I'll Take My Stand*, 152
southern exceptionalism, 13, 18, 32–33, 84–85, 167–68, 180–81
Spanish missionary system, 32, 53, 54, 56, 61, 226n45
spiralic time, 6, 10–11
Sprague, Roger, Sr. (and Oleta Kay Sprague Ham): *Migrant Mother*, 230n4, 231n7
Squint, Kirstin, 140, 240n83
Standing Rock: Dakota Access Pipeline protests, 139–40
Stanyhurst, Richard, translation of Virgil's *Aeneid*, 200–201
Steinbeck, John (*The Grapes of Wrath*), 24, 73; the California flood of 1938 in, 35, 78; Cherokee characters in, 81–82; Cherokee epistemologies and insight into, 11; the Cherokee Nation as starting place of, 24, 78–79, 81–82, 87, 103–4; context of economic and ecological collapse, 79–81; the Dust Bowl as fait accompli in, 79, 88; interstitial chapters, 80–81; and settler anxiety of disappearance, 24, 80, 81; undermining of sympathy for white settler farmers in, 80, 81
Stein, Sally, 71
Stone, Katherine, 35–36
Stryker, Roy, 82–83
subsidence (overall land sinking), 110, 131
sugarcane, 123, 132–33
Sutter, Paul (*Let Us Now Praise Famous Gullies*), 14, 145, 149, 151–52, 165–66, 167–68, 169, 171, 174
Swensen, James, 82
Swift, Earl: *Chesapeake Requiem*, 27, 180, 181, 202–5. *See also* Tangier Island

Tahlequah, Oklahoma, 70, 86
Tangier Island, Virginia: anxiety of disappearance as perennial crisis akin to limbo, 181, 200–203, 205–6, 207; and capitalist need for static land, 181, 205; climate change skepticism in, 27, 180, 202; destroying what they purport to love (eros/ion), 210–11; "erosion" as focus of residents (vs. sea-level rise), 27, 180, 202, 205; as foregone space of loss, 180–81; as Indigenous homelands, 199–200, 201, 205; littoral cell system and, 178–80, 181, 205; managed retreat rejected by residents, 203; and the politics of denying sea-level rise, 202; riprap wall, 203; savior sought by residents, 203–5; sea-level rise and, 202, 205; seawall demanded by residents, 203–5; John Smith as mapping area, 199–200, 205; Trump as widely supported by, 27, 180, 204–5
Taylor, Melanie Benson, 4, 6–7, 20
Taylor, Paul, 83. *See also under* Lange, Dorothea
technocrats, 33–34, 40, 214
temporal optic of retrospective and prospective thinking, 5–6, 8, 10–11, 111–12, 131, 207
terra nullius. See virgin untouched land, as myth

Thailand, 211
theft as generating property, 8; Black life and labor in enslavement, 170; eros and eros/ion and, 210; illustrations of, 80, 122, 125, 127; as imbricating Indigenous peoples, 8, 139; plantation futures and, 159, 160–61; as recursive, 8, 127; title and valuation of land and, 8–9, 122, 127, 158–59. See also Nichols, Robert; property; theft of Indigenous homelands
theft of Indigenous homelands: and Black enslavement, direct relationship of, 26, 150, 157, 159–60, 169–71, 173–74, 175; Confederate Lost Cause and erasure of, 150, 161; and the Dust Bowl, 80, 81, 104; erosion and soil exhaustion of plantation enslavement economy as, 166; erosion of beaches as, 35; erosion of coastal wetlands as, 117–18, 129–30, 140; genocidal campaigns and, 57, 58, 61, 83, 229n103; gold rushes and, 32, 53, 156, 165; intergenerational trauma of, 64, 66, 211; Louisiana Purchase (1803) as, 115, 121, 239n50; return of homelands, 6, 7, 23–24, 65, 66, 67, 231n16; by speculators and land thieves, 64–65; by title and taxation, 104, 117–18, 158–59. See also Allotment policies; Removal
Thomas Aquinas, 200
Thompson, Florence (née Christie, subject of Lange's *Migrant Mother*), 69; birthplace of, 70, 87, 230n4; California presence of, as due to series of removals, 86–87; Cherokee identity of, 70–71, 86–87, 230n4, 231n7, 233n65; as feeling exploited and misrepresented by Lange photograph, 70; marriages and children of, 86
Tidwell, Mike, 14, 237n4
timber extraction, erosion and, 58, 117–18, 124, 149, 150, 155, 157, 161–62
time: spiralic, 6, 10–11. See also kinship time; settler time

timescale: of erosion, 22; of Providence Canyon emergence, 144, 148–49, 166
tobacco-based land exploitation, 165–66, 182–84
Todd, Zoe, 5, 12
topsoil erosion: amount lost in the Dust Bowl, 77–78; media referencing *GWTW* in article on (1940), 141–42; soil scientists on white settler agriculture as cause of, 78, 149–50, 158, 161, 213, 217. See also Dust Bowl
Treaty of Indian Springs (fraudulent), 162
tribalography, 8–9, 11, 16, 128, 131, 207
Tribe, Mark, 33
Trump, Donald: Trumpism, 13, 17–18, 27, 180, 204–5
Tuck, Eve, 11, 14–15, 21, 66
Tunica people, 121
Turner, Frederick Jackson, 180
Tuscaloosa, Alabama, 149
Tuscarora War, 194
Twain, Mark, 133

United Houma Nation. See Houma Nation, United
United Nations: Conference on Trade and Development, 215; Food and Agriculture Organization, 217
United States: all parts of, as Indigenous homelands, 70, 112, 121, 186; as fetishizing the moment colonial feet touched dirt, 187; genocidal military campaigns against Native people, 57, 58, 61, 229n103; government, 77–78, 78, 130, 142, 151, 195, 199; *McGirt v. Oklahoma* (2020), 231n16; North/South regional boundary, debates on, 179; shallowness of civil culture, 43. See also Allotment policies; homestead policy; Removal
US Army Corps of Engineers, 36, 120, 129–30, 190
US Bureau of Indian Affairs (BIA), 121–22

Vaill, Patrick, 94
Vanhoutte, Kristof, 201–2, 205
Venice, Italy: sea-level rise and erosion, 27, 213–15
Verdin, Herbert (father of Monique), 130, 132, 185
Verdin, Matine (grandmother of Monique), 130, 133–34, *134*, 135
Verdin, Monique, 130; as airing out the wound of eros/ion, 211; camera used by, 133; on lack of cultural traction of Bayou Bridge Pipeline protests, 140; as United Houma Nation citizen and activist, 130. Works: chapter in *Unfathomable City* (Solnit and Snedeker), 108; *My Louisiana Love* (Hong, dir.), 130, 240n83
—*Return to Yakni Chitto* (2019), 25, 109, 131; backgrounds of, as adjusting and confounding perspective, 131, 132, 133–39, *134*, *136*, *138*; multiple genealogies utilized in, 132–33; multivocal multimedia storytelling in, 16, 131; and narrative of continuity and change, 134, 135; as recontextualizing the realities of Indigenous claims to homelands, 133; texts accompanying the images, 131, 134, 135, 137–39; womanhood centered as vital perspective, 133, 241n92. Works: *Allison Rodriguez*, 135, *136*, 137; *Graduation*, 137, *138*; *Lil Black Chicken*, 135; *Matine's Map*, 133–34, *134*; *What the Future Holds*, 137–38
vines, 104–5, 127–28, 129
Virgil (*Aeneid*), 200–201
Virgil (character), 201, 247–48n63
Virginia, 184, 199–200. *See also* Tangier Island
virgin untouched land: as myth, 5, 36–37, 39–40, 47, 48, 75–76, 156, 168, 169, 226n35. *See also* Indigenous homelands
visual archive, 9; aerial imagery, 195, *196*, 197; background in composition of, 109, 111, 112; horizon line, 23, 38–39, 40; methodology and, 16–17; *sfumato*, 132, 140

Wadi Gazi wetland, 215, 250n23
walls: in Butler's Earthseed series, as insulation, 52, 65; as isolationism in time of crisis, 203–4; Limbo and, 201, 205; political slogans calling for, 205; as technocratic fantasy, 214–15; Venice and, 214–15; the West Bank Wall and soil erosion, 215. *See also* seawalls and other hardscapes to prevent erosion
Watson, Jay, 152
Watts Island, Virginia, 199–200
Weaver, Jace, 88–89, 90–91, 93
Webb, Elizabeth (*For the Mud Holds What History Refuses* [*Providence in Four Parts*]; 2019): awe and terror as inspiring, 16; erasure of Indigenous homeland thefts, 26, 169–71; Flaherty's documentary *The Land* incorporated into, 168, 169; Flanagan's *The Canyons at Providence* incorporated into, 145, 171; overview of multimedia exhibit, 26, 145, 168–69; as racial reconciliation narrative, 168–71; and the "virgin land" myth, 168, 169; voice of the canyon imagined in, 168–69
Weits, Michal: *Blue Box* (documentary), 249–50n20
Weitz, Yosef, 215, 249–50n20
West Bank, Palestinian territory, 27, 215–16, 249–50n20
West Bank Wall, 215
wheat: and the Dust Bowl, 77–78, 236n123
whelk shells, 192–93
whiteness, 16, 108, 167, 190–93. *See also* white sight
white pathos of loss: in Audubon's writings, 119–20, 129; in Cable's writings, 122, 124–25, 127, 129; corrective narratives of *Migrant Mother* identity and, 71; environmental movements and, 140, 152–53; geological and natural sciences authors as evoking, 14; in *Grapes of Wrath*, 81; Indigenous loss subsumed into, 3, 156–57, 164; Lange and Taylor's

white pathos of loss (continued) work and, 82–84; settlers as agents of their own loss, 80; as the standard of erosion discourse, 129–30. *See also* Confederate Lost Cause pathos; democracy, anxiety about the erosion of; environmentalism and the settler colonial future; lost-cause narratives; white supremacy

white sight, 16, 40, 82

white supremacy: agrarianism and, 152, 161; American exceptionalism as quarantining to the US South, 13, 167–68; continual recursivity of, 145; corrective narratives of *Migrant Mother* identity and, 71; identity codification as buttressing, 164, 190; naturalization of, 113, 145, 148, 167; Wilmington terror plot (1898), 180, 187, 188, 189; and Wrightsville/Shell Beach developers, 180, 187–89. *See also* enslavement; plantation enslavement economy; racism
—anxiety of loss of: apartheid South Africa and, 212–13; Audubon and, 109, 113, 114–15, 117, 140; in Cable, 109, 140; in Caldwell's *Tobacco Road*, 144; concern for the physical earth as lost in debates on, 216–17, 218; contemporary critics trafficking in, 83; in *GWTW*, 158, 159–60; in Lange and Taylor's Dust Bowl documentation, 83–84; Louisiana land loss and, 109–10, 113, 114–15, 117, 140; sharecroppers and, 80, 144; sublimated into environmentalist anxieties of loss, 65; Trump as campaigning on, 204, 205. *See also* Confederate Lost Cause pathos; lost-cause narratives; plantation futures; settler anxiety of disappearance; white pathos of loss

Whyte, Kyle Powys, 10–11, 30–31, 135, 207

Wichita people, 76

Williams, Kim Hester, 4

Williams, Terry Tempest: *Erosion*, 17–19

Wilmeth, Dudley, 151

Wilmington, North Carolina: white terror plot (1898), 180, 187, 188, 189. *See also* Wrightsville Beach

Wilson, Charles Reagan, 152

windstorms, 74, 78

Wiyot people, 57

Womack, Craig, 88–89, 90, 92, 96, 98

Wood, Karenne: *Markings on Earth* (2001), 26–27, 182; call to dynamic Indigenous identity, 184–85; as challenge to settler colonial demand for stasis, 26–27, 186–87; on colorism, 185–86; and complicity vs. resistance to settler colonialism, 182–83; on entanglement of land, race, and belonging, 182, 184–86, 190; form and meaning, 183, 206–7; kinship time and, 206–7; on the land in reciprocal growth, 206–7; on settler moves toward innocence (Jamestown apology), 183–84; on white nonrecognition of Indigenous identity, 184–85. Works: "Colors," 185–86; "First Light," 206–7; "Jamestown Revisited," 183–85; "Oronoco," 182–83, 185

wood-overuse hypothesis, 20–21

Woods, William, 21

work of settler memory: and Black futurism, 64; definition of, 4; elision of Indigenous perspectives, 5; in *GWTW*, 143, 145, 161; possibility as curtailed through, 24; Providence Canyon cause, 167; regional nostalgia, 7; white sight, 16, 40, 82. *See also* Bruyneel, Kevin; erasure of material history of Indigenous loss; settler moves toward innocence

Works Progress Administration, 151, 168

World War I: wheat-growing craze, 77–78

World War II: topsoil loss in the media, 141–42

Worster, Donald, 79

Wrightsville Beach, North Carolina: Ellyn Bache's "Shell Island" set in, 180, 190–93, 202, 203; bridge as patrolled "pass" system for Black

visitors, 191; in Chesnutt's *The Marrow of Tradition*, 180; developers of, as white supremacists, 180, 187; littoral cell system, 178–80, 181; north-end expansion of development, 189, 190, 193; ocean warming and, 190–91; sea wall construction, 178; Shell Island Resort, and erosion, 177–78, *179*, 193, 207–8; and white anxieties of disappearance, 180; white panic about Black proximity on Shell Island, 188, 189; as whites-only beach, 187
—and Shell Island, 187; as Black beach, history of, 187, 188; as Black resort, white development of, 187–89; constructed merger of island into Wrightsville Beach (1965), 190; developer unhappiness with returns on investment, 189; fire devastation, 189; Moore's Inlet as separating, 188; positionality of the author and, 177–78, 207–8; as setting for Bache's "Shell Island," 190–93; whelk shells, 192–93; white panic about Black equal access to leisure, 188, 189
Wright, Thomas, 187

Xia, Rosanna, 29, 31–32, 35, 43, 66, 224n2

Yang, K. Wayne, 11, 14–15, 17, 21, 66, 148, 153, 173; writing as La Paperson, 19, 48, 54
Yeager, Patricia, 153–54, 155, 158
Yokuts people, 37
Yusoff, Kathryn, 4, 32, 153

www.ingramcontent.com/pod-product-compliance
Lightning Source LLC
Chambersburg PA
CBHW021850230426
43671CB00006B/334